T0221438

Computational Intelligence in Telecommunications Networks

Edited by
Witold Pedrycz
Department of Electrical and Computer Engineering
University of Alberta, Edmonton, Canada

Systems Research Institute, Polish Academy of Sciences
Warsaw, Poland

Athanasios Vasilakos
Institute of Computer Science
Foundation for Research and Technology-HELLAS (FORTH), Greece

CRC Press
Boca Raton   London   New York   Washington, D.C.

Library of Congress Cataloging-in-Publication Data

Pedrycz, Witold, 1953–
    Computational intelligence in telecommunications networks / Witold Pedrycz &
Athanasios Vasilakos
        p. cm.
    Includes bibliographical references and index.
    ISBN 0-8493-1075-X (alk.)
    1. Soft computing. 2. Telecommunication systems. I. Vasilakos, Athanasios. II. Title.

QA76.9.S63 P44 2000
006.3—dc21   00-030342

This book contains information obtained from authentic and highly regarded sources. Reprinted material is quoted with permission, and sources are indicated. A wide variety of references are listed. Reasonable efforts have been made to publish reliable data and information, but the author and the publisher cannot assume responsibility for the validity of all materials or for the consequences of their use.

Neither this book nor any part may be reproduced or transmitted in any form or by any means, electronic or mechanical, including photocopying, microfilming, and recording, or by any information storage or retrieval system, without prior permission in writing from the publisher.

All rights reserved. Authorization to photocopy items for internal or personal use, or the personal or internal use of specific clients, may be granted by CRC Press LLC, provided that $.50 per page photocopied is paid directly to Copyright Clearance Center, 222 Rosewood Drive, Danvers, MA 01923 USA. The fee code for users of the Transactional Reporting Service is ISBN 0-8493-1075-X/01/$0.00+$.50. The fee is subject to change without notice. For organizations that have been granted a photocopy license by the CCC, a separate system of payment has been arranged.

The consent of CRC Press LLC does not extend to copying for general distribution, for promotion, for creating new works, or for resale. Specific permission must be obtained in writing from CRC Press LLC for such copying.

Direct all inquiries to CRC Press LLC, 2000 N.W. Corporate Blvd., Boca Raton, Florida 33431.

Trademark Notice: Product or corporate names may be trademarks or registered trademarks, and are used only for identification and explanation, without intent to infringe.

© 2001 by CRC Press LLC

No claim to original U.S. Government works
International Standard Book Number 0-8493-1075-X
Library of Congress Card Number 00-030342

Preface

Today, it is almost a truism to state that telecommunications systems are the symbol of our Information Age. With the rapidly growing traffic and our growing presence in the Cyberspace, telecommunications becomes a fabric of our life. The future challenges are enormous as we anticipate a rapid growth in terms of new services and number of users. What comes with this challenge is a genuine need for more advanced methodology supporting analysis and design of telecommunications architectures. Hopefully, the systems have to evolve to become *intelligent* to some extent and start operating in more autonomous manner.

Computational Intelligence or CI (this term was coined about six years ago and is sometimes referred to as Soft Computing), emerges as a consortium of recent information technologies such as granular computing, neural networks, and evolutionary computing. In this highly synergistic environment, each of these technologies plays a unique and special role. Granular computing (including fuzzy sets) helps develop and manipulate information granules—crucial entities with clearly defined semantics occurring in any problem formulation. Neural networks are a synonym of highly adaptable architectures endowed with significant learning abilities. In the triumvirate of CI, evolutionary computing delivers a badly needed environment of global optimization—a must in the development of complex intelligent systems. The bottom line of CI is straightforward and intuitively appealing: the research agendas of these technologies complement each other very much and the synergy reflects what really occurs when human beings become engaged in complex problem solving. Decomposition, top-down and bottom-up design principles, manipulation at the level of meaningful chunks (granules) of information, coping with inherent uncertainty, and multicriteria decision processes—all are the primordial faculties of intelligent systems.

CI and telecommunications systems have become a reality: CI arises as a solid design and analysis venue of telecommunications systems including ATM networks, mobile networks, and active networks (that emerge when moving from Internet to Activenet). Undoubtedly, this is the first in-depth volume in the field that provides the reader with scope, depth, and details necessary to illustrate the existing and fast-paced progress of CI technology and shows its importance in solving crucial problems of future telecommunications networks. We believe that the incorporation of the CI methodology into the telecommunications networking contributes to the self-sustaining, self-organi   ure, offering contex   ironment.

The contributions in this volume are an excellent testimony to this important trend.

The volume is unique in its coverage. It covers a broad range of topics that are essential to the current telecommunications systems: from networking embracing issues of Call Admission Control (CAC), congestion control, policing, QoS-routing for ATM networks, to network design, network management, optical networks, mobile networks, active networks, and Intelligent Mobile Agents.

We hope that the book will be of indispensable help to a broad audience of readership: researchers, professionals, lecturers, and graduate students. We also trust that the ideas presented in this volume will be a source of inspiration in further theoretical and practical pursuits. More importantly, we hope that this source of inspiration will materialize in a most beneficial and promising avenue of a highly multidisciplinary research in telecommunications and will lead to the development of future computing and networking.

The editors would like to express deep thanks to all the authors who have contributed so generously to the volume by sharing their recent research experience, insights, and in providing an excellent testimony to this rapidly growing area. We owe our debts to numerous reviewers who shared their experience in reviewing the material and provided important and constructive feedback. Our interaction with the CRC Press was superb and we owe our sincere thanks to Bob Stern, Publisher, and Helena Redshaw, Production Manager. These folks provided us with continuous and enthusiastic support, shared professional experience, and made this volume a reality.

Witold Pedrycz and Thanos Vasilakos
Edmonton and Heraklion, 2000

Editors

Witold Pedrycz is a Professor and Director of Computer Engineering and Software Engineering in the Department of Electrical and Computer Engineering, University of Alberta, Edmonton, Canada. He is actively pursuing research in Computational Intelligence, fuzzy modeling, knowledge discovery and data mining, fuzzy control including fuzzy controllers, telecommunications networks, pattern recognition, knowledge-based neural networks, and relational computation. He has published numerous papers in the area of applied fuzzy sets as well as research monographs (Fuzzy Control and Fuzzy Systems, 1988 and 1993; Fuzzy Relation Equations and Their Applications to Knowledge Engineering, 1988; Fuzzy Sets Engineering, 1995; Computational Intelligence: An Introduction, 1997; Fuzzy Sets: Analysis and Design, 1998; Data Mining Techniques, 1998). He is also one of the Editors-in-Chief of the *Handbook of Fuzzy Computation* (Oxford University Press/Institute of Physics, 1998). Dr. Pedrycz is a member of many program committees of international conferences and has been serving on editorial boards of journals on fuzzy set technology and neurocomputing (including *IEEE Transactions on Fuzzy Systems, IEEE Transactions on Systems, Man and Cybernetics, Fuzzy Sets and Systems*), soft computing (Soft Computing Research Journal), intelligent manufacturing (*Journal of Intelligent Manufacturing*), and pattern recognition (*Pattern Recognition Letters*). He is an IEEE Fellow (1999).

Athanasios V. Vasilakos has received B.S. (1983) in Electrical Engineering from the University of Thrace (Greece), M.S. (1986) in Electrical and Computer Engineering from the University of Massachusetts at Amherst (USA), and the Ph.F. (1988) in Computer Engineering from the University of Patras (Greece). From 1988 to 1991 he was with the Computer Engineering Department of the University of Patras and the Computer Technology Institute in Patras. From 1991 to 1995 he was professor of Informatics at the Hellenic Air Force Academy. Since 1995 he has been a Research Professor with the Institute of Computer Science, Foundation for Research and Technology - Hellas (FORTH, Greece), and a consultant to the Greek government. His main interests are in communications networks, B-ISDN, ATM network, Mobile nets, learning theory, and Computational Intelligence. He is a chairman of the Technical Committee of Telecommunications of the EURIDIT Network of Excellence for Fuzzy Logic in Europe. He is a member of numerous program committees of international conferences on telecommunications networks and an editor of the journals *Computer Communications* (Elsevier), *Soft Computing* (Springer), and *ACM Applied Computing Review*. He has authored over

70 papers mostly in the area of applications of Computational Intelligence in the problems of computer networks. Dr. Vasilakos is a member of ACM and ACM SIGCOMM. He has chaired several R&D projects in informatics and telecommunication.

Contributors

Antonios F. Atlasis   National Technical University of Athens, Athens, Greece

Chung-Ju Chang   National Chiao Tung University, Hsinchu, Taiwan, ROC

Ray-Guang Cheng   Industrial Technology Research Institute, Chutung, Hsinchu, Taiwan, ROC

Imrich Chlamtac   The University of Texas at Dallas, Richardson, Texas

N. Davey   University of Hertfordshire, Hertfordshire, United Kingdom

Christos Douligeris   University of Piraeus, Piraeus, Greece

D. Everitt   University of Melbourne, Parkville, Victoria, Australia

R. J. Frank   University of Hertfordshire, Hertfordshire, United Kingdom

Andrea Fumagalli   The University of Texas at Dallas, Richardson, Texas

Sumit Ghosh   Arizona State University, Tempe, Arizona

Kaoru Hirota   Tokyo Institute of Technology, Yokohama, Japan

Cynthia S. Hood   Rensselaer Polytechnic Institute, Troy, New York

S. P. Hunt   University of Hertfordshire, Hertfordshire, United Kingdom

Chuanyi Ji   Rensselaer Polytechnic Institute, Troy, New York

Seong-Soon Joo   Electronics and Telecommunications Research Institute, Taejon, Korea

Abraham Kandel   University of South Florida, Tampa, Florida

Ioannis E. Kassotakis   National Technical University of Athens, Athens, Greece

Yaron Klein   Tel Aviv University, Tel Aviv, Israel

Tony S. Lee   NASA Ames Research Center, Moffett Field, California

Yao-Ching Liu   University of Miami, Coral Gables, Florida

Jon W. Mark   University of Waterloo, Waterloo, Ontario, Canada

Maria E. Markaki   National Technical University of Athens, Athens, Greece

Masaharu Munetomo   Hokkaido University, Sapporo, Japan

Richard G. Ogier   SRI International, Menlo Park, California

M. Palaniswami   University of Melbourne, Parkville, Victoria, Australia

Witold Pedrycz   University of Alberta, Edmonton, Alberta, Canada

Andreas Pitsillides   University of Cyprus, Nicosia, Cyprus

Qutaiba Rasouqi   Arizona State University, Tempe, Arizona

Marios P. Saltouros   National Technical University of Athens, Athens, Greece

Ahmet Sekercioglu   Swinburne University of Technology, Melbourne, Australia

P. Seshasayi   Arizona State University, Tempe, Arizona

Xuemin Shen   University of Waterloo, Waterloo, Ontario, Canada

Leen-Kiat Soh   University of Kansas, Lawrence, Kansas

Nina Taft-Plotkin   SRI International, Menlo Park, California

D. Tissainayagam   University of Melbourne, Parkville, Victoria, Australia

Costas Tsatsoulis   University of Kansas, Lawrence, Kansas

Kiyohiko Uehara   Toshiba Corporation, Kawasaki, Japan

Luca Valcarenghi   The University of Texas at Dallas, Richardson, Texas

Athanasios Vasilakos   Institute of Computer Science, Foundation for Research and Technology–Hellas (FORTH) and Science of Technology Park of Crete, Heraklion, Crete, Greece

P. Venkataram   Indian Institute of Science, Bangalore, India

Contents

1

Computational Intelligence: A Development Environment for Telecommunications Networks

Witold Pedrycz and Athanasios V. Vasilakos

CONTENTS

0-8493-1075-X/01/$0.00+$.50
© 2001 by CRC Press LLC

ABSTRACT In this chapter, we introduce a concept of Computational Intelligence (CI), define its key components—contributing technologies of neurocomputing, granular computing, especially fuzzy sets, and evolutionary methods, and discuss the underlying design methodology. Subsequently, we elaborate on the use of CI as the design and analysis venue of telecommunications systems, especially ATM networks and active networks.

KEY WORDS: *telecommunications networks, Computational Intelligence, learning, knowledge representation, fuzzy sets, granular computing, neurocomputing, ATM networks, active networks, mobile active networks, intelligent mobile agents.*

1.1 Introduction

The objective of this study is to introduce a new paradigm of development and analysis of intelligent systems coming under the name of Computational Intelligence (CI) and discuss its role as a development environment for telecommunications networks. There has been a significant body of research in this realm and our intent is to summarize these findings in a coherent and concise manner.

The material is structured into seven sections. First, we start with a comprehensive introduction to CI being regarded as a new and well-established paradigm supporting the synthesis and analysis of intelligent systems. The main contributing technologies of CI such as neural networks, fuzzy sets, and evolutionary computing are discussed in detail and contrasted in terms of their research agendas and main advantages and shortcomings. Subsequently, we provide the reader with a set of definitions of CI and highlight the facets of synergy forming a cornerstone of CI. As a detailed example of such synergistic links, we elaborate on neurofuzzy computing. Starting from Section 4, there are discussions of various applications of CI to telecommunications systems such as the management of ATM networks (dealing with controls at various levels, say network, call, and cell control). In Section 5, we cover the issue of CI in the management of mobile networks and then, in Section 6, discuss the use of CI in active networks and mobile active networks. Concluding remarks are included in Section 7.

1.2 The Triumvirate of Information Technologies

In this section, we elaborate in a concise way on the three dominant information technologies such as neural networks, granular computing, and evolutionary methods. They constitute a backbone of the new emerging area of

Computational Intelligence. Our intent is to identify the key features of these technologies, elaborate on their complementary character and, finally, identify the general design methodologies present therein.

1.2.1 Neural Networks

Neural networks emerge as powerful and distributed computing architectures that are equipped with significant learning abilities. In a nutshell, neural networks help represent highly nonlinear and multivariable relationships. Starting from pioneering research of Rosenblatt (1961), Minsky and Pappert (1969), neural networks have undergone a significant metamorphosis becoming today an important reservoir of various learning methods subsequently giving rise to the highly adaptable topologies (Rumelhart and McClelland, 1986). Neural networks have already been successfully used in many application areas including system modeling, telecommunications, pattern recognition, robotics, and process control.

From a formal point of view, a computational neuron is regarded as a many-input single-output static processing unit described as

$$y = \phi\left(\sum_{i=0}^{n} w_i x_i \right)$$

where x_i are the inputs of the neuron, w_i are adaptable connections (weights), and "y" denotes the output of the neuron. In essence the above expression represents a weighted sum of the inputs that is followed by a nonlinear function ϕ transforming the result of such weighted summation of the inputs into a single numeric value that describes the activation of the neuron.

The processing abilities of neural networks emerge as a result of a *collective* processing realized by a set of neurons. Traditionally, the neurons are organized into layers giving rise to multilayer neural networks. Neural networks are universal approximators: they can approximate any continuous function to any required degree of accuracy. This important theoretical result is crucial to any application: it states that neural networks are *capable* of representing any nonlinear mapping to the required accuracy. The approximation is realized through learning in which the connections are update (modified) to follow the required (target) values of the mapping.

Neurocomputing comes today with an array of learning algorithms covering various modes of learning such as supervised, unsupervised, and reinforcement learning. Neural networks are inherently distributed, highly plastic, and parallel architectures. This makes them difficult to understand. We commonly refer to them as black-box architectures. This has a certain disadvantage. First, we do not have a general way of incorporating any prior knowledge about the problem. This forces us to train the network from scratch that may lead to a substantially longer training time. Second, the interpretation of the network (that may lead to the global and easy-to-comprehend description of the topology) is not possible.

When it comes to the design methodology, neural networks, we subscribe to the bottom-up design approach. In other words, we proceed with a large set of numeric data that is afterwards captured in the form of the neural network (its connections). In other words, we start from individual data and combine them into a single numeric neural network.

1.2.2 Fuzzy Set Technology

Fuzzy sets or, more generally, granular computing, form a key methodology for representing and processing linguistic or, in general, nonnumeric information. They support a diversity of mechanisms of knowledge representation focusing on a relevant selection of information granularity. Fuzzy sets exploit imprecision in an attempt to a make system complexity manageable. There are two important principles supported by fuzzy sets. The first, formulated by the founder of fuzzy sets (Zadeh, 1973) is called the principle of incompatibility. In brief, this principle promotes fuzzy sets as a basic vehicle useful in overcoming an evident and acute disparity between precision and relevancy when modeling complex phenomena. Similarly, rough sets (Pawlak, 1991), random sets or set theoretic mechanisms contribute to the formation of granular computing (Pedrycz, 1995). The second principle originating within the realm of computer vision (Marr, 1982) alludes to an important idea of least commitment and graceful degradation that advises to postpone any final decision until the point enough evidence has been gathered. In this sense fuzzy sets allow us to quantify uncertainty and take advantage of it rather than blindly discarding it.

In general, fuzzy sets and granular computing promote a top-down approach. In particular, we first capture the skeleton of the problem and subsequently refine the solution by moving into processing smaller, more-detailed information granules. When dealing with fuzzy sets, all processing is carried out at the level of such information granules. The numeric details are hidden on purpose. The fuzzy set-oriented constructs are transparent. They do not exhibit any significant learning abilities. Subsequently, the design exploiting fuzzy sets is intuitive and straightforward but cannot be easily adjusted to the existing numeric data. There are simply no profound learning abilities of the fuzzy systems.

1.2.3 Evolutionary Computing

Evolutionary computing (EC) delivers a suite of tools of *global* optimization. Evolutionary computation embraces genetic algorithms (GA), evolutionary computation, and evolutionary strategies which are biologically-inspired methodologies aimed at global optimization. The pioneering research of Holland (1975) and Fogel et al. (1966) gave rise to a new paradigm of population-driven computing. This global optimization can be perceived both in terms of structural and parametric optimization. Furthermore, some other techniques including tabu search fall under the same category of global optimization. EC is a must when it comes to the optimization of complex systems.

1.3 Computational Intelligence: the Development of Synergy

1.3.1 Defining Computational Intelligence

There are several definitions of the concept of CI. For the first time the term CI has been used by Bezdek (1992, 1994) who put it forward as follows:

> A system is computationally intelligent when it: deals with only numerical (low-level) data, has pattern recognition components, does not use knowledge in the AI sense; and additionally when it (begins to) exhibit (i) computational adaptivity; (ii) computational fault tolerance; (iii) speed approaching human-like turnaround; and (iv) error rates that approximate human performance.

In his attempt to define CI, Marks (1993) focuses on the contributing technologies:

> ... neural networks, genetic algorithms (GAs), fuzzy systems (FSs), evolutionary programming (EP), and artificial life are the building blocks of CI

Another attempt in defining CI was made by Fogel (1995):

> ... These technologies of neural, fuzzy and evolutionary systems were brought together under the rubric of Computational Intelligence, a relatively new term offered to generally describe methods of computation that can be used to adapt solutions to new problems and do not rely on explicit human knowledge

Overall, CI can be regarded as a research endeavor being a home to a number of technologies including genetic, fuzzy, and neural computing. This point of synergy and various synergistic links as well as a need for the formation of a sound methodology have been advocated in Pedrycz (1997).

1.3.2 The Synergy

It is clear that the already identified components of CI encompass neural networks, fuzzy set technology, and evolutionary computation. In this synergistic combination, each of them plays an important, well-defined, and unique role.

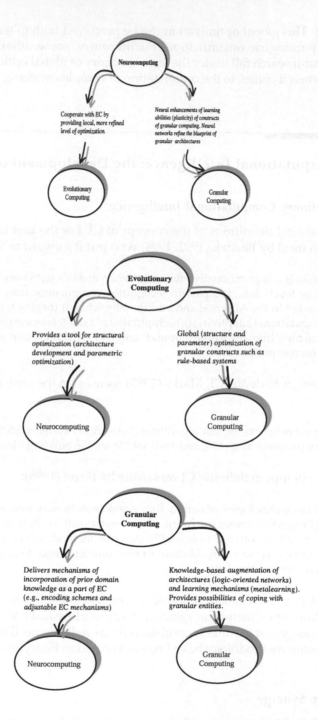

FIGURE 1.1

The web of synergistic links between the contributing information technologies of Computational Intelligence.

As we have already summarized the main technologies, one can easily envision their strong points as well as some deficiencies that require further elimination. The role of synergy between them is to alleviate the existing shortcomings. Figure 1.1 highlights the main links of synergy between EC, granular computing, and neural networks and shows how they support each other.

The bottom line is that the synergy between these key information technologies becomes a necessity. They reflect a way of problem formulation and its ensuing solution. All of these methodologies stem from essential cognitive aspects of fuzzy sets, underlying evolutionary mechanisms of genetic algorithms, and biologically sound foundations of neural networks which provide essential foundations when dealing with engineering problems. With their increasing complexity, it becomes apparent that all of the technologies discussed above should be used concurrently rather than separately.

There are numerous links that give rise to the CI architectures. In the next section, we elaborate on one of them dealing with the synergy between neural networks and fuzzy sets.

1.3.3 Neurofuzzy Computing

There has been an evident trend in research and applications of fuzzy sets and neural networks to develop and deploy highly heterogeneous architectures. They usually come under a banner of neurofuzzy systems. It was John von Neumann who made an interesting observation many years before symbiotic relationships between fuzzy sets and neural networks had started to shape up:

> . . . we have seen how this leads to a lower level of arithmetical precision but to a higher level of logical reliability: a deterioration in arithmetic has been traded for an improvement in logic.

This observation makes a point of emphasizing the need for some trade-off and cooperation—fuzzy sets just attempt to fill out this important computational niche. Fuzzy neural networks are a conceptual and computational vehicle combining the technology of fuzzy sets and neural networks. The term itself carries a number of representations and, in fact, materializes into a series of diverse topologies (and ensuing learning algorithms). Since the inception of fuzzy sets there have been interesting approaches resulting in the form of neurofuzzy structures. The reader may refer to Lee and Lee (1975) as being the first hybrid attempt to the design of neural networks. Some other research pursuits were reported by Buckley and Hayashi (1994) and Keller and Hunt (1985). It is instructive to place all the neurofuzzy (or fuzzy-neuro) architectures in a two-dimensional plane of learning abilities—knowledge transparency, Figure 1.2. The transparency usually implies more efficient training; we avoid learning from scratch by being navigated by some initial pieces

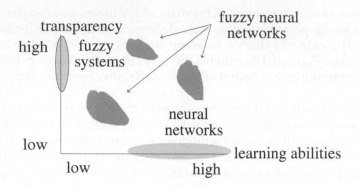

FIGURE 1.2
Fuzzy systems and neural networks: a spectrum of architectures providing a tradeoff between transparency and learning (adaptability).

of domain knowledge (which itself could be of a qualitative nature). Some relationships could be defined in advance (at least at the linguistic level). Some topologies can be also predetermined and fixed by mapping (implanting) existing domain knowledge. As the resulting network is transparent (as opposed to opaque numeric neural networks), this process is highly simplified and significantly augmented. Referring again to Figure 1.2, the points situated along the first axis represent neural networks of different intensity of learning abilities. The elements distributed along the second axis concern fuzzy systems (models). The clusters of elements in this space depict an entire spectrum of hybrid systems; again some of them lean more towards fuzzy structures with some touch of learning abilities whereas some other are very much neural network-dominant.

1.4   Computational Intelligence in the Management of ATM Networks

Asynchronous Transfer Mode (ATM)-based networks are designed to be scalable, high-bandwidth, manageable, and have the flexibility of supporting various classes or multimedia traffic with varying bit rates and Quality of Service (QoS) requirements. Thus, they have the potential to create a unified communications infrastructure that can transport services with widely different demands on the network (such services include real-time video and voice with no tolerance to delays, but some tolerance to loss, and data with some tolerance to delay, but no tolerance to loss).

An important difficulty of exploiting the potential of ATM optimally is the management and control complexity of the scheme itself (the basic concept is simple). Since ATM simultaneously attempts to support voice, data, and

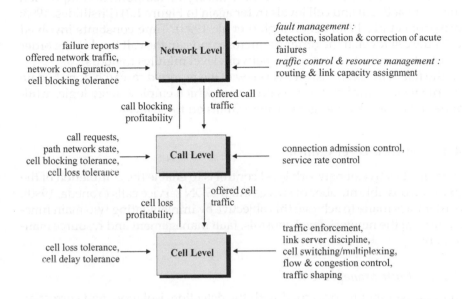

FIGURE 1.3
Multilevel traffic control in ATM networks.

video applications, which all have differing performance and QoS requirements, optimal utilization of the network resources requires complex, nonlinear, and distributed control structures. In order to achieve its potential, ATM networks will need to accommodate several interacting control mechanisms, such as call admission control, flow and congestion control, input rate regulation, routing, bandwidth allocation, queue scheduling, and buffer management.

The complexity of the ATM networks and multidimensionality of the control problems dictate that traffic control in ATM networks be structured. The control structure is most likely to be implemented in a multilevel architecture which partitions the solution into different levels of control with varying temporal decomposition in network, call, and cell levels (see Figure 1.3).

Due to the complex nature of the above-mentioned control issues, some researchers are looking for solutions by application of Computational Intelligence (CI) techniques to design intelligent control systems to various aspects of ATM network management, often supplementing the existing control techniques. Their motivation arises from the reported success of the techniques in various previously unsolvable or difficult control problems in many diverse fields (Habib, 1996; Douligeris and Develekos, 1997; Ghosh et al., 1992).

The complexity of the ATM networks and multidimensionality of the control problems dictate that traffic control in ATM networks be structured and most likely implemented in a multilevel architecture which partitions the solution into different levels of control with varying temporal decomposition in network and call and cell levels (refer again to Figure 1.3) (Pitsillides, 1993; Nordström et al., 1995; Sekercioglou et al., 1999). Time constants involved are: in the cell level in the order of microseconds, in the call level in the order or seconds to minutes, and in the network level minutes to hours.

In the following sections, an overview of the reported research done to date to implement control methods is presented which employ fuzzy logic, artificial neural networks, and genetic algorithms on these levels.

1.4.1   Network Level Control

The main objective of network level control is to enable the completion of the maximum possible number of successful B-ISDN service calls (Yoneda, 1990). An attempt is made to achieve this objective by implementing two main functionalities at the network level controls: fault management and resource management.

1.4.1.1   Fault Management

Fault management is concerned with the detection, isolation, and correction of acute failures that interrupt the availability of network resources. Besides acute failures, some failures may be manifested intermittently, or malfunctions may subtly degrade network facilities to detect degradations in performance caused by such conditions, and respond with appropriate actions in order to minimize the effect on offered services (Chen and Liu, 1994). This responsibility is fulfilled by Operations and Management (OAM) in ATM networks.

In the area of fault management in communication networks, reported research on applications of CI is very limited in scope. Applications of AI in network management have been surveyed and reported by Smith and Fry (1995).

1.4.1.2   Routing and Link Capacity Assignment

In the absence of faults and malfunctions, efficient utilization of network resources is maintained by the traffic control and resource management functions, which involve routing and link capacity assignment. At the network level, route selection and link capacity assignment to virtual paths are performed by using offered call traffic and tolerated call blocking probability information.

The first major function at the network level control in ATM networks is route selection. In B-ISDN networks, links have to be described in terms of multiple metrics, including QoS and policy constraints, which makes routing with requirements a difficult problem to solve.

The second major function of network level control in ATM networks is the optimal allocation of bandwidth to virtual paths (Chen and Liu, 1994;

Pitsillides et al., 1997). Effective implementation of this function is very important for a number of reasons:

1. By reserving capacity in anticipation of the virtual channels, which will belong to a virtual path, the processing effort required to establish individual virtual channels can be minimized.

2. Virtual paths may be used as a means of logically separating traffic types having different QoS requirements while allowing virtual channels to be statistically multiplexed.

3. Virtual paths allow groups of virtual channels to be managed and policed more easily.

4. Dynamic routing control at virtual path level should lead to designing effective network reconfiguration mechanisms.

5. Effective allocation of bandwidth to virtual paths would lead to a more efficient and effective network.

Aboelela and Douligeris (1997) have studied the application of fuzzy control for the route selection in ATM networks. Park and Lee (1995a) have worked on optimized routing using recurrent ANNs, while Atlasis, Saltouros, and Vasilakos (1998) use a Learning Automaton for the route selection in ATM networks in order to maximize the network throughput (call acceptance rate) while guaranteeing the QoS requirements.

The application of combination of evolutionary programming with fuzzy logic could be very beneficial to solve multiobjective optimization problems such as bandwidth to virtual paths. Heuristics have been proposed to reach near-optimal solutions (Vasilakos et al., 1997; Vasilakos et al., 1998). In the study presented in (Vasilakos, Ricudis, Anagnostakis, Pedrycz, and Pitsillides, 1998) the researchers have used evolutionary-fuzzy prediction in interdomain routing of broadband network connections with QoS requirements in the case of an integrated ATM and SDH networking infrastructure. In their method, a subset-interactive autoregressive model is used to predict link utilization levels, based on experience from both static traffic observations as well as dynamic knowledge, acquired during the network's operation. Based on these, the shadow cost of allowing the connection through each feasible path is calculated, which is then used to select the "best" path. The shadow cost is calculated in such a way as to lead the network to states which exhibit the lowest expected blocking probabilities in regard to information about user's demand—thus aiming to match network state with demand at all times.

Early results of a study for VP bandwidth allocation using evolutionary programming techniques has been published by Pitsillides et al. (1997).

Pedrycz and Vasilakos (2000), proposed a new characterization of ATM time series in the setting of information granules and its application to predict network utilization patterns for ATM routing algorithms.

1.4.2   Call Level Control

1.4.2.1   *Connection Admission*

Connection Admission Control (CAC) is defined as a set or actions, performed at connection set-up phase, to determine whether or not the virtual path (VP) or virtual channel (VC) requesting the connection can be accepted. The decision is based on the connection's anticipated traffic characteristics, the requested QoS, and the user terminal declares these source traffic descriptor values to the network when the connection set up is requested. If the request is accepted, network resources are implicitly allocated to the connection.

To attain high utilization of VPs while meeting the QoS requirements, CAC must determine whether to accept a new VC by considering its anticipated traffic characteristics, the QoS requirements of existing VCs, the availability of the network resources, and current utilization of the links. There are many demanding problems waiting to be solved for the development of an effective CAC algorithm, especially

- The statistical behavior of several sources of different types multiplexed on an ATM is difficult to predict. Therefore, deciding how to allocate resources for multiple QoS requirements is hard to solve.
- Developing accurate analytical models to evaluate QoS services could be very difficult.

Application of CI techniques appears to be appropriate, and several researchers have attempted to solve the problem by using artificial neural networks and fuzzy logic control techniques.

Hiramatsu has studied ANN-based ATM CAC (1990) and has published his work on training techniques for ANN application in ATM (1995). Ramalho and Scharf (1994) have used a method for learning the behavior of the traffic in an ATM link. Park and Lee (1995b) also have published their work on adaptive call admission control using feedforward ANNs.

Vasilakos, Atlasis, and Loukas (1995) have proposed a CAC algorithm that uses a Learning Automaton to maximize the call acceptance rate while guaranteeing the QoS requirements. The Learning Automaton receives as feedback a function of a conservative fluid-flow (equivalent bandwidth) approximation and through a learning process attempts to approximate more effectively the optimal call acceptance region than the fluid-flow approximation.

Uehara and Hirota (1997) have proposed a method based on estimation of the possibility distribution of cell loss ratio (CLP). They use fuzzy inference for the estimation of CLR of new connections. The mechanism operates this way: Each call request is placed into a transmission rate class depending on its declared parameters. Then, by using the number of active connections for each class, a CLR estimation is made by the inference engine. If the estimate exceeds the required CLR, the connection request is rejected. The fuzzy sets representing the values of the fuzzy numbers of the rule base are shaped by a learning mechanism and observed CLR data, which gives the scheme its adaptation capability.

Scheffer and Kunicki have studied the application of fuzzy logic techniques for accurate modeling of voice and video sources, and prediction of their behavior (Scheffer et al., 1993, 1994). They have proposed a CAC scheme, which uses a fuzzy logic based traffic prediction algorithm (Scheffer and Kunicki, 1996).

Cheng and Chang (1994, 1996) have devised a fuzzy control system, which combines CAC and a feedback mechanism. The mechanism sends back coding rate control signals to video sources, and congestion control signals to data sources. In this scheme, fuzzy sets representing the linguistic values are selected by evolutionary techniques. Unlike the other schemes mentioned in the previous section, the system does not have the capability of real-time adaptability but, since it has the ability to adjust the cell transmission rate of the sources, and subsequently traffic density at the switches, it can still maintain QoS for the connections.

Early results of a study whose aim is the application of CI techniques to ATM CAC problem and development of a simulation testbed has been published by Czezowski (1998).

1.4.3   Cell Level Control

1.4.3.1   Usage Parameter Control

Usage parameter control (UPC), or in other words, traffic enforcement or policing, is a very important function in ATM networks. Its task is to ensure that traffic sources stay within the limits of the negotiated traffic parameters, which are declared during the call setup phase. Traffic enforcement functions are performed by the network provider at the virtual circuit or virtual path level and corrective measures are taken if a traffic source does not stay within the declared limits. The measures could be as drastic as blocking the traffic source or could be less severe such as selectively discarding the violating cells or tagging violating cells that could be discarded in downstream nodes if necessary.

The ideal UPC mechanism should have desirable characteristics: accurate detection of any traffic situation violating the negotiated values, and separating those connections from the ones that stay within the negotiated limits; fast response to violations; and implementation simplicity and cost effectiveness. Designing a UPC mechanism encompassing these features could be a daunting task. For example, well-studied mechanisms such as leaky bucket and window mechanisms cannot achieve the ideal UPC characteristics but only provide a trade-off between the above requirements.

Catania et al. (1996a, b) and Ascia et al. (1997) have proposed a UPC mechanism based on fuzzy logic control which displays characteristics close to ideal UPC, and have also implemented the algorithm as a VSLI chip. The mechanism they propose ensures that a bursty source conforms to its agreed average cell rate. It is a window-based control mechanism. It allows short term fluctuations of the source cell rate around a negotiated average value, as long as the source respects this value over the long term, and at the same time it is capable of recognizing average value, as long as the source respects this value over the long term, and at the same time it is capable of recognizing a violation immediately. The maximum number of cells, which are considered to be nonviolating in a fixed period, is dynamically updated by a set of fuzzy inference rules. The set of rules is shaped to guarantee transparency to a compliant source by assigning a credit value of allowable cells which is higher than the negotiated value agreed at call set-up phase. This credit value represents the number of cells that the source can send during a particular transmit window. If a source has a high flow of traffic, the UPC intervenes to enforce a reduction of the bit rate or the source. To do so, it reduces the assigned credit to lower the allowable cell rate threshold and identifies any cells that exceed this threshold as violating cells.

The parameters describing the behavior of the source consist of the average number of cell arrivals per window since the start or the connection, the number of cell arrivals in the last transmit window, and current value of the maximum allowable cells that can be transmitted. By using these three parameters, the fuzzy UPC mechanism determines the value of the threshold to be used in the next transmission window.

Atlasis and Vasilakos (1998) proposed a scheme that uses a Learning Automaton to achieve an efficient policing of the source traffic parameters. Specifically, in this methodology the Automaton is used in combination with the well-known Leaky Bucket mechanisms. The Automaton, receiving as feedback the distribution function of the values that the Leaky Bucket counter takes, achieves much better policing of the nonconforming sources. Its effectiveness is based on the fact that this scheme policing the distribution function of the source and not just a single traffic parameter detects faster and smaller deviation from the negotiated values achieving almost the optimal behavior that a UPC mechanism should have.

A UPC mechanism particularly designed for policing voice sources as been proposed by Ndousse (1994). Since voice cells are characterized by a high degree of burstiness, utilization of a classical control approach faces difficulties. Ndousse proposes an intelligent implementation of the Leaky Bucket cell rate control mechanism, which yields a lower cell drop rate than the Leaky Bucket algorithm under similar circumstances.

The fuzzy Leaky Bucket is implemented as a two-level fuzzy logic controller (FLC) by connecting three fuzzy associative memories (FAMs) (Kosko, 1992). In the first level, there are two FAMs, each taking two input variables and generating an output variable which is supplied to the second level FAM. The output of the FAM determines the number of special tokens in the token buffer in the next sampling interval. These special tokens are used to tag the violating cells. Therefore, depending on the availability of the network resources, violating cells are not discarded straight away, but can have a chance of transmission. The FLC dynamically determines the number of special tokens allocated in the token buffer by monitoring the buffer occupancies and buffer growth rates in the token buffer and channel buffer allocated to the voice connection.

1.4.3.2   Flow, Congestion, and Rate Control

In early stages of B-ISDN development, prevailing belief among the research community was that preventive (or, in other words, open-loop) type congestion control at the edge was necessary due to the large bandwidth delay product, and would be sufficient for ATM networks. But, outcomes of subsequent studies have shown that, because of the variety of the traffic to be supported in B-ISDN networks, open-loop congestion and flow control is rendered to be ineffective in ATM networks. Today, the shift is towards closed-loop congestion controls (within the network).

In ATM networks, depending on the nature of the traffic sources, the closed-loop congestion control issue can be approached in two ways:

- For *delay tolerant traffic*, which basically comprises of TCP/IP type traffic, switches can send feedback signals to the sources leading them to reduce the rate at which they release cells to the network. Then excess traffic is queued at the traffic source and consequently delayed.

  The ATM Forum has introduced a service category, called available bit rate (ABR), in order to allocate bandwidth dynamically within an ATM network, while simultaneously minimizing the cell losses, and has selected a feedback control framework to achieve these aims ("Traffic management specification version 4.0," 1996). The framework allows downstream nodes or intermediate ATM switches to periodically send information to the traffic sources relating to maximum cell rates that they can handle. The cell rate information is carried by a stream of resource management (RM) cells generated by the traffic sources and relayed back to the sources by the destination end systems. During their round-trip, while these cells pass through the switching nodes, the cell rate information contents of these cells are dynamically updated by these intermediate systems. For the calculation of rate, several algorithms have been proposed.

- On the other hand, since delay tolerance of *video/voice traffic* is very low, congestion control is performed by sending coding rate signals to these types of sources. In the presence of congestion, the sources can vary their coding rate, and so reduce the frequency or cells generated by using this feedback information. Lower coding rate inevitably reduces the image/sound quality at the receiver but network utilization and quality or offered service rate are maintained at higher levels by minimizing the cell losses and delays due to congestion.

The following sections summarize the research utilizing CI techniques for implementation of congestion and rate control algorithms in ATM networks.

Tarraf, Habib, and Saadawi (1994, 1995a–c) have investigated extensively how ANNs can be used to solve many of the problems encountered in the development of coherent traffic control strategies in ATM networks. In (1995) they present congestion control schemes for ATM networks. Also, they investigate a reinforcement-learning based neural network for congestion control in ATM networks (1995). Liu and Douligeris have published the results of a comparison study on the performance of static and adaptive feedback congestion controllers which uses ANNs (1995).

Huang and Yan (1996) use a recurrent neural network for the dynamic control of communication systems, particularly dynamic congestion control on ATM networks. Mehrvar and Le-Ngoc (1995) apply a neural network scheme for congestion control in packet switch OBP satellite systems.

Pitsillides et al. (1995) have proposed the Fuzzy Explicit Rate Marking (FERM) algorithm, and analyzed its performance regarding fairness, responsiveness, resource utilization, and cell loss in LAN and WAN environments. FERM operates on switching nodes and by periodically monitoring the buffer utilization and queue growth rate, determines a cell rate which is used to update the maximum cell rate information carried by the RM cells passing through the virtual connection.

Douligeris and Develekos (1995) have studied a FLC which is based on the short-term observation of the network status to predict the near future cell discarding behavior of the switching nodes. This prediction is then fed back to traffic shapers in the sources to minimize cell losses.

Jensen (1994) has proposed a three-step FLC for controlling the transmission rate or sources to protect links against overload in the case of connections exceeding their negotiated traffic parameters. The scheme operates as follows: at the call admission stage a service dependent priority is assigned to each connection. This priority is kept as a fixed value for the whole lifetime of the connection. Also, in the switching node, a certain buffer capacity is allocated to the connection. The FLC generates the cell service rate control signals for each buffer. Input parameters of the FLC are: (1) allocated priority level; (2) current buffer occupancy level; (3) bandwidth utilization at the output link of the node; and (4) the difference between the effective bandwidth at which the source is transmitting the cells and the declared bandwidth negotiated during the call set-up stage. Hu et al. (1996) have studied an adaptive traffic controller based on Sugeno's self-tuning fuzzy control methods.

1.4.3.3 Cell Switching and Multiplexing

In an ATM network, cell queuing is required to alleviate congestion at switching nodes. Congestion occurs when multiple cells simultaneously attempt to access an output link in a switch. Cell queuing can be arranged either by placing buffers at input ports (called input queuing), or by placing cell buffers at the output ports (called output queuing). Output queuing yields better performance in terms of cell delay and throughput, but computationally more demanding to operate than input queuing. On the other hand, in input queuing, if the head-of-line blocking problem can be solved, comparable performance can be achieved.

One way of solving this problem is to employ a mechanism called bypass queuing. When bypass queuing is used, a controller module schedules the cells in an optimal fashion to enhance the switch throughput. Additionally, cells can be dispatched optimally if they are assigned priorities, with higher priorities assigned to real-time traffic such as voice and video (due to rigid delay requirements) and lower priorities assigned to data traffic, by an intelligent scheduling mechanism.

Liu and Douligeris (1996) have proposed a fuzzy scheduler to optimize the cell servicing sequences to reduce cell losses. In their mechanism, each traffic class in the switch has its own portion of the dedicated buffer and a fuzzy scheduling algorithm manages the server. Park and Lee (1995a, b) have also worked on optimal scheduling and published their study on application of recurrent ANNs to this problem.

1.5   Computational Intelligence in the Management of Mobile Networks

Modern cellular mobile communications systems are characterized by a high degree of capacity. Consequently, they have to serve the maximum possible number of calls while the number of channels per cell is limited. The objective of bandwidth (channel) allocation is to assign a required number of channels to each cell such that both efficient frequency spectrum utilization is provided and interference effects are minimized.

Kassotakis, Markaki, Vasilakos (to appear early 2000), proposed a Hybrid Genetic Algorithm (HGA) for Isochronous Channel reuse in Multiple Access Metropolitan Networks. HGA can be extended to a variety of resource allocation problems in cellular mobile networks where channel sharing reuse can be used to enhance the channel management.

Sandalidis et al. (1998) and Lai and Coghill (1996) proposed efficient combinatorial evolution algorithms for channel allocation in cellular models. Neural Networks technology (Feed forward NNs, Hopfield NNs) for Fixed Channel Assignment (each cell is allotted a certain number of channels) is proposed by Kunz (1991), Smith and Palaniswami (1997), and Funabaki and Tanefuji (1992) and Dynamic Channel Assignment (channels are not distributed to cells in advance and are assigned, one at a time, on a call-by-call basis in real time) is proposed by Everitt and Manfield (1989), Smith and Palaniswami (1997), and Chan et al. (1994).

Many heuristic techniques have been devised for solving the channel assignment problem (Duque-Anton, Castelino, 1996). Fuzzy Logic Inference Systems to estimate and predict the probability information for direct sequence code division multiple access (DS/CDMA) wireless cellular networks have been proposed by Shen and Mark (1999).

1.6 Computational Intelligence in Active Networks and MobileActive Networks

Traditional networks have the drawback that routers and switches are vertically integrated closed systems whose functions are rigidly built into the embedded software by switch router vendors. Therefore, development and deployment of new protocols in such networks requires a long standardization process rather than pursues rapid introduction of innovative technologies. The range of services provided by the network is also limited as the network cannot anticipate and provide support for future applications. This leads to the fact that the pace of network evolution proceeds too slowly.

Active Networks (AN) (Tennenhouse et al., 1997) offers a different paradigm that enables users to inject customized programs into the models of the network, thereby tailoring the mode processing to be user—and application—specific and modifying the behavior of the network nodes. Active Networks enable a massive increase in the sophistication and customization of the computation that is performed within the network. Therefore they will accelerate the pace of innovation by decoupling network services from the underlying hardware and allowing new services to be loaded into the infrastructure on demand. Contrary to traditional circuit and packet switching nodes, in which a fixed protocol stack is hard coded, Active Network switches allow dynamic protocol composition.

Today's packets are static and subject to a single processing paradigm: routing toward their destination is based on information in the packet header. In Active Networks data transfer is based on smart packets in addition to data and addressing. A smart packet carries "How" information, or "Method"—a set of instructions that can be interpreted consistently by Active Networks modes (as illustrated in Figure 1.4, adapted from Orman, 1997). The network elements become active elements: they apply the methods to the packets in an execution environment, thereby implementing network-based services tailored to the application. The method library opens up the traditional networking system to customized resource allocation, protocol process, and information processing by a range of client control entities.

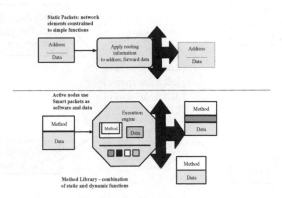

FIGURE 1.4
Smart packet.

1.6.1 MobileActive Networks

In Mobile and wireless networks the combination of large unpredictable variations in the channel, heterogeneous traffic, and changing topology due to mobile nodes poses a significant challenge to the design of a system that can be optimized for the current conditions. To meet the challenge we propose that MobileActive networks (Vasilakos, 2000) that draw on the strength of software radios (Bose, 1999), active networks (Tennenhouse et al., 1997) and intelligent agents (Jennings et al., 1998), a fully programmable software radio can incorporate adaptive link layer techniques and extend the adaptability to the physical layer.

This enables more effective use of the spectrum by dynamically adapting the physical layer of the networks to best meet the current environmental conditions, network traffic conditions and application requirements and provides freedom from worst case design. Hence the physical characteristics of the network can be redesigned as environmental conditions and user requirements change (Vasilakos, 2000).

1.6.2 Enabling Technologies

One of the enabling technologies making active networking a reality will be the intelligent agents (Jennings et al., 1998—Magendanz et al., 1996). Computational Intelligence is an integral part of agent technology as shown in Figure 1.5.

Existing management systems are expensive to maintain and difficult to change, so a major challenge is to devise a management platform using intelligent agent technology that can adapt to the evolving nature of future active networks (Premkumar and Venkataram, 1997).

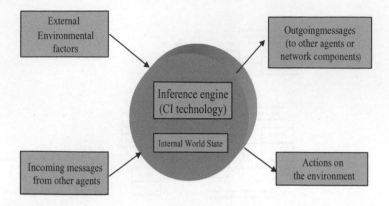

FIGURE 1.5
Intelligent agent interactions with external environment.

Active networking can be applied to the scenario of *IP and ATM integration* (Guarene et al., 1998). One possibility is to embed in each mode of the network an active intelligent agent which can perform different actions (i.e., flow classification) according to the type of the user application and the length of the messages to be sent. Based on the interactions with other agents and some local information, the agent will make decisions on appropriate adaptation strategies (topology-driven, traffic-driven).

Intelligent Agents can be employed in Active Networks to solve the *network management problem* by making managed nodes programmable and implementing intelligent filtering. Also, they can make the network react to *congestion* more promptly by moving the endpoint congestion control algorithms into the network, thereby reducing greatly the duration of each congestion episode.

1.7 Concluding Remarks

CI arises as the comprehensive development environment for intelligent systems. The rationale behind the inception of CI is also fully legitimate. Sooner or later we will be faced with the limitations of neural networks, fuzzy sets, or evolutionary computing when being used as stand-alone approach. To alleviate this deficiency, CI attempts to put the contributing technologies in a certain perspective and develop a sound and coherent methodology as well as come up with a consistent and well-developed algorithmic framework.

The research on applications of CI in telecommunication systems, particularly in ATM networks and cellular mobile networks, has been pursued vigorously for some time resulting in a host of algorithms. However, unlike consumer applications, there are no commercially deployed applications as yet. There could be a number of reasons behind this such as:

- The lack of comprehensive performance comparisons between the best traditional techniques and the ones involving CI applications. The comparisons performed in the research studies usually have been undertaken in simplified networking scenarios, and testing on real hardware has not been undertaken yet except for some partial implementations such as in Ascia et al. (1997). Before the applications of CI techniques to high speed communication networks becomes accepted, it will be necessary to place a greater emphasis on rigorously demonstrating the advantages to be gained, and that is an area we strongly recommend. A common simulative framework (CSF) and a common testbed framework (CTF) may be necessary.

- The reluctance to adopt new technologies by telecommunications companies and equipment manufacturers. This issue is closely related to the lack of comprehensive performance studies mentioned above.

- Inability of CI proponents to socialize and demonstrate, in realistic systems, merits of CI approach and its sound theoretical foundations.

As a final note, we would strongly encourage a thorough study of an integrated control structure implementing a multilevel control strategy spanning network and call and cell levels. The integration can be achieved by appropriate design of each individual strategy in a new multilevel fuzzy logic structure, and/or the integration of existing, or separately designed strategies, with their integration achieved via a fuzzy logic-based supervisor taking care of the overall "goodness" of the network and handling any interactions among the control functions, at the same or different levels.

Active networks and its enabling technology, Intelligent Agents, offer a fertile area for CI applications. Active networking replaces the model of vertically integrated and closed network nodes with open and programmable network nodes, whose primitive functions can be dynamically composed by applications. This capability will provide a foundation for supporting multiple programmable virtual networks and virtual services for different users and services classes (Van der Merwe and Leslie, 1997).

By casting these areas within the framework of CI, we believe that if we inject enough intelligence into telecommunication networking then they will become self-sustaining, self-organizing, self-managing intelligent communities in the future, offering context (or perception)-aware services to users in a ubiquitous computing and networking environment.

Acknowledgment

Support from the Natural Sciences and Engineering Research Council (NSERC) is gratefully appreciated.

References

E. Aboelela, and C. Douligeris, (1997) "Routing in multimetric networks using a fuzzy link cost," in Proceedings of the International Symposium on Computers & Communications ISCC'97, Monash University, Melbourne, Australia, 1163–1168, April.

G. Ascia, V. Catania, V. Ficili, S. Palazzo, and D. Panno, (1997) "A VLSI fuzzy expert system for real-time control in ATM networks," *IEEE Trans. Fuzzy Syst.*, 5, 20–31, February.

A.F. Atlasis, and A.V. Vasilakos, (1998) "LB-SELA: rated-based access control for ATM networks," *Comput. Networks and ISDN Syst.*, 30, 963–980.

A.F. Atlasis, M.P. Saltouros, and A.V. Vasilakos, (1998) "On the use of a stochastic estimator learning algorithm to the ATM routing problem: a methodology," *Comput. Commun.*, 21, 538–546.

J.C. Bezdek, (1992) "On the relationship between neural networks, pattern recognition and intelligence," *Int. J. Approximate Reasoning*, 6, 85–107.

J.C. Bezdek, (1994) "What is computational intelligence?" in *Computational Intelligence Imitating Life*, Zurada, J. M., R. J. M. II, and C. J. Robinson, Eds., IEEE Press, New York, 1–12.

V. Bose, (1999) "Design and implementation of software radios using a general purpose processor," Ph.D. thesis, Massachusetts Institute of Technology.

F. Box, (1978) "A heuristic technique for assigning frequencies to mobile radio nets," *IEEE trans. Vehicular Technol.*, 27, (2), 57–64.

J.Y. Buckley, and Y. Hayashi, (1994) Fuzzy neural networks: a survey, *Fuzzy Sets Syst.*, 66, 1–14.

D.J. Castelino, S. Hurley, and N. Stephens, (1996) "A tabu search algorithm for frequency assignment," *Ann. Oper. Res.*, 63, 301–319.

V. Catania, G. Ficili, S. Palazzo, and D. Panno, (1995) "A fuzzy expert system for usage parameter control in ATM networks," in *ProcGlobecom95*, 1338–1342.

V. Catania, G. Ficili, S. Palazzo, and D. Panno, (1996a) "A comparative analysis of fuzzy versus conventional policing mechanisms for ATM networks," *IEEE/ACM Trans. Networking*, 4, 449–459, June.

V. Catania, G. Ficili, S. Palazzo, and D. Panno, (1996b) "Using fuzzy logic in ATM source traffic control: Lessons and perspectives," *IEEE Commun. Mag.*, 70–81, November.

P.T.H. Chan, M. Palaniswami, D. Everitt, (1994) "NN-based dynamic channel assignment for cellular communication systems," *IEEE Trans. Vehicular Technol.*, 43, 279–288, May.

C. Chang and R. Cheng, (1994) "Traffic control in an ATM network using fuzzy set theory," in Proceedings of the IEEE INFOCOM '94 Conference, IEEE Commun. Soc., Toronto, Canada, 1200–1207, June.

T.M. Chen and S.S. Liu, (1994) "Management and control functions in ATM switching systems," *IEEE Network*, 27–40, July.

R.G. Cheng, and C.J. Chang, (1996) "Design of a fuzzy controller for ATM networks," *IEEE/ACM Trans. Networking*, 4, 460–469, June.

P. Czezowski, (1998) "Computational intelligence in ATM call admission control," in Proceedings of the IEEE Ph.D., Students Conference (GRADCON '98), IEEE, Winnipeg, Canada.

C. Douligeris, and G. Develekos, (1997) "Neuro-fuzzy control in ATM networks," *IEEE Commun. Mag.*, 154–162, May.

C. Douligeris, and G. Develekos, (1995) "A fuzzy logic approach to congestion control in ATM networks," in *ProcIcc95* 36, 1969–1973.

M. Duque-Anton, D. Kunz, and B. Ruber, (1993) "Channel assignment for cellular radio using simulated annealing," *IEEE Trans. Vehicular Technol.*, 42, 14–21, February.

D. Everitt, and D. Manfield, (1989) "Performance analysis of cellular mobile communications systems with dynamic channel assignment," *IEEE Jour. Selected Areas Commun.*, 7, (8), 1172–1180, October.

Z. Fan, and A. Mehaoua, (1998) "Active Networking: a new paradigm for next generation networks?" 2nd IEEE/IFIP International Conference on Management of Multimedia Networks and Services (MMNS 98).

L.J. Fogel, A.J. Owens, and M.J. Walsh, (1966) *Artificial Intelligence through Simulated Evolution*, J. Wiley, Chichester, England.

D. Fogel, (1995) Review of "Computational intelligence: imitating life," *IEEE Trans. Neural Networks*, 6, 1562–1565.

N. Funabaki, and V. Takefuji, (1992) "A neural network parallel algorithm for channel assignment problems in cellular radio networks," *IEEE Trans. Vehicular Technol.*, 41, 430–436, November.

S. Ghosh, Q. Razouqi, H.J. Schumacher, and A. Celmins, (1992) "A survey of recent advances in fuzzy logic in telecommunications networks and new challenges," *IEEE Trans., Fuzzy Systems*, 6, 85–107.

E. Guarene, P. Fasano, and V. Vercellone, (1998) "IP and ATM integration perspectives," *IEEE Comm. Mag.*, 74–80.

I.W. Habib, (1996) "Applications of neurocomputing in traffic management of ATM networks," *Proc. of the IEEE*, 84, 1430–1441, October.

F. Herrera, M. Lozano, and J.L. Verdegay, (1995) The use of fuzzy connectives to design real-coded genetic algorithms, *Mathware and Soft Computing*, 1, 239–251.

E. Herrera and M. Lozano, (1996) Adaptation of genetic algorithm parameters based on fuzzy logic controllers, in *Genetic Algorithms and Soft Computing*, Herrera, F. and Verdegay, J. L., Eds., Physica-Verlag, Heidelberg, Germany, 95–124.

A. Hiramatsu, (1990) "ATM communications network control by neural networks," *IEEE Trans. Neural Networks*, 1, 122–130, March.

A. Hiramatsu, (1995) "Training techniques for neural network applications in ATM," *IEEE Commun. Mag.*, 33, 58, 63–67, October.

J.H. Holland, (1975) *Adaptation in Natural and Artificial Systems*, University of Michigan Press, Ann Arbor.

Q. Hu, D.W. Petr, and C. Braun, (1996) "Self-tuning fuzzy traffic rate control for ATM networks," in *ProcIcc96*, 424–428.

Y. Huang, and W. Yan, (1996) " Dynamic control of communication systems based on simple recurrent neural networks," in Proceedings of the IEEE 1996 National Aerospace and Electronics Conference, IEEE, Communications Society, 1, 254–258.

IEEE Communications Society, (1995) Proceedings of the IEEE Global Telecommunications Conference GLOBECOM'95, Singapore.

IEEE Communications Society, (1996) Proceedings of the 1996 IEEE International Conference on Communications ICC'96, Dallas, TX.

N. Jennings, K. Sycara, and M. Wooldridge, (1998) "A roadmap of agent research and development," *Autonomous Agents and Multi Agent Syst.*, 1, 275–306.

D. Jensen, (1994) "B-ISDN network management by a fuzzy logic controller," in Proceedings of the IEEE Global Telecommunications Conference GLOBE-COM'94, IEEE Communications Society, 799–804.

I. Kassotakis, M. Markaki, and A. Vasilakos, (2000) "A hybrid genetic approach for channel reuse in multiple access telecommunication networks," IEEE JSAC, Special Issue, *Computational Intelligence in High Speed Networks* (to appear early 2000).

M.J. Keller, and D.J. Hunt, (1985) "Incorporating fuzzy membership functions into the perceptron algorithm," *IEEE Trans. Pattern Anal. Machine Intelligence*, PAMI-7, 693–699.

B. Kosko, (1992) *Neural Networks and Fuzzy Systems: A Dynamical Systems Approach to Machine Intelligence*, Prentice-Hall, Englewood Cliffs, NJ.

D. Kunz, (1991) "Channel assignment for cellular radio using neural networks," *IEEE Trans. Vehicular Technol.*, 40, 188–193, February.

K. Lai and E.G. Coghill, (1996) "Channel assignment through evolutionary optimization," *IEEE Trans. Vehicular Technol.*, 45, 91–96, February.

S.C. Lee and E.T. Lee, (1975) "Fuzzy neural networks," *Math. Biosciences*, 23, 151–177.

Y.C. Liu and C. Douligeris, (1995) "Static vs. adaptive feedback congestion controller for ATM networks," in *ProcGlobecom95*.

Y.C. Liu and C. Douligeris, (1996) "Nested threshold cell discarding with dedicated buffers and fuzzy scheduling," in *ProcIcc96*, 429–433.

T. Magedanz, K. Rothermel, and S. Krause, (1996) "Intelligent agents: an emerging technology for next generation telecommunications," in *Proceedings of INFOCOM '96,* San Francisco, California, March 24–28.

R. Marks, (1993) "Intelligence: computational versus artificial," *IEEE Trans. Neural Networks*, 4, 737–739.

D. Marr, (1982) "Vision," W. H. Freeman, San Francisco.

H.R. Mehrvar and T. Le-Ngoc, (1995) "ANN approach for congestion control in packet switch OBP satellite," in *ProcIcc95*, 810–814.

M. Minsky and S. Pappert, (1969) "Perceptrons: An Introduction to Computational Geometry," MIT Press, Cambridge.

T.D. Ndousse, (1994) "Fuzzy neural control of voice cells in ATM networks," *IEEE J. Selected Areas Commun.,* 12, 1488–1494, December.

E. Nordström, J. Carlström, O. Gällmo, and L. Asplund, (1995) "Neural networks for adaptive traffic control in ATM networks," *IEEE Communi. Mag.*, 33, 43–49, October.

H. Orman, (1997) "Active Networks, Technical Report" DARPA ITO, http://www.darpa.mil/ito/research/anets.

OIEEE Communications Society, (1995) Proceedings of the 1995 IEEE International Conference on Communications ICC'95, Washington, USA.

Y.K. Park, and G. Lee, (1995a) "Intelligent congestion control in ATM networks," in Proceedings of the 5th IEEE Computer Society Workshop on Future Trends of Distributed Computing Systems, IEEE Communications Society, 369–375.

Y.K. Park and G. Lee, (1995b) "Applications of neural networks in high-speed communication networks," *IEEE Commun. Mag.*, 33, 68–74, October.

Z. Pawlak, (1991) "Rough Sets. Theoretical Aspects of Reasoning about Data," Kluwer Academic Publishers, Dordrecht.

W. Pedrycz, (1998) "Computational Intelligence An Introduction," CRC Press, Boca Raton, FL.

W. Pedrycz, (1991) "Neurocomputations in relational systems," *IEEE Trans. Pattern Anal. Machine Intelligence*, 13, 289–296.

W. Pedrycz, (1993) "Fuzzy neural networks and neurocomputations," *Fuzzy Sets Syst.*, 56, 1–28.

W. Pedrycz, (1995) "Fuzzy Sets Engineering," CRC Press, Boca Raton, FL.

W. Pedrycz, (1997) "Computational Intelligence: An Introduction," CRC Press, Boca Raton, FL.

W. Pedrycz and F. Gomide, (1998) "An Introduction to Fuzzy Sets," MIT Press, Cambridge.

W. Pedrycz and A. Vasilakos, (2000) "Linguistic models and linguistic modeling: ATM traffic modeling," *IEEE, Trans. Syst., Man and Cybernetics*, February. (to appear)

W. Pedrycz, A. Vasilakos, and A. Gacek, (2000) "Information granulation for concept formation," proc. 15th ATM Symp. Appl. Compt., ATM SAC 2000, Como, Italy, March 19–21.

A. Pitsillides, A. Pattichis, Y.A. Sekercioglou, and A. Vasilakos, (1997) "Bandwidth allocation for virtual paths using evolutionary programming (EP-BAVP)," in Proceedings of the International Conference on Telecommunications ICT'97, Monash University, Melbourne, Australia, 1163–1168, April.

A. Pitsillides, (1993) "Control Strictures and Techniques for Broadband—ISDN Communication Systems," Ph.D. thesis, School of Electrical Engineering, Swinburne University of Technology, Melbourne, Australia.

A. Pitsillides, A. Sekercioglou, and G. Ramamurthy, (1995) "Fuzzy backward congestion notification (FBCN) congestion control in Asynchronous Transfer Mode (ATM)," in *ProcGlobecom95*, 280–285.

A. Pitsillides, A. Sekercioglou, and G. Ramamurthy, (1997) "Effective control of traffic flow in ATM networks using fuzzy logic based explicit rate marking (FERM)," *IEEE J. on Selected Areas Commun.*, 15, 209–225, February.

G. Premkumar and P. Venkataram, (1997) "Artificial Intelligence approaches to network management: recent advances and a survey," *Comput. Commun.*, 20, (4), 1313–1312.

M.F. Ramalho and E. Scharf, (1994) "Fuzzy logic based techniques for connection admission control in ATM networks," in Proc. 11th IEE Teletraffic Symposium, IEE, 12A/1–12A/8, March.

F. Rosenblatt, (1961) "Principles of Neurodynamics: Perceptrons and the Theory of Brain Mechanisms," Spartan Press, Washington.

D.W. Rumelhart and J.J. McClelland (eds.), (1986) "Parallel Distributed Processing," MIT Press, Cambridge.

H. Sandalidis, P. Stavroulakis, and J. Rodriguez, (1998) "An Efficient Evolutionary Algorithm for Channel Resource Management in Cellular Mobile Systems," *IEEE Trans. Evolutionary Computation*, 2, (4), November.

M.F. Scheffer and I.S. Shaw, (1993) "VLSI hardware realization of self-learning recursive fuzzy model for dynamic systems," in Proceedings of IFAC 12th World Congress, Sydney, Australia.

M.F. Scheffer and I.S. Shaw, (1994) "Application using VLSI hardware realization of self-learning recursive fuzzy model," in *Proceedings of SICICA '94*, Budapest, Hungary.

M.F. Scheffer, J.J.P. Beneke, and J.S. Kunicki, (1994) "Fuzzy modeling and prediction of network traffic fluctuations," in *Proceedings of COSIG '94* South Africa.

M.F. Scheffer and J.S. Kunicki, (1996) "Fuzzy adaptive traffic enforcement for ATM networks," in Proceedings of the 4th IEEE AFRICON Conference, IEEE, 1047–1050.

A. Sekercioglou, A. Pitsillides, and A. Vasilakos, (1999) "Computation intelligence in management of ATM Net: a survey of current state of research," SPIE International Conference on Science and Applications of Computational Intelligence, Denver, CO, July 14–20.

X. Shen and J.W. Mark, (1999) "Mobility information for resource management in wireless ATM networks, Computer Networks and ISDN Systems," *Computer Networks*, 31, 1049–1062.

J.A. Smith and M. Fry, (1995) "Artificial intelligence in network management," *Australian J. Intelligent Inf. Process. Syst.* Autumn, 53–62.

K. Smith and M. Palaniswami, (1997) "Static and dynamic channel assignment using neural networks," *IEEE J. Selected Areas Commun.*, 15(2), 238–249, February.

A.A. Tarraf and I.W. Habib, (1994) "A novel neural network traffic enforcement mechanism for ATM networks," *IEEE J. Selected Areas Commun.*, 12, 1088–1096, August.

A.A. Tarraf, I.W. Habib, and T.N. Saadawi, (1995a) "Intelligent traffic control for ATM broadband networks," *IEEE Commun. Mag.*, 33, 76–82, October.

A.A. Tarraf, I.W. Habib, and T.N. Saadawi, (1995b) "Reinforcement learning-based neural network congestion controller," in Proceedings of the Military Communications Conference MILCOM '95, IEEE, 2, 668–672.

A.A. Tarraf, I.W. Habib, and T.N. Saadawi, (1995c) "Congestion control mechanism for ATM networks," in *ProcIcc95*, 206–210.

D. Tennenhouse, J. Smith, W. Sincoskie, D. Wetheral, and J. Mindem, (1997) "A survey of active network research," *IEEE Commun. Mag.*, 35(1), 80–86, http://www.tns.lcs.mit.edu/

D. Tennenhouse and D. Wetherall, (1996) "Towards an active network architecture," SPIE Conference of Multimedia Computing and Networking, 2–16.

"Traffic management specification version 4.0," (1996) Technical Report AF-TM-0056.000, April.

K. Uehara and K. Hirota, (1997) "Fuzzy connection admission control for ATM networks based on possibility distribution of cell loss ratio," *IEEE J. Selected Areas Commun.*, 15, 179–190, February.

A. Vasilakos, A.F. Atlasis, and N.H. Loukas, (1995) "The Use of Learning Algorithms in ATM Networks Call Admission Control Problem: A Methodology," *ICCC '95*.

A. Vasilakos, K. Anagnostakis, and A. Pitsillides, (1997) "An evolutionary fuzzy algorithm for QoS and policy based interdomain routing in heterogeneous ATM and SDH/SONET networks," in Proceedings of the 5th European Congress on Intelligent Techniques and Soft Computing EUFIT'97, Aachen, Germany.

A. Vasilakos, C. Ricudis, K. Anagnostakis, W. Pedrycz, and A. Pitsillides, (1998) "Evolutionary-fuzzy prediction for strategic QoS routing in broadband networks," in Proceedings of the IEEE World Congress on Computational Intelligence WCCCI'98, 1488–1493, Anchorage, Alaska.

A. Vasilakos, (2000) "A roadmap of fuzzy logic in ATM, mobile and active networks: challenges of the new century," *Soft Computing* (to appear).

J. Van der Merwe and I. Leslie, (1997) "Switchlets and dynamic virtual ATM networks," *Integrated Network Management V*, 355–36.

J. Von Neumann, (1958) *The Computer and the Brain*, Yale University Press, New Haven, 66–82.

S. Yoneda, (1990) "Broadband—ISDN ATM layer management: operations, administration, and maintenance considerations," *IEEE Network*, 31–35, May.

L.A. Zadeh, (1973) "Outline of a new approach to the analysis of complex systems and decision processes," *IEEE Trans. Syst., Man, and Cybernetics*, 2, 28–44.

2

Neural Network Methods for Call Admission Control

Richard G. Ogier and Nina Taft-Plotkin

CONTENTS

KEY WORDS: *asymmetric error function, call admission control, call vector compression, equivalent bandwidth, high-speed networks, learning the average,*

0-8493-1075-X/01/$0.00+$.50
© 2001 by CRC Press LLC

modular neural networks, safe-side control, online training, offline training, pattern table, p-percentile delay, power spectral density, quality of service, relative target method, traffic descriptor compression, variance of counts, virtual output buffer.

2.1　Introduction

In this chapter we present the call admission control (CAC) problem and various ways of using neural networks as the key component of a CAC algorithm. Neural networks are essentially used to predict quality of service (QoS), which is the critical information on which CAC decisions are based. The need for CAC algorithms arose in ATM* networks, which were the first to pioneer the idea of providing precise QoS guarantees in a high-speed network. Most networks today, such as IP** networks, cable networks, and others, are also working on infrastructures and mechanisms by which to provide QoS. Thus the need to support CAC for a variety of scenarios has arisen.

After presenting the CAC problem in Section 2.2, we discuss why neural networks (NNs) are a good approach for the CAC problem (Section 2.3). In Section 2.4 we explain how NNs can be designed for CAC by capturing network behaviors (such as buffering schemes and multiplexing) through choices for NN outputs and modular NN units. A detailed discussion of various options for NN inputs and outputs that have been studied or proposed in the literature is presented in Section 2.5. We also present a method for reducing the number of NN inputs through linear compression. Since there can potentially be a large number of such inputs, this compression greatly speeds up the time for a NN to make a decision.

For offline training (Section 2.6) we briefly discuss how to generate training data and how to modify the error to reflect the asymmetric objective of CAC (discussed below). Modern networks anticipate experiencing unpredictable traffic conditions with some regularity. Thus a critical component of a CAC algorithm is its ability to adapt, and so we devote most of the discussion on training to online training (Section 2.7). Towards this end, we present methods for selecting training patterns, for coping with the statistical variability of measurements, for reducing the size of the training set to speed up learning, for coping with rare events, and for gradually phasing out outdated patterns.

2.2　Problem Statement

The call admission control problem is one of deciding whether or not to admit a particular call into the network. A call is admitted if the network

---

* Asynchronous Transfer Mode.
** Internet Protocol.

has sufficient resources to guarantee both the QoS the user requests and the QoS already promised to existing calls. The most common resources to be allocated are bandwidth of transmission links and buffer space in switches or routers. A user requesting a network connection typically supplies two sets of parameters: (1) traffic descriptors and (2) QoS requirements. The traffic descriptors can be any set of parameters that describe the statistical behavior of the traffic the user intends to submit to the network, and that are useful to the network in making the CAC decision. In many cases, applications do not know how to specify their traffic well. Alternatively, the network itself can measure traffic behavior and provide the descriptors (thus it is not always mandatory for users to supply traffic descriptors). Examples of these two types of parameters, which constitute the inputs to a CAC algorithm, are listed below. Note that the traffic descriptors supplied by the network are typically (although not necessarily) different from those provided by the user.

- Sample traffic descriptors:
    - User supplied: peak rate, average rate, maximum burst size, minimum rate, correlation measures, application type (e.g., MPEG video).
    - Network supplied (measurements): counts of cell or packet arrivals, power spectral density parameters, correlation measures, entropy.
- Sample QoS requirements: packet or cell loss rate, maximum or average delay, delay variation.

The CAC problem involves the prediction of the QoS that would be received by the new call and the existing calls, if the new call were admitted. A CAC algorithm uses the above parameters and any other information regarding the existing calls and/or state of the resources to make its decision. The existing calls constitute the aggregate stream, which is the superposition of traffic from the multiple calls. The essence of the CAC problem is to predict whether or not a potential new aggregate stream (composed of the current aggregate stream plus the new call request) will be feasible, where an aggregate stream is defined to be *feasible* if the network can satisfy the QoS constraints for all the calls that compose the stream, and *infeasible* otherwise. The multidimensional space of all possible aggregate streams can be divided into two regions, the feasible region and the infeasible region. The feasible (infeasible) region contains the set of all feasible (infeasible) streams, respectively. An alternate way to state the CAC problem is that the CAC algorithm must learn the decision boundary between the feasible and infeasible regions. The objective of CAC is asymmetric because accepting an infeasible stream is considered a worse mistake than rejecting a feasible one.

The general problem of admitting calls into the network is composed of two parts. First the network needs to find a candidate path for a call—this is usually called the *routing* problem. Calls are rejected if no such path can be found. Second, after a candidate path has been found, each node along the path must decide individually if it can accept the call. The call will be accepted if each of those nodes decides that there are sufficient available resources to guarantee the requested QoS of both the new call and the existing calls. The decision made at these nodes is sometimes referred to as the nodal CAC problem, and is the one we focus on in this chapter.

We now describe the main variations of the CAC problem. Due to the wide variety of traffic characteristics and QoS needs of multimedia traffic, it is useful to classify connections. It is most common to classify them using both traffic descriptors and QoS requirements. For example, calls with the same peak and mean rates with a cell* loss rate (CLR) requirement of 10^{-3} would fall into the same traffic class. However, two calls with the same peak and mean rates but with two different CLR requirements would fall into two different classes. Similarly two calls with the same CLR requirement but different peak rates would fall into two separate traffic classes.

We refer to the problem in which all calls have the same traffic and QoS requirements as the *single-traffic-class CAC problem*. The *multiple-traffic-class CAC problem* refers to the scenario in which different traffic classes coexist. It is difficult to build an efficient admission control system, since the CAC mechanism cannot be tailored to each specific application. The decision boundary between the feasible and infeasible regions is depicted in Figure 2.1 for the case of two traffic classes.

Different networks provide different types (or classes) of services. A service class is defined according to the set of QoS parameters for which the service is willing to make guarantees, and the values of those parameters. Clearly, a telephone service that provides guarantees for delay only, and a video service that provide guarantees for delay, loss, and jitter, are two separate services.

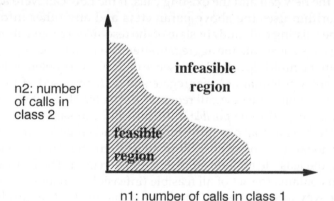

FIGURE 2.1
The decision boundary between feasible and infeasible regions.

---

* A cell is a fixed-length packet. In ATM networks cells are 53 bytes in length.

In addition, two services that guarantee two different levels of delay are also considered distinct services. The CAC problem that allows only a single-service class is called *the single-service-class CAC problem*; similarly the CAC problem that can handle multiple service classes is called *the multi-service CAC problem*.

A single-service class will typically carry multiple-traffic classes. This can occur even if the traffic classes have different QoS requirements, in which case the service guarantees the most stringent QoS among all traffic classes.

The design of the switch or router can affect the nature of the CAC problem. The CAC decision is typically made on a per-link basis, since a single link can carry many calls. In the *single-buffer-per-link CAC problem*, the transmission link is preceded by a single buffer. Since all calls share the same buffer, the CAC needs to predict QoS only for a single buffer. In some networks a link can be partitioned into multiple virtual paths (VP). In such cases CAC can be applied on a per VP basis.

In the *multiple-buffer-per-link CAC problem*, a transmission link is preceded by multiple buffers and a scheduler is used to select the next buffer from which the link transmits. Such switch/router designs may be used by a network to provide separate buffers either for the different classes or for the different services. In this case, the CAC needs to predict QoS for multiple buffers simultaneously.

The following properties are desirable for a good CAC algorithm:

1. *Robustness.* The more traffic classes and services the CAC algorithm can support, the more robust it is considered.

2. *Efficiency.* The CAC should take advantage of statistical multiplexing and minimize the amount of overallocation of bandwidth. Efficient bandwidth allocation will accept more calls, thus lowering the blocking rate and increasing the throughput.

3. *Adaptability.* The CAC should be able to adapt to the appearance of new traffic sources in the network, to sudden behavioral changes in aggregate traffic streams, and to changes in network characteristics.

4. *Minimal Inputs.* Keeping the number of inputs small lowers the burden of the caller to supply such information or lowers the efforts of the network needed for measurements. However the inputs should not be so few as to render it difficult for CAC to make efficient decisions.

5. *Practical.* A useful CAC algorithm must be implementable and able to make decisions fast. An example target decision speed is 1 ms/call so that the CAC can handle 1000 calls per second.

6. *Safe-Side Control.* A CAC algorithm should make "safe" decisions in that it should never accept an infeasible call. This approach may lead to the rejection of a few feasible calls.

Achieving both robustness and efficiency at the same time is a difficult task. In the following section we discuss why neural networks are an attractive approach for this task, and explain how they achieve the above properties.

2.3 Why Use Neural Networks for CAC?

Conventional methods for developing CAC algorithms are based on mathematical and/or simulation modeling. These methods require making assumptions about the traffic processes. QoS prediction is then done using queuing models to reflect the buffering and transmission behavior. Since this approach can quickly become analytically involved, simplifying assumptions need to be made. For example, it is common to assume that traffic sources are Markovian, or stationary, or that cell arrival patterns depend upon some parametric models like Markov Modulated Poisson Processes (MMPP). It has recently become known that high-speed network traffic is more complex, and that none of these assumptions is safe. Exact solutions based on analytical methods exist only for restricted traffic and system models. QoS estimation through analysis can also be inaccurate because declared and actual traffic parameters frequently differ.

Clearly results derived from analyses based on such assumptions have limited applicability and will generate inaccurate QoS estimates. To compensate, such CAC schemes force themselves to err on the side of being conservative and thus typically overallocate resources. This leads to inefficiency.

The easiest method for CAC is to accept a call if there is enough available bandwidth to allocate to the call its peak rate. This is the most inefficient method since it entirely ignores statistical multiplexing. The most well-known analytical result for CAC is the equivalent bandwidth method.[5, 6] This method provides a simple formula to compute the amount of bandwidth needed to meet a call's loss requirement, given its peak rate, mean rate, and average burst duration. The equivalent bandwidth yields a bandwidth that lies between the call's peak and mean rates. This method is exact only asymptotically, as the buffer size approaches infinity and the cell loss probability approaches zero. Although far superior to a CAC scheme that allocates peak bandwidth, the equivalent bandwidth method still overallocates bandwidth in most cases.

Neural networks are attractive for solving CAC problems because they are a class of approximators that are well suited for learning nonlinear functions. A neural network represents a multiple-input multiple-output nonlinear mapping. A NN can learn this mapping from a set of sample data. Feedforward neural networks can approximate any piecewise-continuous function with arbitrary accuracy, given enough hidden neurons.[7]

Some advantages of using neural networks for CAC are:

1. Neural networks do not require an accurate mathematical model of either the traffic or the system. No assumptions need to be made since the neural network is trained on observed data. Not assuming a specific traffic behavior *a priori* is a preferable approach because multimedia traffic is not well understood and continuously changing. NNs are also not affected by mistakes in declared traffic descriptors. These features allow a NN to yield more accurate QoS estimation, which leads to greater efficiency and robustness.

2. When NNs are properly trained, they can generalize and extrapolate additional details of the function mapping the inputs to outputs. If the training set is sufficiently large, a NN will generalize accurately and will produce accurate outputs for inputs not in the training set. This also contributes to robustness of the CAC scheme.

3. NNs are adaptable since they can be retrained in real time using the latest measurements.

2.4 How to Implement CAC with Neural Networks

Neural networks can be used to solve the CAC problem and all its variants. First we describe a model using a NN for the single-buffer-per-link case, with a FIFO (first-in first-out) buffer. The NN has n inputs $u_1, \dots, u_n$ and a single output $Y = f(u)$, where $u = (u_1, \dots, u_n)$ and $f(\,)$ denotes the input-output transfer function. The output can be a QoS estimate, which can be an estimate of the buffer delay, the buffer's loss rate, or the amount of bandwidth needed on the link to carry all the calls whose traffic enters the FIFO buffer. This version of this model, with a single output representing a particular QoS estimate, only supports a service that makes guarantees on one QoS parameter. For services that provide guarantees on m QoS parameters, a version of this model with m outputs, one per QoS parameter, could be used.

In another version of this model, a single output is used that takes on binary values that represent the accept or reject decision. In this version, particular QoS estimates are internal to the NN. Such a NN is called a classifier. With this choice for the output, the NN can be used to represent any definition of a feasible stream, including definitions that involve multiple QoS constraints (e.g., delay and CLR).

In the case of a single output representing a QoS estimate, a call is admitted to the network if the QoS estimate for the new candidate aggregate stream is below the most stringent QoS requirement (i.e., the decision threshold) for all calls in the stream. In the case of multiple outputs, each QoS estimate needs to be compared with the relevant decision threshold before a call is accepted. In the case of a single binary output (e.g., trained to learn 0 for accept and 1 for reject), the output is compared to a threshold such as 1/2. (We explain in Section 2.6 that it may be desirable to modify this threshold.)

The choice of output influences which CAC problem is being modeled. Consider the case in which the output is a queue-related parameter, such as delay or loss. When FIFO queuing is used, there is only a single loss rate or distribution of delay associated with that queue. The NN predicts the QoS of the aggregate stream of superposed traffic sources. When many sessions that have different QoS requirements are multiplexed together, the switch ensures the most stringent of all the delay and/or loss requirements. (Hence, some sessions will experience better QoS than they requested.) Thus this model can support multiple traffic classes, but they will all receive the same QoS. If a scheduling mechanism (e.g., earliest deadline first) is used to prioritize among traffic classes, then multiple loss rates (one for each class) could be computed for a single buffer. A NN used for this scenario would require multiple outputs, one for each class.

We now consider a second model for the multiple-buffers-per-link problem. In this case, there is at least one QoS estimate per buffer. If there are b buffers, and x QoS parameters per buffer, then this NN model could have bx outputs. Alternatively, a single binary output (representing an accept or reject decision for a given call) could still be used in this case, since all call types and services share a single transmission link.

A third way of modeling CAC problems is to use a modular design in which multiple NN units are organized in a hierarchical fashion. Modularity is a means to solve a complex computational task by decomposing it into simpler subtasks and then combining the individual solutions. This approach is attractive if the functional relationship between the neural network inputs and outputs is very complex, and if parts of this function can naturally be separated. Consequently, the modules of the network tend to specialize by learning different regions of the input space. The decomposition should be designed so as to facilitate this.

An example, based on the work in,[20] is given in Figure 2.2. There is a bank of NN units in level 1 of this model. Each NN unit is associated with a single traffic class, and its inputs correspond to descriptors for that class. Let's assume that one virtual path is assigned for each traffic class. The output gives the bandwidth estimate for all the calls in that class, i.e., for the VP assigned to that class. The NN unit at level 2 takes as inputs the bandwidth per class for each VP and outputs the link bandwidth needed to support the VPs. The attraction of this model is that it naturally separates out the two levels of statistical multiplexing that occur in a switch that supports VPs. The NNs in level 1 determine the amount of multiplexing gain for mixing calls from the same class onto a single VP, while the NN in level 2 determines the amount of multiplexing gain for mixing VPs onto a link. Another example of a modular NN design can be found in.[19]

Modular networks offer several advantages over a single neural network.[7] First, the training is faster, which allows the NN units to be more adaptable. Second, the representation of input data developed by a modular network

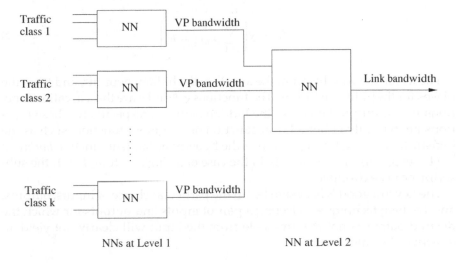

FIGURE 2.2
Sample modular NN design for CAC.

tends to be easier to understand than in the case of an ordinary multilayer NN. Third, this type of design should lead to more accurate estimates since the input-output mapping that each NN unit has to learn is simpler. It has been proven[1] that the number of input-output patterns that a NN can deterministically learn is equal to twice the number of its weights. Fourth, a useful feature of a modular approach is that it also provides a better fit to a discontinuous input-output mapping.[7]

2.5 Architecture

A common architecture for all the models described above is a 3-layer feedforward neural network, with a layer of inputs, a single hidden layer of neurons, and an output layer. We establish the following notation for the neural network elements.

For a given set of inputs $u_1, \ldots, u_I$, the kth output of the NN is given by

$$Y_k = g^{out}\left(\sum_{j=1}^{J} W_{jk} V_j + b_k^{out}\right) \tag{2.1}$$

where W_{jk} is the connection weight from the jth hidden neuron to the kth output, and b_k^{out} is the bias for the kth output. V_j, $j = 1, \ldots, J$, is the output of the jth hidden neuron, and is given by

$$V_j = g\left(\sum_{i=1}^{I} w_{ij}u_i + b_j \right) \tag{2.2}$$

The w_{ij} are the weights from the inputs to the hidden neurons, and b_j are the biases for the hidden neurons. The functions g^{out} and g are the activation functions for the output layer and the hidden neurons, respectively. These functions are typically either a linear function or a sigmoid function, such as the logistic function $1/(1 + \exp(-av))$ or the hyperbolic tangent function $\tanh(v/2) = (1 - \exp(-v))/(1 + \exp(-v))$. In the case of a single output, $k = 1$, the subscript k can be dropped.

The key to a good NN design lies in the particular choice of inputs, outputs, and training technique. Selecting a pair of inputs and outputs for which the desired output is not determinable from the input will clearly not yield an effective CAC algorithm.

2.5.1   NN Inputs

The NN inputs can be any input that helps the NN to predict the QoS of an aggregate traffic stream. The NN inputs typically include either traffic descriptors or system state parameters or some combination of both. Examples of traffic descriptors, either provided by the user or measured by the network, were given in Section 2.2. In this section, we now expand the discussion of these types of inputs. The advantage of having the users supply traffic descriptors is that the network need not spend any resources or time to measure the traffic. However the disadvantage comes from the fact that there is usually a difference between the declared traffic parameters and the actual traffic parameters since most applications today do not understand well the traffic they generate. When the number of connections in a network becomes large, the difference between the declared and actual traffic parameters can be quite large. The advantages and disadvantages of using measurements as NN inputs are exactly the reverse of the advantages and disadvantages of the user supplied approach.

Examples of system state parameters that can be used as inputs include the number of existing calls for each traffic class, the buffer loss rate, and the buffer level.

It is desirable for the neural network inputs to have the following properties:

1. *Capture key elements of traffic behavior that influence queuing.* Many researchers believe that traffic descriptors that capture the correlation and burstiness properties of a traffic stream will be successful for CAC. In order to avoid overallocating or underallocating resources, it is necessary to estimate QoS well, which in turn requires proper traffic and system characterization.

2. *Additive.* When the NN input vector is additive, the current input vector can be updated efficiently by simply adding the traffic descriptors of the new call to that of the aggregate call (the current input vector). This additive property of traffic descriptors greatly speeds up the decision process of accepting or rejecting a call.

3. *Support a large number of traffic classes.* This will make the algorithm robust.

4. *Keep the number of inputs reasonably small.* This will make the algorithm practical since the forward calculation speed is proportional to the number of weights.

We now give examples of NN inputs that have been proposed in recent research efforts.

2.5.1.1 Number of Calls per Traffic Class

In this case, the NN input is a vector $s = (s_k)$ whose kth component gives the number of calls in the stream that belong to the kth traffic class. This input has been used by Hiramatsu[10, 11] and Tham and Soh.[19] In Reference 15, they use this input coupled with the link load level. We refer to this particular input as the *call vector* in subsequent sections of this chapter. The advantage of this approach is that the user need not supply any traffic descriptors at all, and the decision boundary is determinable from this input. The disadvantage of this approach is that it does not scale well in the number of traffic classes. There could be a very large number of traffic classes in any network. It has been suggested[11, 19] that a practical number of classes is less than 100.

2.5.1.2 Counts of Arrivals

In Reference 20 they used online traffic measurements for the NN inputs. In each interval q, the number of cell arrivals, $N_S(q)$, is counted for each stream S. If one keeps track of the arrival process over consecutive intervals, and also provides this data to the NN, then the NN can be trained to capture the correlations that exist among cell arrivals. This approach requires a careful choice of the measurement interval. The advantages of this input are that this information entirely characterizes the input stream and that the user need not supply any traffic descriptors. The disadvantage of this choice for NN inputs is that it does not scale in the number of calls supported simultaneously. This choice of inputs is not considered very practical since the measurement intervals typically need to be very large.

2.5.1.3 Variance of Counts

In Reference 16, we used the *variance of counts* (VOCs) as NN inputs. To calculate or measure VOCs, time is divided into intervals of equal length. As above, $N_S(q)$ denotes the number of cells or packets arriving in interval q for stream S. Let $N_S(q, h)$ denote the number of cells arriving in the interval consisting of intervals q through h. Let λ_S denote the mean of $N_S(q)$. The variance of counts for an interval of length m is defined by

$$VOC_S(m) = \frac{\text{var}\{N_S(q + 1, q + m)\}}{m},\qquad(2.3)$$

where q is an arbitrary time slot. For a given stream S, the NN inputs are scaled versions of λ_S and $VOC_S(m)$ for $m = 1, 2, 4, \ldots, 2^M$, where $M + 1$ is the number of VOCs used. To limit the number of NN inputs while considering a representative set of VOCs, we used VOCs over intervals of exponentially increasing length.

The VOC traffic descriptor is an unnormalized IDC as described in Reference 8, that is, $VOC_S(m) = IDC_S(m)l_S$. Unnormalized IDCs are preferable because then the VOCs are additive, i.e., if S is the sum of statistically independent streams S_i, then $VOC_S(m)$ is the sum of $VOC_{Si}(m)$ over the S_i.

The advantages of these NN inputs include: 1) VOCs characterize all second-order statistics of the stream, 2) they are additive, 3) moments of interval counts have been shown to accurately predict queuing delay for some models,[8] and 4) this method is independent of the number of traffic classes.

It is known[8] that $IDC_S(m)$ converges to $\text{var}(X) / E^2(X)$, the squared coefficient of variation of the interarrival time X, and so $VOC_S(m)$ also converges to a constant if the interarrival time has finite variance. One can thus choose the number $M + 1$ of VOCs to be large enough so that $VOC_S(2^M)$ is close to the limit for most streams. This limits the number of NN inputs to $M + 2$, which is typically much smaller than the number of traffic classes as used in Section 2.5.1.1.

These NN inputs can be measured by either the user or the network. They could be calculated by the user when the user's traffic is a bursty on-off process. VOCs can be precomputed for a set of traffic classes and stored in a table, in which case a user only needs to indicate a traffic class in the call request.

2.5.1.4   *Power Spectral Density Parameters*

All of the inputs discussed above focus on the time domain. Another type of input that can capture correlation and burstiness properties of traffic streams is the power-spectral-density (PSD) function in the frequency domain. The PSD is the Fourier transform of the autocorrelation function of the input process. Using the PSD to derive the NN inputs for CAC was first proposed in Reference 2. In this method, a MMPP traffic source is characterized by three power spectrum parameters: the DC component (u), the half-power bandwidth (v), and the average power (w). As u increases, the traffic load increases; as v decreases, the input power in the low frequency band increases; and as w increases, the variance of the input rate increases.

There are two advantages to this approach. First, it has been shown in Reference 12 that the low frequency band of the input PSD has a dominant impact on the queuing performance while the high frequency band can usually be neglected. This is because the low frequency component of the PSD contains the correlation component. The larger the low frequency component, the burstier the traffic source. Second, since the PSD has the additive property, so do these three parameters.

In Reference 2, the CAC controller is designed so that the user can input three simple parameters: its peak rate, mean rate, and peak cell rate duration. The controller applies the FFT* to these inputs and outputs the three PSD parameters (u, v, w), which are in turn fed as inputs to a NN.

2.5.1.5 Entropy

Entropy has been proposed as a traffic descriptor in References 4 and 17. The entropy of traffic streams is attractive as a descriptor because it can capture the behavior of correlations over many time scales. The entropy has been used as an input for CAC in References 4 and 17 but it has not yet been tried in a CAC algorithm based on neural networks.

2.5.2 NN Outputs

A NN output variable can represent any of the following:

1. *Accept/Reject Decision.* With this output, the NN has to learn the boundary between the feasible and infeasible performance regions for a given input space.

2. *Loss Rate.* In this case, the NN predicts the average buffer overflow rate. Since the loss rate can have an exponentially wide range, from 10^{-1} to 10^{-9}, it is common to use log(*loss*) instead.

3. *Delay.* In this case, the NN predicts the average buffer delay, or average buffer occupancy.

4. *Jitter.* In this case, the NN predicts the variation of the buffer delay.

5. *Bandwidth.* In this case, the NN predicts the amount of bandwidth needed to achieve a specific QoS level for the given input stream.

6. *p-th percentile delay.* In this case, the NN predicts the value D such that the probability that a cell or packet experiences a delay less than or equal to D is $p\%$. The percentile is typically chosen to be around 90%. In this approach, all the calls need to have the same percentile requirement.

7. *Probability distribution of delay.* In this case, the NN output represents the probability that the delay experienced will be less than the requested delay (which can be one of the NN inputs), conditioned on the system state (represented by the other NN inputs). This method, proposed in Reference 18, works well when the probability is con-

---

* Fast Fourier Transform.

ditioned on the number of active connections. This method allows calls to have different delay requirements.

2.5.3 Compression of Neural Network Inputs

In Section 2.5.1, we discussed the mapping of the call vector to a vector of parameters related to second order statistics of the aggregate traffic process (based on VOCs or the PSD). Such a mapping can be considered a compression of the call vector (whose dimension is the number of traffic classes, which can be in the hundreds) to a smaller vector whose dimension is independent of the number of traffic classes. One benefit of reducing the number of NN inputs is that the NN output can be computed in less time, thus allowing more calls to be processed per second. The best compression is one that maps the call vector to the fewest parameters without reducing performance significantly.

In this section, we present a method for finding a linear compression of the call vector to I parameters (where I is any positive integer) that is optimal in the sense of minimizing the output error function. Linear compression is chosen so that the compressed parameters are additive. A similar method was presented in Reference 16 to compress vectors of VOCs, where it was shown that compression to three parameters did not result in a significant reduction of performance. (These results are also summarized in Section 2.5.5.) An advantage of compressing call vectors instead of VOCs is that the statistics of calls need not be known or measured. Although we focus on the call vector in this section, the method can be applied to any choice of NN inputs that are additive.

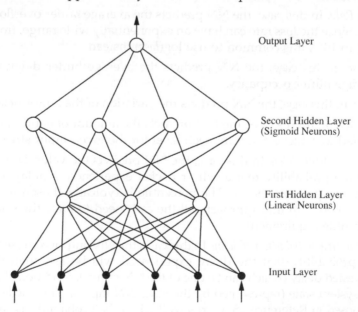

FIGURE 2.3
Neural network with additional hidden layer for compressing inputs.

The compression method involves adding a new hidden layer to the NN, between the input layer and the original hidden layer, as shown in Figure 2.3. The neurons in the new hidden layer have linear activation functions, and there is one such neuron for each compressed parameter. For this design, Equation 2.2 is replaced by the following two equations:

$$V_j = g\left(\sum_{i=1}^{I} w_{ij} v_i + b_j\right).$$ (2.4)

$$v_i = \sum_{l=1}^{L} \alpha_{li} s_l.$$ (2.5)

The outputs v_i of the hidden linear neurons form the compressed vector corresponding to the call vector s, and the weights α_{li} from the lth input to the first hidden layer form the compressed vector corresponding to the lth traffic class. Equation 2.5 can be expressed in vector form as

$$v = \sum_{l=1}^{L} \alpha_l s_l,$$ (2.6)

where v is the compressed vector for call vector s and α_l is the compressed vector for a single call of traffic class l.

This method was motivated by the well-known technique of using a hidden layer of linear neurons for image or data compression.[7, 9, 13] Assuming that NN weights (including the α_{li}) are found that minimize the output error function, the compressed parameters v_i are optimal by definition. Once this NN is trained and thus the matrix $\{\alpha_{li}\}$ is obtained, the original input layer is no longer required. The first hidden layer becomes the new input layer; i.e., the compressed parameters v_i are now used as the NN inputs.

Since the compressed parameters are additive, when a new call of class l arrives, the compressed vector for the new call vector is updated simply by adding the compressed vector α_l for the new call. To perform this operation quickly, the compressed vector α_l corresponding to each traffic class l can be stored in a look-up table.

Consider the special case in which only a single compressed parameter is used. Thus, a single parameter α_l is computed for each traffic class l, and the compressed parameter for a given call vector is equal to the sum of α_l over all calls represented by the call vector. Moreover, it is possible to scale the α_l so that a given call vector falls into the call acceptance region (learned by the neural network) if and only if this sum is less than the link bandwidth. Therefore, compression to a single parameter corresponds to learning the equivalent bandwidth for each traffic class.

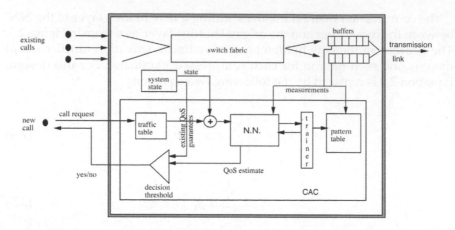

FIGURE 2.4
Design of CAC controller using a neural network.

2.5.4 Design of CAC Controller

Figure 2.4 depicts a method of incorporating a NN into the design of a CAC algorithm. This figure is general in that it illustrates potential CAC inputs, and not required inputs. The CAC inputs can come from either the user, the system state or measurements, or some combination thereof. A traffic table can be used to convert user supplied traffic descriptors into VOCs or PSDs, if needed. If both the system state and the user's traffic descriptors (or the mapped version of the user's descriptors) are used, we assume they have the additive property (indicated by the adder in Figure 2.4). The system state is updated each time new connections are established or torn down. The pattern table and trainer are present only when the system is designed for online training (discussed in Section 2.7).

2.5.5 Performance Evaluation for an Example NN CAC

One way to evaluate the performance of a CAC method is to measure how well the method approximates the feasible region. The *acceptance region* is defined as the set of aggregate streams that are accepted (classified as feasible) by a particular CAC method. We define a performance measure, called the "percent of feasibles accepted", to be the ratio of the size of the acceptance region to the size of the feasible region. To measure this ratio, we randomly generate a large number of streams, and then divide the number of streams that fall within the acceptance region by the number that fall within the feasible region. A natural way to randomly generate a stream is to select the number of calls according to a uniform distribution between 1 and the maximum number of calls allowed, and then select the parameters for each call independently, according to the probability distribution assumed for these parameters.

Note that the above measure does not depend on the call arrival model. One can also evaluate a CAC method by measuring the throughput (average number of packets successfully transmitted per second) or the number of calls accepted for a particular call arrival model.

Simulation experiments were conducted[16] to compare NNs using VOCs as inputs with a conventional CAC method based on equivalent bandwidth (EB).[6] Different degrees of input compression were also compared for the VOC NNs. In these experiments, multiple heterogeneous on-off traffic sources (calls) were fed into a single FIFO queue with a constant service rate. A stream was defined to be feasible if the CLR was less than 10^{-3}. The NNs were trained using the asymmetric error function described in the next section.

Table 2.1 shows the percent of feasibles accepted for the NN method with no compression (NN-0) and with 1–3 compressed parameters (NN-1 through NN-3), and for the EB method. The table shows that NN-3 performs about the same as NN-0 and accepts about 150% more feasibles than the EB method. This shows that three compressed parameters are sufficient for the NN model tested.

Figure 2.5 shows, for the NN-1, NN-3, and EB methods, the average throughput (normalized by the link capacity) as a function of the call arrival rate, for a buffer size of 2000 cells. (A Poisson call arrival process was used.) The NN method accepted approximately 80% more calls than EB (averaged over all arrival rates) while guaranteeing the CLR constraint. Smaller improvements were observed for larger buffer sizes, which is to be expected since the EB method approaches optimality as the buffer size tends toward infinity.

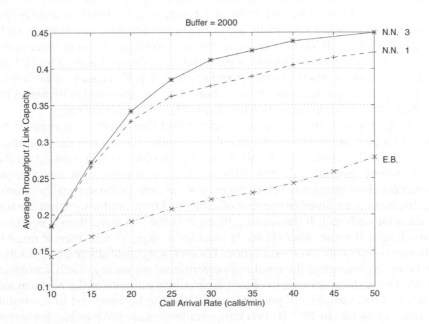

FIGURE 2.5
Comparison of throughput for NN and EB methods.

TABLE 2.1

Comparison of Percent Feasibles
Accepted for NN and EB Methods

Method	% Feasibles Accepted
NN-0	89.8
NN-3	90.2
NN-2	81.4
NN-1	79.6
EB	36.0

2.6 Offline Training

The feedforward NNs described above can be trained using standard back-propagation algorithms and their variations (e.g., Reference 7). In this section we briefly discuss methods for generating the training set (the set of input-output patterns) used for backpropagation and present a nonstandard error function that helps to achieve the asymmetric goal of the CAC problem. Methods for the more difficult problem of online training are presented in the next section.

When a NN is trained offline for CAC, the training set consists of a large number (typically 1000 or more) of input-output patterns (u^k, y_k) that are usually generated by simulating the traffic and queuing processes. The traffic can be modeled as an on-off Markov chain, a Markov-modulated Poisson process, or a self-similar traffic process. More complex traffic models or traces of traffic from actual networks can also be used. A fixed number of traffic classes can be defined by specifying the parameters for each class, or an infinite number of classes can be obtained by allowing any choice of parameters within some range.

Aggregate traffic streams must be generated that cover all regions of the space of input patterns u^k. If the number of traffic classes and the maximum number of calls per class are small, it may be possible to use a training set that includes every possible call state. Otherwise, one way to generate an aggregate stream is to first randomly select the number of calls in the stream (between 1 and the maximum possible number of calls), and then randomly select the traffic class for each call. If the input pattern u^k is the call state, then it is known immediately. If some other choice is used for u^k (e.g., VOCs), then u^k must be obtained using analysis or simulation. For each aggregate stream, a simulation can be run to determine the resulting performance measure y_k. Each simulation should be run long enough to obtain an accurate estimate of the performance measure. For example, 10^8 packets would need to be observed to accurately estimate a loss rate of 10^{-6}. If such long simulations are not feasible, the virtual output buffer method discussed in the next section can be used to improve the estimate through extrapolation.

Once the training set is generated, the NN can be trained using a version of backpropagation. A commonly used error function for training is

$$J = \sum_{k=1}^{K} |f(u^k) - y_k|^N,$$ (2.7)

where usually $N = 2$ (equivalent to the mean squared error).

Recall that the CAC objective is asymmetric: to accept as many feasible input patterns as possible while rejecting *all* infeasible input patterns. After the NN is trained, the decision threshold can be adjusted so that all infeasible input patterns in the training set are rejected. That is, assuming that a NN output greater than the decision threshold corresponds to a reject decision, the threshold is chosen so that it is slightly less than the smallest NN output for an infeasible input pattern in the training set.

In order to accept as many feasible streams as possible, the maximum error over all infeasible training patterns u should be minimized, so that the decision threshold can be chosen as large as possible without accepting any infeasible patterns. One way to achieve this objective is to use the following asymmetric error function:[16]

$$J = \sum_{\text{feasible } k} [f(u^k) - y_k]^2 + \sum_{\text{infeasible } k} |f(u^k) - y_k|^N$$ (2.8)

where $N > 2$. For large N, minimizing the second sum will tend to minimize the maximum error $|f(u^k) - y_k|$ over all infeasible patterns.

In simulation experiments,[16] significantly more feasible streams were correctly classified using the asymmetric error function with $N = 4$, than using the error function in Equation 2.7 with $N = 2$ or $N = 4$.

In Reference 16 we compared the performance of three different error functions with respect to the "percent feasibles accepted" measure defined in Section 2.5.5. These results are summarized in Table 2.2. The results show that, for the NN model tested, the asymmetric error function with $N = 4$ accepted about 7% more feasibles than the symmetric error function (7) with $N = 4$, and more than twice as many feasibles as the symmetric error function with $N = 2$.

TABLE 2.2

Comparison of Different Error Functions

Error Function	% Feasibles Accepted
Asym ($N = 4$)	89%
Sym ($N = 4$)	83%
Sym ($N = 2$)	42%

2.7   Online Training

In online training, the NN is trained continuously or frequently, based on actual measurements obtained while the NN is being used for CAC. Online training is useful if the CAC needs to adapt to changing network characteristics or new traffic classes, or to fine-tune an offline-trained NN using more accurate measurements.

Online training is more difficult than offline training in part because the call state changes frequently. Thus, the performance measurement (NN output) obtained for a given training pattern is based on fewer packets (or cells) and is therefore subject to more statistical variability (i.e., noise). In online training, unlike offline training, one cannot fix the call state and observe the performance of a large number of packets in order to obtain a good estimate of the resulting loss rate or average delay. This difficulty can be reduced by exploiting the ability of NNs to learn the *average* of several different measurements associated with the same call state, as explained below.

However, online training can be slow because a large number of packets must be observed before a small loss rate can be estimated with any accuracy. For example, more than a billion packets must be observed to accurately estimate a loss rate on the order of 10^{-7}. In this case, a packet loss can be considered a rare event. The online training can be made faster by using the ability of a neural network to extrapolate from estimates of measures that are based on more common events. For example, in the virtual output buffer method,[10, 11] discussed below, the NN learns the loss rate that would occur if the packets were fed into imaginary queues with smaller service rates than the actual queue, and extrapolates that knowledge to improve the estimate of the actual loss rate.

For a NN CAC to be adaptive, old measurements for a given input pattern must eventually be "forgotten" and replaced by new measurements for similar input patterns. There is a trade-off between adaptability and accuracy: if the NN is trained using past measurements made over a large time window, it can achieve good accuracy but will not adapt quickly to network changes. If a small time window is used (so that the NN quickly forgets past measurements), faster adaptation is achieved at the cost of less accuracy.

Another reason online training is more difficult than offline training is that input patterns that are marginally unacceptable occur rarely or never, assuming the CAC is performing well. If the NN remembers that these patterns are unacceptable, it may not be able to adapt to changing network conditions that cause these patterns to become acceptable. If the NN eventually "forgets" these patterns, it will start to accept infeasible call vectors and thus perform poorly for some period of time while it is relearning the decision boundary. Such behavior should be avoided, since it is more important to reject a given infeasible call vector than to accept a given feasible call vector, i.e., the NN should achieve "safe-side" control. One way to help achieve this goal is to start with a NN CAC that has been trained offline to perform a conservative version of CAC, such as one based on peak rate or equivalent bandwidth, and then use online training with a slow learning rate so that the CAC gradually learns to accept more calls. The virtual buffer method also helps to achieve safe-side control by learning more quickly that the call vector is approaching the decision boundary.

Other problems that will be addressed in this section include how to decide which patterns to store, given a bounded storage capacity, and summarizing different measurements for the same input pattern, in order to reduce the training set and thus reduce the time required to train the NN. In the following subsections, we assume that the NN input is the call vector and that the NN has a single output. However, the methods are applicable to other choices for the NN input and can easily be extended to multiple NN outputs.

2.7.1 Learning the Average

If a feedforward NN is trained using backpropagation with the mean-square error function, and the training set consists of several input-output patterns having the same input pattern but several different training outputs, the NN will learn the *average* of the training outputs. In reality, the training set will include several different input patterns, and the NN will perform nonlinear regression, so that the learned output for a given input pattern is also affected by training patterns whose input patterns are "close to" the given input pattern.

For example, a NN can learn a loss rate of 10^{-6} from 999 training patterns with an output of 0 and one pattern with an output of 10^{-3}, all patterns having the same input pattern. However, the learned average is the actual packet loss rate only if each pattern corresponds to the same number of packets. A common mistake is to measure the loss rate (or average delay) over intervals of equal length, and use the resulting input-output patterns to train the NN. Since the number of packets arriving in each such interval can vary greatly, this method does not give the average measure per packet, but gives the average measure per *interval*. It therefore does not give the correct packet loss rate (or average packet delay). For example, suppose that 10 packets arrive in one interval, none of which are lost, and 20 packets arrive in the next interval, 10 of which are lost. Then the correct loss rate is 1/3, whereas the above method would give 1/4 (the average of 0 and 1/2).

The correct average performance per packet (where the performance can be loss rate, average delay, or another measure) can be obtained by modifying the mean-square error function by multiplying each term by the number of packets used to arrive at the corresponding measurement. Thus we apply backpropagation with the following error function:

$$J = \sum_{k=1}^{K} m_k [f(s^k) - y_k]^2,$$ (2.9)

where K is the number of patterns, s^k and y_k are the call vector and output for the kth training pattern, m_k is the number of packets used to obtain the measurement y_k, and $f(s^k)$ is the NN output for input s^k. The measurements need not be over intervals of equal length. In the special case where $m_k = 1$ for all k, each measurement corresponds to a single packet, and the above function becomes the standard mean-square error function.

If the performance to be estimated is average delay or packet loss rate, the training output y_k is simply the average delay or packet loss rate for the kth measurement. To estimate p-percentile delay, y_k can be defined as follows:

$$y_k = f(s^k) + \frac{1 - \phi'_k}{1 - p} - \frac{\phi'_k}{p}, \tag{2.10}$$

where ϕ'_k is the fraction of the m_k observed packets whose delays were less than $f(s^k)$. This is an example of the relative target method presented in Reference 11, where a similar example is presented for the case in which each observation is for a single packet ($m_k = 1$). Since, for a given call vector s, $f(s)$ will converge to the average (weighted by m_k) of all training outputs y_k corresponding to state s, by taking the average of each side of the above equation over these training outputs, it follows that the corresponding averages of the last two terms on the right-hand side will be equal, implying that the fraction of all packets observed in state s having delay less than $f(s)$ will be p. That is, $f(s)$ will converge to the p-percentile delay for call vector s.

If the training set contains several measurements y_k for the same call vector s, these measurements can be summarized by a single measurement y^s, obtained by taking the weighted average of the individual measurements. This reduces the training set so that it contains at most one training pattern per call vector, thus allowing the training to be faster. If each summarized measurement is based on a sufficient number of packets to ensure good accuracy, the standard mean-square error function can be used. However, if the space of call vectors is very large, the number of packets observed for a given call vector is likely to be insufficient to ensure good accuracy. In this case, the error function should be of the form

$$J = \sum_{\text{observed states } s} m^s [f(s) - y^s]^2, \tag{2.11}$$

where m^s is the total number of packets used to obtain the summarized measurement y^s. Using this error function will improve the accuracy of the estimate through nonlinear regression (interpolation and extrapolation). Another method for summarizing multiple measurements for the same input pattern, where each measurement is for a single packet, is given in Reference 14.

As discussed previously, when the NN is used to estimate packet loss rate r, it is common to use $\log r$ instead of r as the target output. As pointed out by Hiramatsu,[11] if the usual backpropagation method is used, the NN will learn the average of $\log r$, not the log of the average packet loss rate. This can be corrected, so that the NN learns the log of the average loss rate, by replacing the error function (2.9) by the error function

$$J = \sum_{k=1}^{K} m_k [\exp(f(s^k)) - r_k]^2 \tag{2.12}$$

where r_k is the measured packet loss rate for the kth training pattern. As a result, $\exp(f(s^k))$ will converge to the average loss rate for input pattern s^k, so that the NN output $f(s^k)$ will converge to the log of this rate. This modification of the error function is equivalent to adding a neuron to the NN whose output is the exponential of the old NN output (which estimates $\log r$).

2.7.2 Procedures for Online Training

In online training, we cannot assume that the queuing system reaches equilibrium between changes in the call vector. Therefore, the neural network can only learn, for each possible state of the call vector, the average performance of packets that are admitted when the call vector is in that state. A training pattern will therefore consist of a pair (s^k, y_k) and a size m_k, where y_k is a measurement of the average performance of m_k packets that were admitted when the call vector was s^k. Since the call vector can change at arbitrary times, the intervals over which these measurements are performed need not have equal length. In an extreme case, each measurement can correspond to a single packet, as in the method of Reference 14.

The simplest method for online training is to perform one step of gradient descent (backpropagation) per measurement, in the same order that the measurements are observed. However, this method does not provide the benefit of storing a large window of past measurements, including measurements for call vectors that rarely occur, and using each pattern repeatedly for training, so that better convergence is achieved and past measurements are not quickly forgotten.

Therefore, training patterns (each consisting of a call state and a corresponding measurement) should be stored in a *pattern table* (see Figure 2.4). The time window over which past measurements are stored can be selected depending on the desired degree of adaptability. If the network characteristics are not expected to change, and good estimation accuracy is desired, then a very large time window can be used. In addition, the total number of patterns that are stored should be limited, either because of storage limitations or to limit the time required to train the NN. To achieve this, a circular buffer can be used, so that the newest pattern replaces the oldest pattern.[14]

Given a pattern table, the NN can be trained (using a version of backpropagation) either in *pattern mode* or *batch mode*.[9] In pattern mode, one pattern at a time is selected from the table (possibly in random order), and one gradient-descent step is performed with respect to the single term of the error function (2.9) corresponding to that pattern. In batch mode, all patterns are used simultaneously, and an improvement is made to the entire error function (2.9), using either gradient descent or a searching technique that helps to avoid getting trapped in local minima (e.g., Reference 14). Because of the long time required to fully train a NN, it is more practical to perform one improvement step at a time (using either pattern mode or batch mode). A gradual change in the NN weights (and thus the admission policy) also helps to prevent a sudden worsening of performance when an important pattern moves out of the time window (and is thus forgotten). The frequency with which the NN is trained depends on the time required to perform one step of training, and is independent of the frequency with which new measurements are made.

Other rules can be used in place of the circular buffer mentioned above, to decide which old measurements to discard when new measurements are added to the pattern table. Since some input patterns (e.g., marginally unacceptable call states) occur much less often than others, it may be better to discard a measurement for a commonly occurring input pattern than one for an older, rarely occurring input pattern. Put another way, it would be good to use a rule that results in a pattern table that is as evenly distributed as possible over all portions of the input space. This can be accomplished by dividing the pattern table into blocks, each block corresponding to a different region of the input space.[3] For example, an m-dimensional input space of call vectors $(s_1, \ldots, s_m)$ can be divided into blocks defined by specifying ranges for the number of calls in each class. The above goal can then be achieved by limiting the number of measurements per block, and discarding the oldest measurement in the block when this limit is exceeded.

2.7.3 Using Exponential Moving Averages

The online training methods described above may require storing several different measurements for the same input pattern. To achieve good accuracy, a very large number of these measurements may be required. We now present a new method that uses an exponential moving average (EMA) to summarize the measurements for each input pattern, thus reducing the number of stored measurements to the number of input patterns that were observed within a given time window (or even less, using the block method above). Another benefit of this method is that the effect of old patterns on the NN weights diminishes gradually, i.e., these patterns are gradually forgotten, thus avoiding sudden changes in the admission policy when an old pattern suddenly moves out of the time window.

The following numbers are stored for each observed input pattern s: the summarized measurement y^s, the last time t^s that this pattern was observed, and an exponentially decayed count m^s of the number of packets contributing to the measurement. When a new measurement y_k is made, based on m_k packets admitted during call state s, m^s and y^s are updated as follows:

$$m^s(t) = \exp(-\beta(t - t^s))m^s(t^s) + m_k \tag{2.13}$$

$$y^s(t) = [\exp(-\beta(t - t^s))m^s(t^s)y^s(t^s) + m_k y_k] / m^s(t), \tag{2.14}$$

where t is the current time and β is the rate at which old patterns are "forgotten." Notice that the degree to which each observation y_k contributes to the summarized measurement $y^s(t)$ depends on both the age of the measurement and the number of packets on which it was based. In this method, the error function (2.9) is replaced with

$$J = \sum_{\substack{\text{observed states } s}} \exp(-\beta(t - t^s))m^s[f(s) - y^s]^2. \tag{2.15}$$

Notice that the weight for a given call state s decays exponentially as a function of the time that has passed since it was last observed. Thus, input patterns that are no longer observed are gradually forgotten. The block method discussed above can be applied to the EMA method by limiting the number of input patterns per block, and discarding the oldest pattern in the block (the one that has not been observed for the longest time) when this limit is exceeded.

Even if the EMA method is used, the number of patterns that must be stored can be overwhelming if the input space is huge. For example, a space of 100-dimensional call vectors where up to 10 calls per class are allowed has 10^{100} possible input patterns. In this case, the method given in Section 2.5.3 for the compression of NN inputs can be used to reduce the input space to a reasonable size.

2.7.4 Virtual Output Buffer Method

To accurately measure a very small packet loss rate (e.g., 10^{-10}), a large number of packets must be observed over a large time interval. The virtual output buffer (VOB) method[10] allows a neural network to accurately estimate a very small packet loss rate with fewer observations. In this method, the NN is trained to learn the loss rates that would occur if the same packets were fed into a number of imaginary buffers (VOBs) that are connected to imaginary links having smaller bandwidths than the actual link. Each VOB therefore has a higher packet loss rate than the actual buffer, and can thus be estimated accurately with fewer observations. The NN then extrapolates this data to form an improved estimate of the actual loss rate. For example, if the actual buffer is connected to a 100 Mb/s link, three VOBs connected to 70, 80, and 90 Mb/s links, respectively, can be used.

The packet loss rate for each VOB is calculated using a set of counters that simulates the packet queuing process. Each VOB has the same packet arrivals as the actual buffer. Every time a measurement is performed on the actual packet loss rate, a corresponding measurement is performed for each of the VOBs, and all of these measurements are stored in the pattern table. In addition to the call vector s, the NN has a new input b, which is the bandwidth of the actual link or one of the imaginary links. The stored input-output patterns therefore have the form (s, b, y), where y is the measured packet loss rate for either the real output buffer or one of the VOBs. The NN actually extrapolates the estimation function along both the link-bandwidth axis (b) and the call-state axis (s). The VOB method can help to achieve safe-side control because, by extrapolating the virtual packet loss data, the neural network can better estimate the decision boundary before the call state reaches the boundary.

2.8   Conclusions

In this chapter, we have described how the CAC problem can be solved using neural networks, and have presented several existing methods, as well as some new methods and ideas. The success of the neural network approach to CAC has been demonstrated in various research efforts.[2, 10, 14, 16, 19, 20] These works and others have demonstrated through simulations that NN-based CAC methods are significantly more efficient and robust than conventional methods for admission control.

References

1. W. Bernard. 30 years of adaptive neural networks: perception, madaline, and back propagation. *Proceed. of the IEEE.*, 78(9), September 1990.
2. C. J. Chang, S. Y. Lin, R. G. Cheng, and Y. R. Shiue. PSD-based neural-net connection admission control. *IEEE Infocom Proc.*, April 1997.
3. A. Diaz-Esterlla, A. Jurado, and F. Sandoval. New training pattern selection method for ATM call admission neural control. *Electronic Lett.*, 300, March 1994.
4. N. G. Duffield, J. T. Lewing, N. O'Connell, R. Russell, and F. Toomey. Entropy of ATM traffic streams: a tool for estimating QoS parameters. *IEEE J. Selected Areas Commun.*, August 1995.
5. A. Elwalid and D. Mitra. Effective bandwidth of general markovian traffic sources and admission control of high speed networks. *IEEE Infocom Proc.*, June 1994.
6. R. Guerin, H. Ahmadi, and M. Naghshineh. Equivalent capacity and its application to bandwith allocation in high-speed networks. *IEEE J. Selected Areas Commun.*, September 1991.
7. S. Haykin. *Neural Networks, A Comprehensive Foundation.* Macmillan Publishing Company, New York, 1994.
8. H. Heffes and D. Lucantoni. A Markov modulated characterization of packetized voice and data traffic and related statistical multiplexer performance. *IEEE J. Selected Areas Commun.*, September 1986.
9. J. Hertz, A. Krogh, and R. Palmer. *Introduction to the Theory of Neural Computation.* Addison-Wesley Publishing Company, Reading, MA, 1991.

10. A. Hiramatsu. ATM call admission control using a neural network trained with virtual output buffer method. *IEEE Int. Conf. Neural Networks*, 6, 1994.
11. A. Hiramatsu. Training techniques for neural network applications in ATM. *IEEE Commun. Mag.*, October 1995.
12. S. Q. Li and C. L. Hwang. Queue response to input correlation functions: discrete spectral analysis, *IEEE/ACT Trans. Networking*, October 1993.
13. T. Masters. *Practical Neural Network Recipes in C++*. Academic Press, New York, 1993.
14. R. Morris and B. Samadi. Neural network control of communications systems. *IEEE Trans. Neural Networks*, 1994.
15. E. Nordstrom, J. Carlstrom, O. Gallmo, and L. Asplund. Neural networks for adaptive traffic control in ATM networks. *IEEE Comm. Mag.*, October 1995.
16. R. Ogier, N. T. Plotkin, and I. Khan. Neural network methods with traffic descriptor compression for call admission control. *IEEE Infocom Proc.*, March 1996.
17. N. T. Plotkin and C. Roche. The entropy of cells streams as a traffic descriptor in ATM networks. *IFIP Performance Commun. Syst.*, October 1995.
18. A. Sarajedini and P. M. Chau. Quality of service prediction using neural networks. *MILCOM Proc.*, 2, 1996.
19. C. K. Tham and W.-S. Soh. Multi-service connection admission control using modular neural networks, *IEEE Infocom Proc.*, March 1998.
20. S. Youssef, I. Habib, and T. Saadawi. A neurocomputing controller for bandwidth allocation in ATM networks. *IEEE J. Selected Areas in Commun.*, February 1997.

Problems

1. Assume your goal is to design a NN-based CAC with two services. The first service makes QoS guarantees to users on bandwidth and delay parameters (i.e., it guarantees a minimum bandwidth B_1 and maximum delay D_1). The second service makes QoS guarantees to users on bandwidth, delay, and jitter parameters (i.e., it guarantees a minimum bandwidth B_2, maximum delay D_2, and maximum jitter J_2). Assume that there are 6 traffic classes and that the call vector is used as the NN input.

   (a) How many inputs and outputs are there, and what are they?
   (b) Describe the CAC decision based on these outputs.
   (c) Draw a diagram of the NN with all inputs and outputs and any hidden neurons.

2. Suppose you want to design a neural network that gives, for a fixed delay requirement D and for each call vector s, the probability that the delay of a packet is at most D, conditioned on the packet arriving when the call vector is s. Online training is used, so you want to use the ability of a NN to learn the average.

(a) What should the training output y_k of equation (2.9) be if each observation is a single packet delay?

(b) What should the training output y_k be if each observation is for multiple packets?

3. Suppose that you want to design a NN CAC to accept a call only if the probability that the packet delay exceeds 100 ms is at most 10^{-6}. Assume the call vector is used to describe the system state. Online training is used, so you want to use the ability of a NN to learn the average. Also, since a packet delay that exceeds 10^{-6} is a rare event, you want to use the ability of a NN to extrapolate from measurements based on more common events (as does the virtual output buffer method). Explain how this can be done by specifying the NN inputs and NN output, and the input-output training patterns.

4. Consider a neural network whose input is a 5-dimensional call vector $s = (s_1, \ldots, s_5)$ corresponding to 5 traffic classes, and suppose that the maximum number of calls for each class is 16. Describe how the block method discussed in Section 2.7.2 can be used to reduce the maximum number of training patterns to exactly 2048, such that all blocks have the same size and the maximum number of training patterns allowed per block is the same for all blocks.

5. Suppose you wish to train a neural network to give the average packet delay y^s for each possible call vector s. Assume that the kth training pattern has the form (s^k, y_k), where s^k is a call vector and y_k is the delay of a single packet, which can be any positive real number. The output neuron has a sigmoid activation function $h()$ whose values lie between -1 and 1, so you decide to train the neural network to learn $h(y^s)$ instead of y^s. What error function (expressed as a sum over k) should you use?

6. Consider the following pattern table method for online training, where the desired NN output is the average packet delay, which should not exceed some maximum value. Since feasible patterns occur much more frequently than infeasible patterns, feasible patterns and infeasible patterns are stored in two different tables. Each table is limited to 1000 input-output patterns, and the oldest pattern in the table is dropped when this limit is exceeded. Note that due to noisy measurements, the same input pattern can occur in both tables. Explain why this method does not work.

7. Assume the following. Each traffic source (call) i is characterized by leaky bucket parameters r_i, b_i, so that the total number of cells $N_i(t, u)$ arriving within a time interval $[t, u]$ is bounded by $r_i(u - t) + b_i$. The corresponding parameter r_S (respectively, b_S) for an aggregate stream S is defined by summing r_i (respectively, b_i) over all calls i in the stream. The parameters r_i and b_i are known for each call but can differ for different calls, so that the number of traffic classes is

unlimited. A given stream S is feasible if $r_S \le R$ and $b_S \le B$, where R and B are the *unknown* link rate and buffer size (R and B must be learned via training).

(a) Design a NN with a single hidden layer and a *single* output that correctly performs CAC under the above assumptions. Assume that the hidden and output neurons have sigmoid activation functions that approximate a step function that is 0 for negative input and 1 for positive input. An output less than 0.5 should result in acceptance. Specify the NN inputs, describe the outputs of the hidden and output neurons, and explain which NN weights correspond to R and B, which should be the only two weights that are variable.

(b) How would you modify the NN if the condition for feasibility becomes $g(r_S, b_S) \le 0$, where g is an unknown continuous function?

(c) Suppose that there are 5 traffic classes such that all calls in a given class have the same values for (r_i, b_i), and assume that feasibility is determined as in part (b). Show how the parameters r_i, b_i for each class can be determined using the linear compression method of Section 2.5.3, by drawing the NN used for training.

Answers to Problems

1. (a) There are 5 outputs, for estimating the delay of service class 1, bandwidth of service class 1, delay of service class 2, bandwidth of service class 2, and jitter of service class 2.

   (b) Let D_1, B_1, D_2, B_2, J_2 denote the guarantees made. The CAC decision is as follows. If the requesting stream asks for service type 1, then the call is admitted if the potential new aggregate stream receiving service type 1 has estimated delay less than D_1 and estimated bandwidth requirement less than B_1. If the requesting stream asks for service type 2, then the call is admitted if the potential new aggregate traffic receiving service type 2 has estimated delay less than D_2, estimated bandwidth requirement less than B_2, and estimated jitter less than J_2.

2. (a) The training output y_k should be 1 if the packet delay does not exceed D, and should be 0 otherwise.

   (b) The training output y_k should be the fraction of the packets in the observation whose delays do not exceed D (not the average delay).

3. The NN inputs should be the call vector and the specified packet delay D. The output should be the probability that the packet delay is greater than D. (The log of this probability can also be used if the

error function (12) is used.) The values used for the specified delay should be 100 ms and a few smaller delays, e.g., 90 ms, 80 ms, and 70 ms. In this case, each observation would generate four input-output training patterns of the form (s, D, y), where D is one of the four specified delays and y is the fraction of packets in the observation whose delays exceed D.

4. The number of calls in each class should be divided into four ranges of equal size, to obtain $4^5 = 1024$ blocks, and the number of training patterns allowed per block should be 2.

5. The error function is $J = \Sigma_k m_k [h^{-1}(f(s^k)) - y_k]^2$.

6. The NN output for a given NN input pattern converges to the average of the training outputs for that input pattern. This is equal to the average observed delay only if the training outputs have the same distribution as the observed delays, which is not the case in the method described.

7. (a) The NN inputs are r_S and b_S. There are two hidden neurons, whose outputs are $I(r_S > R)$ and $I(b_S > B)$, respectively, where $I(x)$ is 1 if x is true and 0 otherwise. The NN output is $I(r_S > R \text{ or } b_S > B)$. The biases of the two hidden neurons are $-R$ and $-B$, which are learned by the NN.

(b) The two hidden neurons should be replaced by a larger number of hidden neurons, and all NN weights should be variable.

(c) The NN used for training has 5 inputs corresponding to the call vector, a first hidden layer consisting of two linear neurons (representing the compressed parameters r_S and b_S), and a second hidden layer consisting of nonlinear neurons. After the NN is trained, r_i and b_i are obtained as the weights from the ith input to the two linear neurons.

3

CAC and Computational Intelligence

Ray-Guang Cheng and Chung-Ju Chang

CONTENTS

ABSTRACT In this chapter, we address the application of intelligent techniques such as fuzzy logics, neural networks, and neural fuzzy networks on connection admission control (CAC) in multimedia high-speed networks. Conventional CAC schemes using analytic approaches can provide good solutions under stationary conditions. However, because of the unpredictable traffic fluctuations in services, multimedia high-speed networks would be nonstationary dynamic systems, and thus solutions from mathematical analysis based on parametric models would be uncertain. Fuzzy logic systems have been successfully applied to deal with traffic-control-related problems and have provided a robust mathematical framework for dealing with real-world imprecision. Neural networks have learning and adaptive capabilities that can be used to construct intelligent computational algorithms for traffic control. However, there is no clear and general technique to map domain knowledge on traffic control onto the parameters of a fuzzy logic system; the

0-8493-1075-X/01/$0.00+$.50
© 2001 by CRC Press LLC

knowledge embodied in conventional methods is difficult to incorporate into the design of neural networks. Neural fuzzy networks combine the linguistic control capabilities of a fuzzy logic controller and the learning abilities of a neural network. They provide a robust framework to mimic experts' knowledge embodied in existing traffic control techniques and can construct efficient and intelligent computation algorithms for traffic control. Performance comparisons among the conventional CAC, the fuzzy logic CAC, the neural network CAC, and the neural fuzzy CAC are performed, from the aspects of cell loss ratio, the system utilization, and the training time under the QoS guaranteed constraint.

KEY WORDS: *fuzzy logic controller (FLC), neural network, neural fuzzy network, connection admission control (CAC), genetic algorithm (GA), congestion control, equivalent capacity, traffic control, multimedia network, quality of service (QoS), fuzzy set.*

3.1 Introduction

Multimedia high-speed networks have to be capable of handling bursty traffic and satisfying various quality-of-service (QoS) and bandwidth requirements. From network providers' point of view, to increase the network utilization as much as possible is their major purpose. However, the major constraint while maximizing the utilization is that the QoS of each individual connection should be fulfilled. Therefore, effective control mechanisms are required to guarantee QoSs for existing connections and to achieve high network utilization. Among the control mechanisms, the connection admission control (CAC) is one of the most important issues that should be considered.

CAC is defined as a set of actions taken by a network during the connection setup phase to determine whether a new connection request can be accepted or rejected. A request will be accepted only if there are sufficient resources available to fulfill the QoS requirements of the new connection as well as the existing connections. In order to meet the above mentioned requirements, the network has to monitor the usage of the network resource and predict the impact of adding a new connection on the QoS of existing connections. Different schemes were proposed to find the relationship between the number of connections of each service class and their respective QoS. The cell loss ratio, cell delay, and cell delay variation are three of the widely used QoS metrics.

Conventional CAC schemes[1-6] applied a parametric model of the traffic being offered, either by requiring each connection to provide an accurate description of its traffic behavior (via traffic parameters), or by measuring the observed traffic and fitting it to a model, and then inferred the cell loss ratio and/or other network performance measures via mathematical analysis. For this scheme, when a new connection is requested, the network examines the required bandwidth and the QoS requirements to decide whether to accept the new connection or not. In most of the approaches disclosed in the literature, complicated mathematical equations were derived and approximations were required to meet the real-time operation requirement for CAC. An "equivalent capacity" method for individual and multiplexed connections was proposed based on their statistical characteristics defined by traffic parameters and the desired QoS.[1,3] A unified metric was then obtained to represent the effective bandwidth (capacity) used by connections and the corresponding effective load on network links. Although the paper can provide an exact approach to the computation of the equivalent capacity, the associated complexity makes it infeasible for real-time calculation. Hence, an approximation is introduced and it results in the degradation in utilization. Saito[5] proposed a connection admission scheme by inferring the upper bound of cell loss probability from the traffic parameters specified by users (i.e., the maximum number of cells arriving during a fixed interval, and the average and variance of the number of cells arriving during a fixed interval). The QoS requirement is guaranteed under this control without assumptions of a cell arrival process.

In these conventional algorithmic/analytical approaches, researchers inferred equations to characterize a decision boundary of CAC. Some of the approaches are simple but lack flexibility, and the others may be too complicated to be calculated in real-time fashion. These conventional CAC approaches, based on mathematical analysis, can provide robust solutions for different kinds of traffic environments but suffer from estimation error (due to modelling) and approximation error (due to the need to complete calculations in real time); it is not suitable for dynamic environments. Also, these approaches are confined in a restricted number of traffic types and may not hold at all for some traffic sources. One limitation of these traditional approaches is the difficulty of finding equations, which can incorporate all of the possible information, for the CAC decision. For example, buffer size, QoS measures, and congestion degree indication, which provide rich information of network resource and network behavior, are hardly to be incorporated into parametric model for the equations of CAC.

Nowadays, intelligent techniques such as fuzzy logics, neural networks, and neural fuzzy networks have successfully been applied to traffic control problems in high-speed networks.[7] Researchers have been shown that the intelligent techniques have better performance than the analytical methods.[8-16]

Fuzzy set theory appears to provide a robust mathematical framework for dealing with real-world imprecision, and the fuzzy logic approach exhibits a soft behavior which means to have a greater ability to adapt itself to dynamic, imprecise, and bursty environments.[8-10] Bonde and Ghosh[8] used fuzzy mathematics to provide a flexible, high-performance solution to queue management in ATM networks. Ndousse[9] proposed a fuzzy logic implementation of the leaky bucket mechanism that used a channel utilization feedback to improve performance. In Reference 10, a fuzzy traffic controller which simultaneously incorporates CAC and congestion control was proposed. Fuzzy-logic-based CAC is excellent in dealing with real-world imprecision and has a greater ability to adapt itself to dynamic, imprecise, and bursty environments. However, it lacks the learning capability needed to automatically construct its rule structure and membership functions so as to achieve optimal performance.

The self-learning capability of neural networks has been applied to characterize the relationship between input traffic and system performance.[11–15] Hiramatsu[11] proposed a neural-net-based connection admission controller. The proposed controller used a back-propagation neural network. In the proposed method, the declared traffic parameters were used only to divide connections into several bit-rate classes. The neural network in the controller learns the relationship between the numbers of existing connections in each bit-rate class and their corresponding QoSs according to the statistical characteristics of each bit-rate class. In Reference 12, Tran-Gia and Gropp investigated the possible use of a neural network to perform CAC. Youssef, et al.[13] proposed a CAC controller for ATM networks. A neural network is trained to compute the effective bandwidth required to support connections with different QoS requirements. They showed that the adaptability of the neural network controller to new traffic situations had been achieved by adopting a hierarchical approach to the design. In Reference 15, a neural-network-based connection admission control (NNCAC) method for ATM networks was proposed. The NNCAC method uses three preprocessed input parameters to simplify the training process and to improve the controlled performance. In general, neural-net-based CAC provides learning and adaptation capabilities to reduce the estimation error of conventional CAC and achieve performance similar to that of a fuzzy logic controller. However, the knowledge embodied in conventional methods is difficult to incorporate into the design of a neural network.

A neural fuzzy CAC (NFCAC)[16] scheme that absorbs benefits of the three approaches while minimizing their drawbacks was proposed for multimedia high-speed networks. The NFCAC scheme utilizes the learning capability of the neural network to reduce decision errors of conventional CAC policies resulted from modeling, approximation, and unpredictable traffic fluctuations of the system. It also employs the rule structure of the fuzzy logic controller to prevent operating errors, due to incorrect learning, and to decrease training time. Furthermore, the neural fuzzy network is a simple structured network. Input variables and the rule structure could be properly chosen and designed for the NFCAC scheme so that it not only provides a robust framework to mimic experts' knowledge embodied in existing traffic control techniques but also constructs intelligent computational algorithm for traffic control.

The rest of the chapter is organized as follows. An architecture of an intelligent traffic controller including an intelligent CAC controller is introduced in Section 3.2. CAC approaches using intelligent techniques such as fuzzy logics, neural networks, and neural fuzzy networks will be designed and discussed in Sections 3.3, 3.4, and 3.5, respectively. The performance comparison of these approaches via simulations will be presented in Section 3.6. Finally, Section 3.7 gives the summary.

3.2 Intelligent Traffic Controller

This section introduces an implementation of an intelligent traffic controller. As shown in Figure 3.1, the intelligent traffic controller is composed of an intelligent CAC controller, a fuzzy congestion controller, a fuzzy bandwidth estimator, and a network resource estimator. The *fuzzy congestion controller* generates a congestion indicator y according to the measured system statistics, such as the queue length q, the change rate of the queue length Δq, and the cell loss ratio p_l. The design of the fuzzy congestion controller is based on the knowledge obtained from a buffer-threshold method proposed in the literature. The *fuzzy bandwidth estimator* estimates the required capacity C_e for a

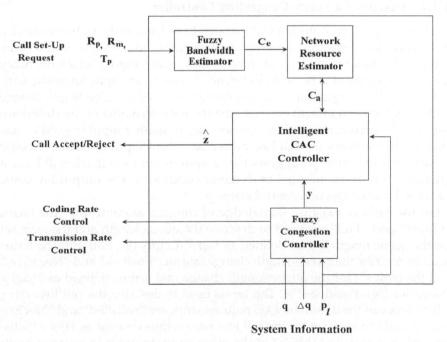

FIGURE 3.1
The functional diagram of an intelligent traffic controller.

new connection from its traffic description parameters such as the peak cell rate, sustainable cell rate, and peak cell rate duration, denoted by R_p, R_m, and T_p, respectively. It employs the equivalent-capacity-based algorithm proposed in the literature.

The *network resource estimator* does the accounting for system-resource usage. When a new connection with bandwidth C_e is accepted, the value of C_a is updated by subtracting C_e from the original value of C_a. Conversely, when an existing connection with bandwidth C_e is disconnected, the value of C_a is updated by adding C_e to the original value of C_a. C_a is initially set to 1. *The intelligent CAC controller* takes the available capacity C_a, the congestion indicator y, and the system performance feedback of cell loss ratio p_l as input linguistic variables to handle the CAC procedure and sends a decision signal $\hat{z}$ back to the new connection to indicate acceptance or rejection of the new connection request. Based on the architecture mentioned above, we can construct a traffic controller which can simultaneously handle congestion control and CAC. This architecture is a closed-loop control system capable of adjusting itself to provide stable, robust operation and avert congestion not only at the cell level but also at the connection level.

In the following subsection, we will describe the design of the fuzzy congestion controller and the fuzzy bandwidth estimator.

3.2.1 Design of a Fuzzy Congestion Controller

A fuzzy congestion controller is illustrated to be used in the traffic controller. It is a fuzzy implementation of the two-threshold congestion control scheme proposed in Reference 17. In addition to the queue length q, which the fuzzy queue management scheme in Reference 8 uses as an input linguistic variable, the fuzzy congestion controller further selects the queue-length change rate Δq, which can effectively describe the local dynamic of the difference between the arrival rate and the service rate, as another input linguistic variable. It also employs the cell loss ratio p_l as a third input linguistic variable because control theory suggests that a system with feedback will have a greater ability to be adapted to dynamic conditions. The output linguistic variable is the congestion control action y.

On the basis of existing knowledge of congestion control,[8, 17] the terms "Empty" and "Full" are used to describe the queue length and the term set for the queue length is thus defined as $T(q)$ = {Empty (E), Full (F)}. The terms used to describe the queue-length change rate are "Positive" and "Negative," and the term set for the queue-length change rate is thus defined as $T(\Delta q)$ = {Negative (N), Positive (P)}. The terms used to describe the cell loss ratio, which is one of the dominant QoS requirements, are "Satisfied" and "Not Satisfied," and the term set for the cell loss ratio is thus defined as $T(p_l)$ = {Satisfied (S), Not Satisfied (NS)}. On the other hand, in order to provide a soft congestion control, not only the control action of "No Change" is considered, but the two-threshold control actions of "Increase" and "Decrease" are also

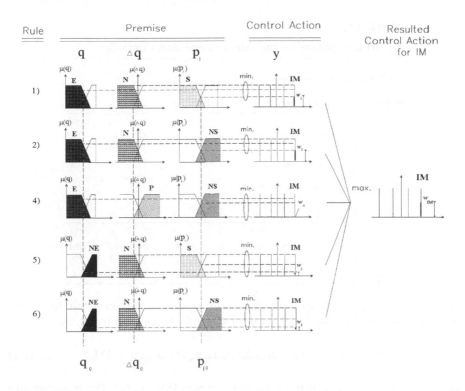

FIGURE 3.2
The max-min inference method.

further divided into four levels, and the term set for the output control action is defined as T(*y*) = {Decrease More (DM), Decrease Slightly (DS), No Change (NC), Increase Slightly (IS), Increase More (IM)}. The triangular and trapezoidal functions are chosen as the membership function because they are suitable for real-time application.

The parameters and rules of the fuzzy congestion controller are initially set, induced from many analytical results of the two-threshold congestion control method[17] and then further calibrated through simulations using a genetic algorithm (GA) optimization technique.[18] Details of the setting of the membership functions and the fuzzy control rules could be referred to.[10] The optimization procedure using GA could be referred to.[19] The initial and optimal rule structures are shown in Table 3.1 (a) and (b), respectively.

The fuzzy congestion controller adopts the max-min inference method for the inference engine because it is designed for real-time operation. Figure 3.2 shows an example of the max-min inference method for the fuzzy congestion controller, where rules 1, 2, 4, 5, and 6, which have the same control action, "*y* is IM," are depicted. In this figure, we assumed that the performance measures estimator measures q, Δq, and p_i and yields q_0, Δq_0, and p_{i0}. The membership values of q_0, Δq_0, and p_{i0} corresponding to the premise of rule 1 (for example) of Table 3.1—q is Empty, Δq is Negative, and p_i is Safe—are given by $\mu_E(q_0)$, $\mu_N(\Delta q_0)$,

TABLE 3.1

The Rule Structure for the Fuzzy Congestion Controller

(a) The initial rule structure

Rule	q	Δq	p_l	y	Rule	q	Δq	p_l	y
1	E	N	S	IM	5	F	N	S	DM
2	E	N	NS	IM	6	F	N	NS	DM
3	E	P	S	IM	7	F	P	S	DM
4	E	P	NS	IM	8	F	P	NS	DM

(b) The optimal rule structure

Rule	q	Δq	p_l	y	Rule	q	Δq	p_l	y
1	E	N	S	IM	5	F	N	S	IM
2	E	N	NS	IM	6	F	N	NS	IM
3	E	P	S	IS	7	F	P	S	DS
4	E	P	NS	IM	8	F	P	NS	NC

and $\mu_s(p_{l0})$, respectively. Applying the "min" operator, we can obtain the membership value of the control action $y =$ IM of rule 1, denoted by w_1, by

$$w_1 = \min(\mu_E(q_0), \mu_N(\Delta q_0), \mu_S(p_{l0})) \cdot \mu_{IM}(y = IM_c), \qquad (3.1)$$

where IM_c is the center of the membership function $\mu_{IM}(y)$. The membership values of rules 2, 4, 5, and 6, denoted by w_2, w_4, w_5, and w_6, respectively, can be obtained in the same manner. Subsequently, applying the "max" operator yields the overall membership value of the control action $y =$ IM, denoted by w_{IM}, as follows:

$$w_{IM} = \max(w_1, w_2, w_4, w_5, w_6). \qquad (3.2)$$

The overall membership values of the control actions IS, NC, DS, and DM, denoted by w_{IS}, w_{NC}, w_{DS}, and w_{DM}, respectively, can be calculated in a similar way.

The fuzzy congestion controller uses Tsukamoto's defuzzification method for the defuzzifier because of its simplicity in computation. This defuzzification method obtains a crisp value y_0, of the control action y by combining w_{IM}, w_{IS}, w_{NC}, w_{DS}, and w_{DM} as follows:

$$y_0 = \frac{IM_c \times w_{IM} + IS_c \times w_{IS} + NC_c \times w_{NC} + DS_c \times w_{DS} + DM_c \times w_{DM}}{w_{IM} + w_{IS} + w_{NC} + w_{DS} + w_{DM}}. \qquad (3.3)$$

On the basis of crisp value y_0, the coding rate manager (the transmission rate manager) sends a coding (transmission) rate control command to throttle the incoming traffic. Low priority cells of type-1 traffic are then selectively discarded, and the transmission rate of type-2 traffic is then altered.

TABLE 3.2

The Rule Structure for the Fuzzy Bandwidth Predictor

Rule	R_p	R_m	T_p	C_e	Rule	R_p	R_m	T_p	C_e	Rule	R_p	R_m	T_p	C_e
1	S	Lo	Sh	C_1	7	M	Lo	Sh	C_1	13	L	Lo	Sh	C_4
2	S	Lo	Me	C_2	8	M	Lo	Me	C_3	14	L	Lo	Me	C_6
3	S	Lo	Lg	C_5	9	M	Lo	Lg	C_6	15	L	Lo	Lg	C_6
4	S	Hi	Sh	C_1	10	M	Hi	Sh	C_1	16	L	Hi	Sh	C_3
5	S	Hi	Me	C_1	11	M	Hi	Me	C_2	17	L	Hi	Me	C_5
6	S	Hi	Lg	C_4	12	M	Hi	Lg	C_5	18	L	Hi	Lg	C_6

3.2.2 Fuzzy Bandwidth Estimator

The fuzzy bandwidth estimator, which is a fuzzy implementation of the equivalent capacity method proposed by Guègrin et al.[1] is illustrated to be used in the traffic controller. In accordance with expert knowledge,[1–3] the fuzzy bandwidth estimator selects R_p, R_m, and T_p as the input linguistic (traffic) variables and uses the estimated equivalent capacity, denoted by C_e, as the output linguistic variable.

In order to extract knowledge of the equivalent capacity method, a total of 10^5 items of numerical data were acquired by extensively calculating the equivalent capacity in Reference 1, Eq. (3.2) for different combinations of R_p, R_m, and T_p. The numerical results show that three terms are good enough for R_p, two terms are good enough for R_m, and three terms are needed to describe T_p. Accordingly, the term sets for R_p, R_m, and T_p are respectively defined as $T(R_p) = \{Small\ (S),\ Medium\ (M),\ Large\ (L)\}$, $T(R_m) = \{Low\ (Lo),\ High\ (Hi)\}$, and $T(T_p) = \{Short\ (Sh),\ Medium\ (Me),\ Long\ (Lg)\}$. Besides, it is known that the estimated equivalent capacity for a connection should fall between its R_m and R_p. The fuzzy bandwidth estimator illustrated in this section quantizes the range between R_m and R_p into six levels. Let C_i ($i = 1, \ldots, 6$) denote the ith level of the capacity estimation and the term set for the estimated capacity is defined as $T(C_e) = \{C_1, C_2, C_3, C_4, C_5, C_6\}$. The positions of membership functions and structure of control rules can be induced from numerous numerical data. The rule structure of the fuzzy bandwidth estimator is shown in Table 3.2. Details of the design of the fuzzy bandwidth estimator could be referred to.[10]

In the following, three intelligent approaches to the design of a CAC controller will be illustrated.

3.3 Fuzzy Logic Connection Admission Control (FLCAC)

Fuzzy-logic-based control systems have demonstrated the ability to make intelligent decisions by deploying "soft" thresholds, performing calculations based on imprecise quantities, and modeling linguistic rules. They emulate the way in which an expert operator makes decisions and have been particularly useful when precise mathematical models are impractical or unavailable.

In this section, we use the CAC controller proposed in Reference 10 as an example to illustrate the way to design an intelligent CAC controller using a fuzzy logic controller. Unlike the equivalent capacity CAC method proposed in Reference 1, which uses only the available capacity C_a as a variable for connection set-up decisions, the FLCAC controller considers the network congestion control action y, which contains information on congestion within the network, and the cell loss ratio p_l, which is used as a channel utilization feedback,[11] as two further input linguistic variables. The accept/reject decision z is the output linguistic variable.

In the CAC methods proposed in the literature, the terms used to describe available capacity for a new connection are "Not Enough" and "Enough," and the term set for the available capacity is thus defined as $T(C_a) = \{$Not Enough (NE), Enough (E)$\}$. The system is either in a congestion state ("y is Negative") or a congestion-free state ("y is Positive"), and the term set for the congestion control action is defined as $T(y) = \{$Negative (N), Positive (P)$\}$. The term set for the cell loss ratio is defined the same as $T(p_l)$ in the fuzzy congestion controller. In order to provide a soft admission decision, not only "Accept" and "Reject" but also "Weak Accept" and "Weak Reject" are employed to describe the accept/reject decision. Thus, the term set of the output linguistic variable is defined as $T(z) = \{$Accept (A), Weak Accept (WA), Weak Reject (WR), Reject (R)$\}$.

The membership functions for $T(C_a)$, $T(y)$, and $T(p_l)$ are defined as

$$\mu_{NE}(C_a) = g(C_a; 0, NE_e, 0, NE_w), \qquad (3.4)$$

$$\mu_{E}(C_a) = g(C_a; E_e, C, E_w, 0), \qquad (3.5)$$

$$\mu_{N}(y) = g(y; -y_{max}, N_e, 0, N_w), \qquad (3.6)$$

$$\mu_{P}(y) = g(y; P_e, y_{max}, P_w, 0), \qquad (3.7)$$

$$\mu_{S}(P_l) = g(p_l; 0, S_e, 0, S_w), \qquad (3.8)$$

$$\mu_{NS}(p_l) = g(p_l; NS_e, 1, NS_w, 0), \qquad (3.9)$$

where C denotes the total network capacity provided for type-i traffic (C is C_r for type-1 traffic and is $(1 - C_r)$ for type-2 traffic); y_{max} denotes the maximum percentage of cells that are prohibited from entering the network; E_e is designed to be a fraction of C and is used to tolerate the estimation uncertainty resulting from the dynamic traffic characteristics as well as the fuzzy implementation of the equivalent capacity method; NE_e is smaller than E_e and is used to indicate an emergency due to lack of capacity; and N_e and P_e are values properly set by monitoring the control action y of the fuzzy congestion controller during congestion and congestion-free periods, respectively. N_e and P_e are calibrated via simulation. In the simulations, a number of short-term congestion was stimulated by overloading the system. The values of control action y were sampled and clustered according to congestion or congestion-free states. We set N_e and P_e to be the mean values of the control action y in the congestion and the congestion-free states.

Similarly, the membership functions for $T(z)$ are defined as

$$\mu_R(z) = f(z; R_c, 0, 0),   \tag{3.10}$$

$$\mu_{WR}(z) = f(z; WR_c, 0, 0),   \tag{3.11}$$

$$\mu_{WA}(z) = f(z; WA_c, 0, 0),   \tag{3.12}$$

$$\mu_A(z) = f(z; A_c, 0, 0).   \tag{3.13}$$

A new connection request can be accepted if the output of the FLCAC controller z is greater than an acceptance threshold z_a, $R_c \le z_a \le A_c$. Without loss of generality, we set $R_c = 0$, $A_c = 1$, $WR_c = (R_c + z_a)/2$, and $WA_c = (A_c + z_a)/2$.

The order of significance of the input linguistic variables for the FLCAC controller would be the cell loss ratio p_l, the congestion control action y, and then the available system capacity C_a. The control rules of FLCAC are designed in Table 3.3. A new connection will have no chance of entering the network and will be rejected if the cell loss ratio is not satisfied ($p_l = NS$), except that a new connection will be weakly rejected to improve the network utilization when the network is congestion-free ($y = P$) and has enough capacity ($C_a = E$). On the other hand, a new connection has a good chance of entering the network and will be accepted or weakly accepted if the cell loss ratio is satisfied ($p_l = S$), except that a new connection will be weakly rejected if the network is experiencing congestion ($y = N$) and the available capacity is insufficient ($C_a = NE$).

Simulation results[10] show that the FLCAC controller improves system utilization by 11%, while maintaining a QoS contract compared with that of a conventional equivalent capacity method. At the same time, the performance

TABLE 3.3

The Rule Structure for the FLCAC Controller

Rule	p_l	y	C_a	z	Rule	p_l	y	C_a	z
1	S	P	E	A	5	NS	P	E	WR
2	S	P	NE	WA	6	NS	P	NE	R
3	S	N	E	WA	7	NS	N	E	R
4	S	N	NE	WR	8	NS	N	NE	R

of the fuzzy congestion control method is also 4% better than that of the conventional two-threshold congestion control method during congestion periods. FLCAC controller is effective because it employs input variables that provide much more information than conventional methods. Moreover, the linguistic capability of fuzzy logic handles the traffic complexity and provides soft control. It can also be implemented, using fuzzy logic chips, for real-time application.[14]

3.4   Neural Network Connection Admission Control (NNCAC)

The CAC is used to determine whether to accept a new connection or not, and can be categorized as a pattern recognition problem. Therefore, the neural network can be used to classify uses into acceptable and rejection regions upon observing information of the users and the network. In this section, we will describe two kinds of neural network approaches to design an intelligent CAC controller. The first approach, named NNCAC, used a multi-layer feedforward neural network while the second approach, named RBFCAC, employed a radial-basis function network (RBFN). In this section, both of them used 30 hidden nodes.

3.4.1 The NNCAC Approach

The multi-layer feedforward neural network is widely used to cope with pattern recognition problems. A neural-network-based CAC (NNCAC) controller uses a multi-layer feedforward neural network to approximate a perfect connection acceptance decision function from input/output data pairs $\{X, Z\}$. That is, upon recognition of the input pattern X, a yes/no decision must be made by NNCAC controller to accept/reject a connection request.

A back-propagation learning algorithm[15] is used here to train the NNCAC controller. Let $X(i)$ denote the sampled input vector of the NNCAC controller at time instant t_i, $NNCAC(X(i), W) = \hat{z}(i)$ denote the corresponding decision of the NNCAC controller, and $f(X(i)) = z(i)$ denote the desired decision. The objective of the back-propagation learning algorithm is to minimize decision error E by recursively adjusting its weight W in each layer, where E is defined as

$$E = \frac{1}{2}\|N\ N\ C\ AC(X(i), W) - f(X(i))\|^2$$
$$= \frac{1}{2}(\hat{z}(i) - z(i))^2.$$

(3.14)

Details of the weight-adjusting equations could be referred to[15].

3.4.2 The RBFN Approach

Recently, a radial-basis function network (RBFN) has been employed in the applications such as signal processing, pattern recognition, control, and function approximation. Generally, RBFNs cannot achieve the same accuracy but require less training time than the multi-layer feedforward neural network does. Moreover, RBFNs have a great potential to relax the size growing and learning difficulty encountered in multi-layer feedforward neural networks and have powerful adaptive and learning capabilities. The hidden nodes in the RBFN perform the normalized Gaussian activation function:

$$z_q \equiv \frac{\exp[-|\vec{x} - \vec{m}_q|^2 / 2\sigma_q^2]}{\sum_{l=1}^{k} \exp[-|\vec{x} - \vec{m}_l|^2 / 2\sigma_l^2]}, \quad 1 \leq q \leq k, \qquad (3.15)$$

where $\vec{x}$ is the input vector, k is the number of hidden nodes, and $\vec{m}_q$ (σ_q) is the mean (the standard deviation) of the qth Gaussian function. For an input vector $\vec{x}$ lying somewhere in the input space, the receptive fields which are close to it will be properly activated. The output of RBFN is simply a mapping of the weighted sum of the outputs of hidden nodes:

$$y_i = a_i \left(\sum_{q=1}^{k} w_{iq} z_q \right), \qquad (3.16)$$

where w_{iq} is the weight of the link and $a_i(\cdot)$ is the output activation function.

Similar to the NNCAC approach, an RBFN-based CAC (RBFCAC) controller employs the same input vector and generates the desired decision. The only difference between NNCAC and RBFCAC is that different neural networks and learning algorithms were selected. Generally, RBFN can be trained by either the hybrid learning rule or the error back-propagation rule. The former uses unsupervised learning in the input layer and supervised learning in the output layer. The latter is a purely supervised learning rule. If RBFN is trained by the error back-propagation learning rule, it does not learn appreciably faster than the back-propagation network and it may encounter large-width problem. In this section, the hybrid learning rule is chosen.

3.4.3 Training Data Collection

The training data may greatly affect the performance of the neural network. The procedure for constructing the set of training data is described below:

[*Construction of Training Data:*]

For a new connection request with traffic parameters of R_m, R_p, and T_p

Estimate the required capacity C_e by using fuzzy bandwidth estimator

Count the available capacity C_a by using network resource estimator

Generate a congestion indicator y by using fuzzy congestion controller

Get the cell loss ratio p_l measured from system information statistics

If $p_l > $ QoS

Then

    Reject the request and set the desired output $z = 0$

Else

    Accept the request and set the desired output $z = 1$

[*Verification of the Acceptance Decision:*]

    Continue the simulation for a pre-defined time interval, without accepting any new connection requests

    Obtain the statistics of cell loss ratio p_l'

If $p_l' > $ QoS (acceptance decision is failed),

then

    Set $z = 0$

    EndIf

EndIf

    Store training data of C_a, y, p_l, and z

According to these training data, the NNCAC and RBFCAC controllers can be offline trained to approximate the complicated connection acceptance decision function.

Simulation results[15, 16] showed that the NNCAC controller provided an amount of 30% system utilization improvement over Hiramatsu's neural-net based CAC scheme, 10% system utilization improvement over the FLCAC controller, while maintaining QoS contracts. The RBFCAC achieves similar utilization but requires less training time than NNCAC does.

3.5 Neural Fuzzy Connection Admission Control (NFCAC)

Fuzzy logic controllers provide us the capabilities of simultaneously achieving several objectives and providing smoother changes in the connection acceptance decision boundary. Expert knowledge can be easily incorporated in the design of the controller. Neural networks provide us with adaptive learning capability, high computation rate, generation of learning, and a high degree of robustness or fault tolerance. Neural networks act as a black box that performs as expected in situations where there may be no prior knowledge or experience, while fuzzy-logic-based systems employ expert knowledge and experience to control the network. A combination of fuzzy logics and neural networks, named neural fuzzy networks, can retain their benefits but reduce their drawbacks. It can also be employed to as an intelligent CAC controller.

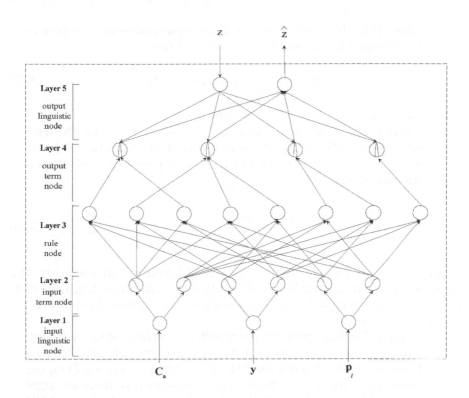

FIGURE 3.3
The architecture of the NFCAC controller.

An NFCAC controller is a five-layered neural fuzzy network.[20] As shown in Figure 3.3, the NFCAC controller has nodes in layer one as input linguistic nodes. It has two kinds of output linguistic nodes in layer five. One is for feeding training data (desired output) into the net and the other is for pumping decision signals (actual output) out of the net. The nodes in layer two and layer four are term nodes which act as membership functions of the respective linguistic variables. The nodes in layer three are rule nodes; each node represents one fuzzy rule and all nodes form a fuzzy rule base. The links in layer three and layer four function act as an inference engine—layer-three links define preconditions of the rule nodes and layer-four links define consequences of the rule nodes. The links in layer two and layer five are fully connected between the linguistic nodes and their corresponding term nodes.

The NFCAC controller adopts three linguistic inputs of an available capacity C_a, a congestion indicator y, and a cell loss ratio p_l and outputs a decision signal $\hat{z}$ to indicate acceptance or rejection of the new connection request. In general, the NFCAC controller (shown in Figure 3.3) has a net input function $f_i^{(k)}(u_{ij}^{(k)})$ and an activation output function $a_i^{(k)}(f_i^{(k)})$ for node i in layer k, where $u_{ij}^{(k)}$ denotes a possible input to node i in layer k from node j in layer $(k-1)$. These layers are described below.

Layer 1: In this layer, there are three input nodes with respective input linguistic variables C_a, y, and p_l. Define

$$f_i^{(1)}(u_{ij}^{(1)}) = u_{1i}^{(1)} \text{ and } a_i^{(1)} = f_i^{(1)},\qquad(3.17)$$

where $u_{11}^{(1)} = C_a$, $u_{22}^{(1)} = y$, $u_{33}^{(1)} = p_l$, and $1 \le i \le 3$.

Layer 2: The nodes in this layer are used as the fuzzifier. The term sets used in NFCAC are the same as that defined in FLCAC. In all, we have six nodes in this layer. Each node performs a bell-shaped function defined as

$$f_i^{(2)}(u_{ij}^{(2)}) = -\frac{(u_{ij}^{(2)} - m_{jn}^{(l)})^2}{\sigma_{jn}^{(l)2}} \quad \text{and} \quad a_i^{(2)} = e^{f_i^{(2)}},\qquad(3.18)$$

where $u_{ij}^{(2)} = a_j^{(1)}$, $1 \le i \le 6$, $j = \lceil (i+1)/2 \rceil$, and $m_{jn}^{(l)}$ and $\sigma_{jn}^{(l)}$ are the mean and the standard deviation of the nth term of the input linguistic variable from node j in input layer, respectively. $n = 1$ if i is the odd node and $n = 2$ if i is the even node.

Layer 3: The links perform precondition matching of fuzzy control rules. According to fuzzy set theory, the fuzzy rule base forms a fuzzy set with dimensions $|T(C_a)| \times |T(y)| \times |T(p_l)|$($|T(x)|$ denotes the number of terms in $T(x)$). Consequently, there are eight rule nodes in this layer. Each rule node performs the fuzzy AND operation defined as

$$f_i^{(3)}(u_{ij}^{(3)}) = \min(u_{ij}^{(3)}; \forall j \in P_i) \quad \text{and} \quad a_i^{(3)} = f_i^{(3)},\qquad(3.19)$$

where $u_{ij}^{(3)} = a_j^{(2)}$ and $P_i = \{j|$ all j that are precondition nodes of the ith rule$\}$, $1 \le i \le 8$.

Layer 4: The nodes in this layer have two operating modes: *down-up* and *up-down*. In the down-up operating mode, the links perform consequence matching of fuzzy control rules. The term set of the output linguistic variable $\hat{z}$ is the same as that used in FLCAC. Therefore, there are four nodes in this layer. Each node performs a fuzzy OR operation to integrate the fired strength of rules that have the same consequence. Thus, we define

$$f_i^{(4)}(u_{ij}^{(4)}) = \max(u_{ij}^{(4)}; \forall j \in C_i) \quad \text{and} \quad a_i^{(4)} = f_i^{(4)},\qquad(3.20)$$

where $u_{ij}^{(4)} = a_j^{(3)}$ and $C_i = \{j|$ all j that have the same consequence of the ith term in the term set of $\hat{z}\}$, $1 \le i \le 4$. The up-down operating mode is used during the training period. The nodes in this layer and the links in layer five have functions similar to those in layer two. Each node performs a bell-shaped function defined as

$$f_i^{(4)}(u_{ij}^{(4)}) = -\frac{(u_{ij}^{(4)} - m_j^{(O)})}{\sigma_j^{(O)2}} \quad \text{and} \quad a_i^{(4)} = e^{f_i^{(4)}}, \quad (3.21)$$

where $u_{ij}^{(4)}$ is set to be $a_j^{(5)}$ obtained from the up-down operating nodes in layer five, and $m_j^{(O)}$ and $\sigma_j^{(O)}$ are the mean and the standard deviation of the jth term of $\hat{z}$, respectively, $1 \le i \le 4$, $j = 1$.

Layer 5: There are two nodes in this layer. One node performs the down-up operation for the actual decision signal $\hat{z}$. The node and its links act as the defuzzifier. The function used to simulate a center-of-area defuzzification method is approximated by

$$f_i^{(5)}(u_{ij}^{(5)}) = \sum_{j=1}^{4} m_j^{(O)}\sigma_j^{(O)}u_{ij}^{(5)} \quad \text{and} \quad a_i^{(5)} = U(\frac{f_i^{(5)}}{\sum_{j=1}^{4}\sigma_{ij}^{(O)}} - z_a), \quad (3.22)$$

where $u_{ij}^{(5)} = a_j^{(4)}$, $i = 1$, z_a is the decision threshold, and

$$U(x) = \begin{cases} 1 & \text{if } x \ge 0, \\ 0 & \text{otherwise.} \end{cases} \quad (3.23)$$

Clearly, $\hat{z} = a_i^{(5)}$ and a new connection will be accepted only if $\hat{z} = 1$. The other node performs the up-down operation during the training period. It feeds the desired decision signal z into the controller to adjust the link weights optimally. For this kind of node,

$$f_i^{(5)}(u_{ij}^{(5)}) = u_{ij}^{(5)}, \quad \text{and} \quad a_i^{(5)} = f_i^{(5)} \quad (3.24)$$

where $i = j = 1$ and $u_{11}^{(5)} = z$.

A hybrid learning algorithm is applied in the design of the NFCAC controller. The algorithm is a two-phase learning method. In phase one, a self-organized learning scheme is used to construct the rules and to locate the initial membership functions. In phase two, a supervised learning scheme is adopted to optimally adjust the membership functions for desired outputs. Training data must be provided for the learning process, in addition to the size of the term set for each input/output linguistic variable and the fuzzy control rules. The procedure for constructing the set of training data is the same as that described in the previous section. Given the input training data C_a, y, p_l, the desired output z, the fuzzy partitions $|C_a|$, $|y|$, $|p_l|$, $|\hat{z}|$, and the desired shape of the membership functions, the self-organized training would locate the membership functions and find the fuzzy control rules. If an initial knowledge base is employed (for example, FLCAC is used in this section) to help constructing an initial structure of the fuzzy control rules, a number of possible rule structures can be formed by slight

TABLE 3.4

The Rule Structure for the NFCAC

Rule	C_a	y	p_l	$\hat{z}$	Rule	C_a	y	p_l	$\hat{z}$
1	NE	N	NS	R	5	E	N	NS	WR
2	NE	N	S	WR	6	E	N	S	WA
3	NE	P	NS	R	7	E	P	NS	WA
4	NE	P	S	WR	8	E	P	S	A

modification of rules. Among all of the possible structures, the one that yields the minimum square error E for the training data is selected. E is defined as

$$E = \frac{1}{2}\sum_{j=1}^{N}[z(t_j) - \hat{z}(t_j)]^2, \tag{3.25}$$

where N is the number of training data, $z(t_j)$ and $\hat{z}(t_j)$ are the desired output and the actual output obtained at time t_j, respectively.

If an initial knowledge base is not provided, the initial locations of membership functions are estimated by using Kohonen's self organizing feature-maps algorithm and the *N-nearest-neighbors* scheme,[21] and the initial rule structure is constructed via GAs.[22] Based on the initial membership functions, an optimal rule structure shown in Table 3.4 was obtained by using GA in the self-organized learning phase. When the fuzzy logic rules were found, the NFCAC controller entered the supervised learning phase, in which the membership functions were adjusted optimally. Details of the hybrid learning algorithm could be referred to.[16]

Simulation results[16] show that the NFCAC scheme provides system utilization about 32% and 11% higher than the EBCAC and FLCAC schemes, respectively, and the NFCAC scheme requires only a fraction of the 10^3 order and the 10^1 order of training cycles, consumed by the NNCAC scheme and RBFCAC scheme, respectively.

3.6 Simulation Results and Discussion

Simulations were performed to test the effectiveness of the FLCAC, NNCAC, and NFCAC schemes. An ATM network is chosen to be the high-speed network supporting multimedia services. The input traffic is categorized into two types: real-time (type-1) and nonreal-time (type-2) traffic. Video and voice services are examples of type-1 traffic, while data services are examples of type-2 traffic. The network system provides two separate finite buffers with size K_i, in order to support different QoS requirements for type-i traffic, $i = 1$ and 2. When the buffer is full, incoming cells are blocked and lost.

The system reserves C_r portion of its capacity for type-1 traffic and the remaining $(1 - C_r)$ portion for type-2 traffic. When there is unused type-1 or type-2 capacity, it is used by the other type of traffic. In the simulations described here, $K_1 = K_2 = 100$ cells and $C_r = 0.8$. Also, the QoS requirement for type-1 traffic $QoS_1 = 10^{-5}$ and that for type-2 traffic $QoS_2 = 10^{-6}$. Details about the cell generation process for voice, video, and data sources as well as their respective parameters will not be addressed here and could be referred to.[16]

The performance of the CAC is evaluated via simulation by concerning the aspects of the cell loss ratio, the system utilization, and/or the training time under the constraint of QoS guarantee. The effective bandwidth CAC (EBCAC) scheme proposed in Reference 3, the fuzzy logic CAC (FLCAC) scheme, the neural-network-based CAC (NNCAC) scheme, the radial-basis-function CAC (RBFCAC) scheme, and the neural fuzzy CAC (NFCAC) scheme are compared and discussed. In the simulations, the FLCAC, NNCAC, RBFCAC, and NFCAC controllers are all equipped with the same peripheral processors. The sizes of training set and test set are all equal to 200, the number of repeated experiments is 20, and the standard deviation is less than 5%.

In order to elaborate more on the design process, the NFCAC scheme is used as an example to illustrate the way for the setting of membership functions and to show the eventual modifications with respect to the change of the traffic. The membership functions of the linguistic variables for type-1 and type-2 traffic are initially specified in the left-hand side of Figure 3.4(a) and Figure 3.4(b), respectively. As we know, the available capacity C_a, deduced from the quivalent capacity C_e of the existing calls, may possess estimation errors. Making network utilization as much as possible, we may employ an idea of "budget deficit" to over-assign the capacity. Thus, the mean value $m_{11}^{(l)}$ of the membership function of NE is set to be a negative value and the mean value $m_{12}^{(l)}$ of the membership function of E is set to be a value close to zero.

The behavior of the congestion indicator y could be monitored from the congestion and congestion-free states during a long-term simulation of the network operation. Thus, the membership functions of y could be initially optimized based on the obtained information. The mean value $m_{22}^{(l)}$ of the membership function of P would be set to be the mean value of the queue-length change rate during congestion-free periods, the mean value $m_{21}^{(l)}$ of the membership function of N would be set to be the mean value of the queue-length change rate during congestion periods, and let $\sigma_{21}^{(l)} = \sigma_{22}^{(l)} = m_{22}^{(l)} - m_{21}^{(l)}$. These parameters could be further optimized via GA by simulation, within which these parameters are coded as genes and the resulted cell loss ratio is used as a cost function.

The initial membership functions of the cell loss ratio p_l are set according to the QoS requirement. The mean value $m_{32}^{(l)}$ of the membership function of NS would be set to be the QoS requirement, the mean value $m_{31}^{(l)}$ of the member-ship function of S would be set to be a fraction of the QoS requirement, and the standard deviations would be set to be $\sigma_{31}^{(l)} = \sigma_{32}^{(l)} = m_{32}^{(l)} - m_{31}^{(l)}$. As a result, there exists a safety margin between the membership functions of

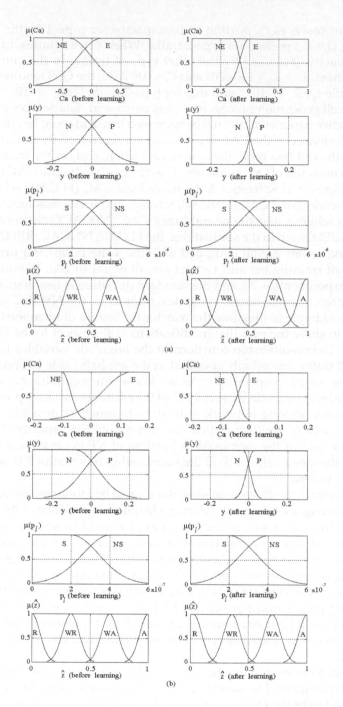

FIGURE 3.4
Membership functions of C_a, y, pl, and $\hat{z}$ for (a) type-1 traffic; (b) type-2 traffic.

terms S and NS provided to tolerate the dynamic behavior of the network operation and insure the QoS requirement.

Here, little information about the setting of initial values for the mean $m_j^{(O)}$ of the term set $T(\hat{z})$ could be employed; therefore, the values of $m_j^{(O)}$ are set to be equally spaced in the range of [0,1]. Based on the initial membership functions, an optimal rule structure was obtained by using GA in the self-organized learning phase. When the fuzzy logic rules are found, the NFCAC controller enters the supervised learning phase, in which the membership functions are adjusted optimally.

In the design of the NFCAC scheme, three different values of the learning constant η are used for the variables C_a, y, p_l, and $\hat{z}$. η is set to zero for p_l because the membership functions are specified by the QoS constraint and should not be modified. $\eta = 0.001$ is used for y because the membership functions of y are initially optimized. As for C_a and $\hat{z}$, their initial membership functions are heuristically set and required further optimization in the supervised learning phase. Thus, $\eta = 0.01$ is used. The use of different η may drastically reduce the training time required in the supervised learning phase. The learned membership functions of the linguistic variables for type-1 and type-2 traffic are shown in the right-hand side of Figure 3.4(a), and Figure 3.4(b), respectively.

For type-1 traffic in Figure 3.4(a), it can be found that the difference of the membership functions before and after learning are: For the membership functions of C_a, the mean value $m_{11}^{(I)}$ of the membership function of NE is properly modified from −0.4 to −0.27. Similarly, the mean value $m_{12}^{(I)}$ of the membership function of E is properly modified from 0.16 to −0.02. There is a drastic change for membership functions of C_a because their initial values are heuristically set. It can also be found that the membership functions of y are more compressed because a little change on y may have a great impact on the CAC decision. Since η for p_l is chosen to be zero, its membership functions are not changed. Finally, it could be found that the position of the decision $\hat{z}$ for WR and WA are accordingly modified. The mean $m_1^{(O)}$ of the membership function of R is slightly increased from 0 to 0.05, representing that the effect of "Reject" is decreased (i.e., the final decision is a weighted sum of R, WR, WA, and A). Also, the mean $m_3^{(O)}$ of the membership function of WA is slightly increased from 0.67 to 0.72, representing that the effect of "Weak Accept" is increased. The changes of membership functions of $\hat{z}$ imply that the NFCAC controller prefers to accept new calls. This phenomenon demonstrates that the NFCAC controller intends to recover some system bandwidth which the equivalent capacity method wastes due to over-estimation, while keeping the QoS contract. It may be the reason for the utilization improvement of the proposed NFCAC controller, which will be shown below. Similar results could be found for type-2 traffic in Figure 3.4(b).

Figure 3.5 shows the cell loss ratios of an ATM traffic controller employing the NFCAC scheme, and the EBCAC, FLCAC, NNCAC, and RBFCAC schemes. It is found that the QoSs for both types of traffic are indeed guaranteed

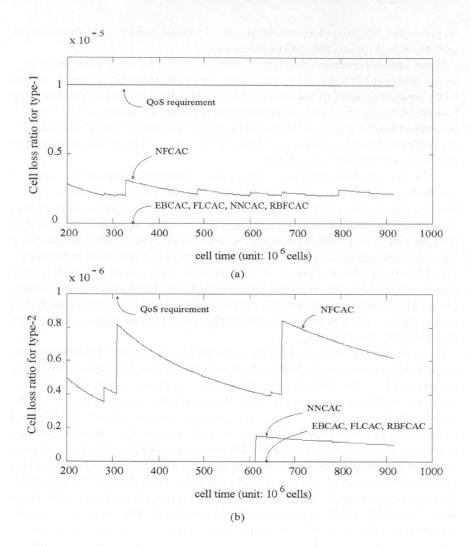

FIGURE 3.5
Cell loss ratio for (a) type-1 traffic; (b) type-2 traffic.

for all of these control schemes. Figure 3.6 shows the system utilization of the NFCAC scheme and the four schemes. We can find that the utilization of the NFCAC scheme is slightly greater than that of the NNCAC and the RBFCAC schemes; the system utilizations of NFCAC, NNCAC, and RBFCAC are 91%, 90.5%, and 89%, respectively; and the NFCAC scheme offers about 32% and 11% greater system utilization than the EBCAC scheme and the FLCAC scheme. It is because NFCAC can incorporate the domain knowledge obtained from both the analytical-based method (the equivalent capacity scheme[1] is employed in the bandwidth estimator) and the measurement-based method (the system statistics of the queue length, the change rate of the queue length, and the cell loss ratio are considered in the congestion controller).

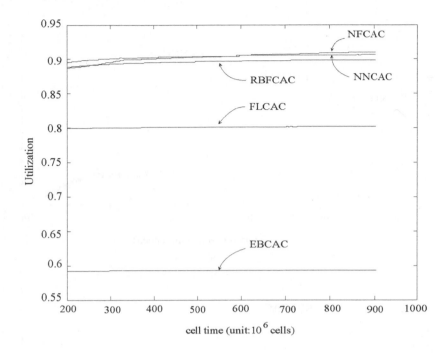

FIGURE 3.6
System utilization.

Also, the reason for the performance improvement is that NFCAC possesses the learning capability of the neural network.

Figure 3.7 shows the training time required for the NFCAC scheme and the NNCAC, RBFCAC schemes. Here, a widely used back-propagation learning algorithm was employed to adjust the membership functions (i.e., represented in terms of weights) of the multi-layer neural fuzzy network and neural network for the NFCAC and NNCAC schemes, while the RBFCAC scheme is basically trained by the hybrid learning rule: unsupervised learning in the input layer and supervised learning in the output layer. It is found that NFCAC has training time of 7 (4) epochs, while RBFCAC and NNCAC have training time of 103 (40) and 5×10^4 (6×10^2), respectively, for type-1 (type-2) traffic. The NFCAC has higher learning speed than the RBFCAC and NNCAC. One reason is that the neural fuzzy network is a structured network, thus the NFCAC controller can easily adopt the domain knowledge of conventional control methods to construct the initial rule structure and the parameters of the membership functions, providing an excellent initial guess in adjusting its weights; on the contrast, the neural network is a nonstructured network, which cannot incorporate domain knowledge about system. The other reason is that the neural fuzzy network has simpler structure than the neural network; the number of tuning parameters used in the neural fuzzy network is quite small, as compared to the neural network such as

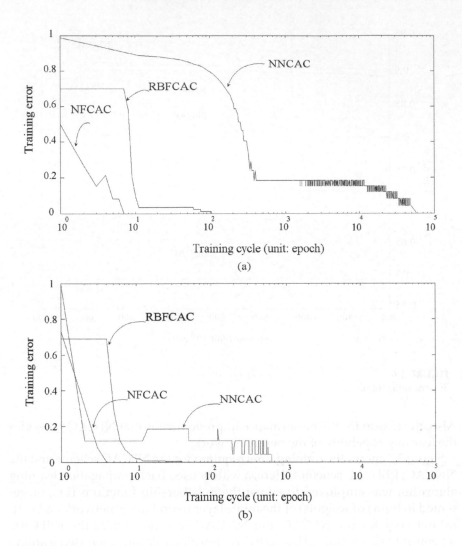

FIGURE 3.7
Training cycles needed for (a) type-1 traffic; (b) type-2 traffic.

multilayer feedforward network and RBFN considered here. In this section, there are only 16 weighting parameters used in NFCAC, while there are 150 and 480 weighting parameters required for the RBFCAC and NNCAC, respectively. It is also noted that the RBFCAC scheme has less learning time than the NNCAC scheme. It is because the RBFCAC scheme can have the proper initial setting of means and variances for the Gaussian activation functions during unsupervised learning according to the prior knowledge, and it has only one layer of connection needed to be trained by supervised learning.

As usually noted, RBFCAC can have faster training speed than NNCAC but cannot achieve the same accuracy as NNCAC does. In the simulations, we first adopted the same set of data used to train NFCAC and NNCAC for RBFCAC. However, it was found that RBFCAC finally violated the QoS contracts due to its error decision of accepting more users than it should be. In order to provide QoS guarantee for RBFCAC, we have to prepare much more training data, especially those around the acceptance/rejection boundary. This will increase the training time of RBFCAC in each epoch than those required by NFCAC and NNCAC. Moreover, the overall processing time of RBFCAC is greater than that needed by either NFCAC or NNCAC because RBFCAC uses more nodes (compared with NFCAC) and a more complicated activation function (compared with NNCAC). All these would degrade the performance of RBFCAC in real application.

3.7 Summary

This chapter discusses intelligent computational approaches for CAC in high-speed multimedia networks. Some of the possible application of fuzzy logics, neural networks, and neural fuzzy networks in traffic control related issues for high-speed multimedia networks are designed and illustrated.

For a skilled engineer, expert knowledge about traffic control can be easily employed in the design of FLCAC. New input linguistic variables which affect the controlling performance can be introduced to adjust the controller to optimum operation. On the other hand, the NNCAC method is suitable for designers who are not familiar with fuzzy logic control schemes, or lack requisite knowledge of CAC. For designers who intend to provide an adaptive controller under dynamic environment or require to monitor the operation of the controller, the NFCAC scheme will be the best choice. NFCAC can automatically construct a rule structure and can self-calibrate the parameters of the membership functions by learning from examples. It is extremely effective in dealing with real-world imprecision, and in showing insight information for the way it works. It can be easily trained and enhance system utilization.

Generally, most of the intelligent computational approaches proposed in the literature are a combination of analytical methods and simulation experimentations. From the survey presented in this chapter, it is apparent that fuzzy logics, neural networks, and neural fuzzy networks can play an important role in the CAC of high-speed multimedia networks because they can provide real-time, model-free, and adaptive control. However, some issues in system design, stability, convergence, and compatibility with proposed standards should be extensively evaluated to prove the adequacy of the proposed schemes in the aspects of speed of operation, sensitivity to design parameters, robustness, and integration with existing network peripheral devices.

3.8 Problems

1. Write a program to simulate the M/M/1 queueing system.

2. If the fuzzy partition of X, denoted by $|T(X_i)|$, is $|T(X_i)| = 3$, for $i = 1, \ldots, 3$, what is the maximum number of fuzzy control rules in the fuzzy logic controller?

3. Minimize the function $f(x, y, z) = x^2 + y^2 + z^2$ using GA, where $-512 \leq x, y, z \leq 512$. Use 10-bit coding for each substring.

4. Use a 3-layered back-propagation neural network to approximate the function $f(x) = x^2 + y^2$.

5. Derive the gradient decent learning algorithm for the neural fuzzy system with triangular membership functions.

6. Derive the gradient decent learning algorithm for the neural fuzzy system with trapezoidal membership functions.

7. Identify the following system using various neural fuzzy systems: $y = \sin(\pi x_1) + 2 \times \cos(\pi x_2)$. The input-output data pairs for training are generated by sampling (x_1, x_2) within $(-1, 1]$ and the desired output y^d is normalized within $[0, 1]$. Twenty fuzzy logic rules are used to express the input-output relationship. A total of 20 training pairs are used. A neural fuzzy system with triangular membership functions are used and the learning will stop until training error $E < 0.02$, where

$$E = \sum_{k=1}^{20} (y_k - y_k^d)^2.$$

8. Repeat Problem 7 and use a neural fuzzy system with trapezoidal membership functions.

References

1. R. Guèrin, H. Ahmadi, and M. Naghshineh, "Equivalent capacity and its application to bandwidth allocation in high-speed networks," *IEEE J. Select. Areas Commun.*, vol. 9. no. 7, pp. 968–981, September 1991.

2. A.I. Elwalid and D. Mitra, "Effective bandwidth of general Markovian traffic sources and admission control of high speed network," *IEEE/ACM Trans. Networking*, vol. 1, no. 3, pp. 329–343, June 1993.

3. G. Kesidis, J. Walrand, and C.S. Chang, "Effective bandwidths for multiclass Markov fluids and other ATM sources," *IEEE/ACM Trans. Networking*, vol. 1, no. 4, pp. 424–428, August 1993.

4. A.I. Elwalid, D. Heyman, T. V. Lakshman, D. Mitra, and A. Weiss, "Fundamental bounds and approximations for ATM multiplexers with applications to video teleconferencing," *IEEE J. Select. Areas Commun.*, vol. 13, no. 6, pp. 1004–1016, August 1995.

5. H. Saito, "Call admission control in an ATM network using upper bound of cell loss probability," *IEEE Trans. Commun.*, vol. 40, no. 9, pp. 1512–1521, September 1992.

6. D. Tse and M. Grossglauser, "Measurement-based call admission control: analysis and simulation," *Infocom'97*, 983–991, 1997.

7. C. Douligeris and G. Develekos, "Neural-fuzzy control in ATM networks," *IEEE Commun. Mag.*, vol. 35, no. 5, pp. 154–162, May 1997.

8. A.R. Bonde and S. Ghosh, "A comparative study of fuzzy versus 'fixed' thresholds for robust queue management in cell-switching networks," *IEEE/ACM Trans. Networking*, vol. 2, August, 1994.

9. T.D. Ndousse "Fuzzy neural control of voice cells in ATM networks," *IEEE J. Select. Areas Commun.*, vol. 12, pp. 1488–1494, December 1994.

10. R.G. Cheng and C.J. Chang, "Design of a fuzzy traffic controller for ATM networks," *IEEE/ACM Trans. Networking*, vol. 4, no. 3, pp. 460–469, June 1996.

11. A. Hiramatsu, "ATM communications network control by neural networks," *IEEE Trans. Neural Networks*, vol. 1, no. 1, pp. 122–130, March 1990.

12. P. Tran-Gia and O. Gropp, "Performance of a neural net used as admission controller in ATM systems," *IEEE Globecom '92*, 1303–1309, 1992.

13. S.Y. Youssef, I.W. Habib, and T.N. Saadawi, "A neural network control for effective admission control in ATM networks," *IEEE ICC '96*, 434–438, June 1996.

14. L.F. Lin, Z.S. Eul, R.G. Cheng, and C.J. Chang, "Implementation of an admission controller for high-speed multimedia networks," *IEEE ICCE '98*, 254–255, May 1998.

15. R.G. Cheng and C.J. Chang, "Neural network connection admission control for ATM networks," *IEE Proc. Commun.*, 144(2) 93–98, April 1997.

16. R.G. Cheng, C.J. Chang, and L. F. Lin, "A QoS-provisioning neural fuzzy connection admission controller for multimedia high-speed networks," *IEEE/ACM Trans. Networking*, 7(1), 111–121, February 1999.

17. N. Yin, S.Q. Li, and T. E. Stern, "Congestion control for packet voice by selective packet discarding," *IEEE Trans. Commun.*, 38, 674–683, May 1990.

18. D.E. Goldberg, *"Genetic algorithms,"* Addison-Wesley, Reading, MA, 1989.

19. R.G. Cheng, "The traffic control strategies for B-ISDN using fuzzy sets theory," Master thesis, Department of Commun. Eng., National Chiao Tung Univ., 1993.

20. C.T. Lin and C.S.G. Lee, "Neural-network-based fuzzy logic control and decision system," *IEEE Trans. Computers*, 40(12), 1320–1336, December 1991.

21. S.T. Welstead, *Neural Network and Fuzzy Logic Applications in C++*, John Wiley & Sons, New York, 1994.

22. Y. Lin and G.A. Cunningham III, "A new approach to fuzzy-neural system," *IEEE Trans. Fuzzy Systems*, 190–198, May 1995.

4

Fuzzy Connection Admission Control for ATM Networks

Kiyohiko Uehara and Kaoru Hirota

CONTENTS

4.1   Introduction

Asynchronous transfer mode (ATM) is the most promising technology for supporting broadband multimedia communication services.[1] In ATM networks, user information is divided into fixed length units, called *cells*. Excessive input of cells to the networks causes congestion and some cells may be lost. In order

Portions reprinted, with permission, from *IEEE Journal on Selected Areas in Communications;* 15(2), 179–190; February 1997. © 1997 IEEE.

0-8493-1075-X/01/$0.00+$.50
© 2001 by CRC Press LLC

to avoid this situation, connection admission control (CAC) is performed in which cell loss ratio (CLR) is estimated in advance of the transmission and an incoming call is judged as to whether it can be accepted or not.

Conventionally, the estimation of CLR is often performed on the basis of analytical models of the cell generation process in terminals and ATM switch architectures.[5-9] However, a wide variety of cell generation patterns[1-4, 16, 17] and increasingly complex ATM switches make the construction of the analytical models difficult. Thus, approximation is often required in constructing the analytical models. It results in the CLR estimation at excessively high values and therefore the multiplexing gain is degraded.

In order to solve the problems, learning systems have been applied to the CLR estimation.[10-15] The learning systems extract the relation between the CLR and the number of connections in each transmission rate class by using observed CLR data. Thus, such observation-based methods do not require the analytical models and provide adaptive mechanisms to cope with dynamic situations in ATM switches. The conventional observation-based methods, however, cannot always perform CAC guaranteeing the allowed CLR because they estimate the average of CLRs. In order to guarantee the allowed CLR, the upper bound of CLR has to be estimated.[18]

The fuzzy inference approach is feasible for estimating the upper bound of CLR because the then-parts can provide the possibility distribution of CLR for the conditions given by the if-parts. The upper bound of CLR can be obtained from the possibility distribution. The conventional fuzzy inference methods, however, often estimate the upper bound of CLR at excessively high values. It stems from the fuzziness increase and specificity decrease in inference consequences. The conventional fuzzy-inference methods do not have the scheme for controlling the fuzziness and specificity. In contrast, the inference method based on a weighted mean* of fuzzy sets,[19, 20] has the scheme for controlling the fuzziness and specificity and can avoid the estimation at excessively high values. Therefore, in this chapter, the fuzzy inference based on a weighted mean of fuzzy sets is applied to the CLR estimation.[18] Moreover, the CAC method is proposed which takes advantage of the effective properties in this inference method.[18]

In this chapter, first, ATM networks and their CAC are introduced and the problems in conventional methods for CLR estimation are described. Then, the feasibility of the fuzzy inference approach is discussed in order to solve the problems. Second, the method is proposed for the CLR estimation by using the fuzzy inference based on a weighted mean of fuzzy sets. The estimation scheme is provided with a learning mechanism. The energy functions are considered for effectively estimating the upper bound of CLR. Moreover, the real-time compensation method for the estimation is proposed, taking advantage of the inference method based on a weighted mean of fuzzy sets. The simulation results are also shown in which the

---

* In this chapter, a weighted mean is used for a weighted arithmetic mean.

upper bound of the CLR is well extracted. Third, a CAC method is proposed which applies the fuzzy inference based on the weighted mean of fuzzy sets. In this method, fuzzy rules, in the area with no observed CLR data, are generated successively by using adjacent fuzzy rules. Simulation studies are conducted to show the feasibility of the proposed CAC method. Finally, this chapter concludes with brief discussions.

4.2 ATM Networks and Their Control by Fuzzy Inference

In this section, first, ATM networks and their connection admission control are introduced. Then, the problems in conventional methods for this control are described. In order to solve the problems, the feasibility of the fuzzy inference approach is considered.

4.2.1 Connection Admission Control in ATM Networks

In ATM networks, transmission data from source terminals are divided into 48-byte length units and a header including destination address is added to each unit. Each transmission unit with its header, called a *cell*, can be sent at the allowed rate on demand. The cells from the terminals are multiplexed *asynchronously* in the networks. Thus, ATM networks can support a wide variety of transmission rates and provide high transmission efficiency by asynchronous multiplexing.

Although ATM networks have the advantages mentioned above, cells might be lost in ATM switches if cells are excessively fed into the networks. In order to avoid this situation, the terminals are required to declare their transmission rates as traffic parameters, e.g., peak cell rate (PCR) and sustainable cell rate (SCR), in advance of transmission. According to these declarations of transmission rates, ATM switches judge whether the required quality-of-service (QOS), evaluated by *cell loss ratio* (CLR) in this chapter, can be achieved or not. Although it must be guaranteed that the QOS objectives are satisfied under the specified *cell delay variation* (CDV), this chapter assumes for simplicity as a first step that the traffic parameters are transformed by taking account of CDV or that CDV is removed by shaping.[1]

The transmission rates are often classified into a number of classes on the basis of the transmission rates such as PCR and SCR. Namely, terminals select one of the transmission rate classes in advance of transmission. When a call request comes from the terminals, ATM switches have to predict whether the required quality in CLR can be achieved or not if the call is accepted. If ATM switches judge the required quality can be achieved, they accept the call. Otherwise, they reject the call. This process is called *connection admission control* (CAC). Therefore, CAC requires the estimation of CLR from the number of connections in each transmission rate class.

In conventional methods, the estimation of CLR is often performed on the basis of analytical models of the cell generation process in terminals and ATM switch architectures.[5-9] The cell generation process, however, has a wide variety of patterns[1-4, 16, 17] and ATM switches have become increasingly complex in order to attain higher performance. It makes the construction of the analytical models difficult. Moreover, the analysis often requires approximation with excessive estimation of CLR.

In order to solve the problems, learning systems have been applied to the estimation of CLR.[10-15] In the methods, the CLRs are observed in ATM switches, and the learning systems extract the relation between the CLR and the number of connections in transmission rate classes. Such observation-based methods are quite effective because analytical models do not have to be constructed and ATM switches can be equipped with adaptive mechanisms to cope with dynamic situations. The conventional observation-based methods, however, cannot always perform CAC guaranteeing the allowed CLR as described in the following.

4.2.2 Fuzzy Inference Approach to Connection Admission Control

The relation between CLR and the number of connections is often nonlinear. Thus, in the observation-based CAC, artificial neural networks (ANNs) have often been adopted.[10-15] The ANNs, however, learn only the average of observed CLR data.[13] The observed CLR data are expected to disperse because a wide variety of cell generation patterns exists even in a transmission rate class. The CAC must guarantee the allowed CLR for any cell arrival process satisfying the traffic parameters of the transmission class. This means that average learning may not guarantee the allowed CLR in CAC.[18] Even if the maximum value of observed CLR at each number of connections is used, it cannot guarantee the allowed CLR. This is because the maximum values also disperse and this average learning provides the average of these dispersing maximum values. From this point of view, the estimation of the possibility distribution of CLR is needed in order to guarantee the allowed CLR in CAC.[18] That is, if possibility distribution can be obtained, its upper bound can make possible the CAC guaranteeing the allowed CLR. Figure 4.1 depicts the comparison between average estimation and upper-bound estimation, where the symbols "+" show the observed CLR data.

The nonparametric approach has also been studied[1-4] for estimating the upper bound of CLR because the cell arrival process in practice is not specified. Although this approach is effective to guarantee the allowed CLR, it tends to estimate excessively high CLR and result in lower multiplexing

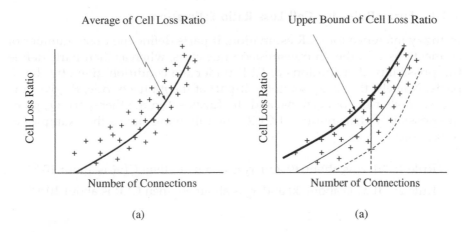

FIGURE 4.1
Average estimation and upper-bound estimation.

gain in a practical environment. Even in constructing the analytical models without use of the nonparametric approach, approximations have been often performed so as to guarantee the allowed CLR and make the multiplexing gain lower in practice.

From the discussion above, the fuzzy inference approach is feasible.[18] This is because the then-part of each fuzzy rule can give the possibility distribution of CLR for the number of connections covered with the if-part in the fuzzy rule. The inference consequence therefore provides the estimated possibility distribution of CLR for the input of the number of connections. The conventional fuzzy inference,[22] however, tends to estimate CLRs having excessively large values because the fuzziness and specificity of inference consequences often become larger and smaller than those of then-parts, respectively. Moreover, it is difficult to derive a learning algorithm for the conventional inference scheme with the fuzzy consequences which give the estimated possibility distribution of CLR. Therefore, we adopt the fuzzy inference method based on a weighted mean of fuzzy sets[19, 20] in order to solve these problems.[18] It provides the useful properties in its application to CAC, as described later.

4.3 Estimation of Upper Bound of Cell Loss Ratio by Fuzzy Inference

By using the properties of the inference method based on a weighted mean of fuzzy sets, effective CLR estimation can be achieved. This section describes the scheme for CLR estimation based on the inference method.

4.3.1 Fuzzy Rules for Cell Loss Ratio Estimation

In fuzzy inference for CLR estimation, if-parts define the fuzzy number of connections x_i in the ith transmission rate class, whereas then-parts define the possibility distributions of CLR under the condition given by the if-parts. Namely, the fuzzy set in the if-part of the jth fuzzy rule, P_j, gives the fuzzy number of connections, and the fuzzy set in the then-part, Q_j, gives the possibility distribution of CLR. The following shows the examples of fuzzy rules:

Rule 1: "If x_1 is about 10 and x_2 is about 20, then CLR is about 10^{-9}."

Rule 2: "If x_1 is about 30 and x_2 is about 50, then CLR is about 10^{-8}."

.

.

.

Rule n_c: "If x_1 is about 100 and x_2 is about 30, then CLR is about 10^{-5}."

In these examples, the number of the transmission rate classes is two and thus P_j is defined by a two-dimensional fuzzy set. In this chapter, the transmission rate is classified into a number of classes, which also means that the other parameters obtained by the declared transmission rates, like burstiness, are considered to be quantized into the classes taking the transmission rates into account. This decreases the complexity in the CLR estimation and makes it possible without specific analytical models.

The fuzzy sets in each fuzzy rule are automatically extracted and tuned by the learning algorithm explained later. In this inference, the fact $\tilde{P}$, namely the input to the inference engine, is the number of connections in each transmission rate class. Thus, $\tilde{P}$ is given by a singleton. The inference consequence $\tilde{Q}$, namely the output of the inference engine, gives the possibility distribution of CLR, reflecting the effect of the inference method based on a weighted mean of fuzzy sets and the learning algorithm described later.

In estimating CLR, this chapter adopted the simplified form of the operations for fuzzy inference based on a weighted mean of fuzzy sets. The membership functions are defined for this simplified form as follows:

$$P_j(x) = f(x; x^\circ_{P_j}, \eta_{P_j}) \tag{4.1}$$

$$Q_j(y) = g(y; y^\circ_{Q_j}, \eta_{Q_j}) \tag{4.2}$$

$$\tilde{Q}(y) = g(y; y^\circ_{\tilde{Q}}, \eta_{\tilde{Q}}) \tag{4.3}$$

where

$$f(x; x^°, \eta) = \begin{cases} \max\left[\frac{2}{\eta}(x - x^°) + 1, 0\right], & x \le x^° \\ \max\left[-\frac{2}{\eta}(x - x^°) + 1, 0\right], & x > x^°, \end{cases} \tag{4.4}$$

$$g(y; y^°, \eta) = \exp\left[-\frac{(y - y^°)^2}{2\eta^2}\right]. \tag{4.5}$$

The parameters $x^°$ and $y^°$ give central positions of membership functions while η is the parameter for their widths which is closely related with fuzziness and specificity. Each of the membership functions is represented by only these two parameters namely positional and width parameters. It reduces the computational complexity in both the inference mode and the learning mode. In this simplification, the fuzziness and specificity are increased and decreased at the same time, respectively, as the value of the width parameter η is larger. By using this simplification, the inference consequence $\tilde{Q}$ is obtained by

$$y_{\tilde{Q}}^° = \frac{\sum_j \tilde{p}_j y_{\tilde{Q}_j}^°}{\sum_j \tilde{p}_j} \tag{4.6}$$

$$\eta_{\tilde{Q}} = \frac{\sum_j \tilde{p}_j \eta_{Q_j}}{\sum_j \tilde{p}_j}. \tag{4.7}$$

4.3.2 Learning Algorithm for Adjusting Fuzzy Rules

The inference method based on a weighted mean of fuzzy sets can control the fuzziness and specificity of its final inference consequence by tuning the width parameters of membership functions in then-parts. Here, the learning algorithm is considered in order to tune the width parameters of membership functions, together with their positional parameters. Thereby, the possibility distribution of CLR can be obtained without excessively high value in its estimation.

In the learning algorithm, energy functions play an important role for the estimation of the possibility distribution. The center of the distribution is given by the average of the observed data. Thus, the parameter $y_{\tilde{Q}}^°$ can be obtained by minimizing the energy function below:[18]

$$E^° = \frac{1}{2}\sum_s (y_{\tilde{Q}}^° - \hat{y}_s)^2. \tag{4.8}$$

Here, $\hat{y}_s$, $(s = 1, 2, \ldots, S)$, denote observed CLRs which are given as target data. In the learning with the energy function defined in Eq. (4.8), the parameters of fuzzy sets in each fuzzy rule are tuned by the error back-propagation algorithm.[23]

In terms of adjusting the width of the membership functions in then-parts, an energy function E_η is proposed as follows:[18]

$$E_\eta = \frac{1}{2}\sum_s (\tilde{Q}(\hat{y}_s) - \hat{\alpha})^2 \qquad (4.9a)$$

$$\hat{\alpha} = \begin{cases} 1, & \tilde{Q}(\hat{y}_s) \le \bar{\alpha} \\ 0, & \tilde{Q}(\hat{y}_s) > \bar{\alpha} \end{cases} \qquad (4.9b)$$

$$\bar{\alpha} = h(\bar{\alpha}_{old} + \delta), \quad \begin{cases} \delta > 0, & \alpha_{min} < \alpha_0 \\ \delta = 0, & \alpha_{min} = \alpha_0 \\ \delta < 0, & \alpha_{min} > \alpha_0 \end{cases} \qquad (4.9c)$$

$$\alpha_{min} = \min_s \tilde{Q}(\hat{y}_s) \qquad (4.9d)$$

where α_0 denotes the threshold for obtaining the upper bound of CLR and $\bar{\alpha}_{old}$ is the value of $\bar{\alpha}$ at one epoch before in adjusting. The function h is a nondecreasing function with its range in [0, 1]. Moreover, the height of the membership function $Q_j(y)$ is fixed to one like Eq. (4.2). In the learning with the energy function defined in Eqs. (4.9), η_{Q_j} is tuned by the error back-propagation algorithm.

In the energy function defined by Eqs. (4.9), the target data $\hat{\alpha}$ is dynamically changed as follows. If $\tilde{Q}(\hat{y}_s)$ is smaller than or equal to $\bar{\alpha}$, the target data $\hat{\alpha}$ is set to one. Thus, the membership function of inference consequence is forced to the highest value of membership grade. As a result, it makes the width of inference consequence larger because the height of the inference consequence is fixed to one. On the other hand, if $\tilde{Q}(\hat{y}_s)$ is larger than $\bar{\alpha}$, the target data $\hat{\alpha}$ is set to zero. Thus, the membership function of inference consequence is forced to the lowest value of membership grade. As a result, it makes the width of inference consequence smaller. Figure 4.2 illustrates these processes mentioned above. The width of the membership function $\tilde{Q}_j(y)$ is adjusted in this way and then the widths of inference consequences are balanced in their certain values. The balanced point depends on the value of $\bar{\alpha}$.

If the value of $\bar{\alpha}$ is set to a smaller number, the width of $\tilde{Q}(y)$ is going to be smaller. Conversely, if the value of $\bar{\alpha}$ is set to a larger number, the width is going to be larger. The dynamics in Eqs. (4.9) controls the value of $\bar{\alpha}$ so that the

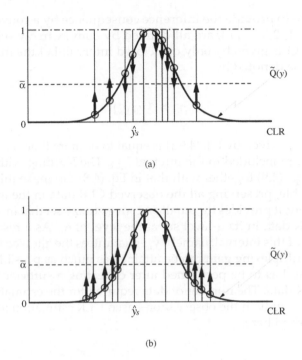

FIGURE 4.2
Learning process for adjusting the widths of inference consequences.

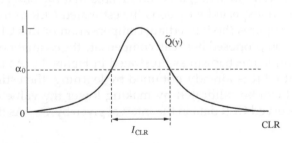

FIGURE 4.3
Estimated interval of cell loss ratio.

observed data are included in the interval given by the inference consequence at a threshold α_0. This interval, I_{CLR}, is the α-level set of $\tilde{Q}$ at the level of α_0. That is,

$$
\begin{aligned}
I_{CLR} &= \{y \mid \tilde{Q}(y) \geq \alpha_0\} \\
&= \tilde{Q}_{\alpha_0}.
\end{aligned}
\tag{4.10}
$$

Figure 4.3 explains how to obtain the interval I_{CLR} from α_0 and a final inference consequence. The inference method based on a weighted mean of fuzzy sets can prove to provide the inference consequence by a convex fuzzy set when Q_j, $(j = 1, 2, \ldots, n)$, are all defined by convex fuzzy sets. Thus, an α-level set of $\tilde{Q}$ is given by only one closed interval. Let the interval given by Eq. (4.10) be denoted by

$$I_{CLR} = [\tilde{y}_l, \tilde{y}_u]. \tag{4.11}$$

As long as α_{min} given by Eq. (4.9d) is equal to or more than α_0, all observed CLR data $\hat{y}_s$ are included in the interval I_{CLR}. The learning with the energy function in Eqs. (4.9) together with that in Eq. (4.8) can make this interval as small as possible, preserving all the observed CLR data in the interval completely. Namely, it provides the smallest convex fuzzy set that includes all the observed CLR data in its α-level set at the level of α_0. As a result, the least upper bound of this interval, namely $\tilde{y}_u$, guarantees the allowed CLR in CAC and higher multiplexing gain can be attained as much as possible.

The learning has to be performed after collecting a sufficient number of observed CLR data. The number of data required for the estimation is a subject for further study in the observation-based CLR estimation including the proposed method here.

4.3.3 Real-Time Compensation for CLR Estimation Errors

In operation of ATM switches, it is conceivable that the observed CLR after accepting a call unexpectedly exceeds the estimated CLR in the CAC stage. This situation requires the immediate compensation of this CLR estimation error. The method proposed here can compensate the estimation errors in real time with a simple mechanism as illustrated in Figure 4.4. As the possibility distribution of CLR is already obtained by learning, the estimated upper bound of CLR can be calibrated by making lower the value of α_0 so as to include the observed CLR data in the interval given by $\tilde{Q}_{\alpha_0}$. As the value of α_0

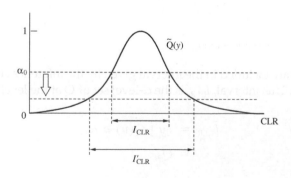

FIGURE 4.4
Self-compensation mechanism for estimation errors.

is lower, the width of the interval is wider. In Figure 4.4, this newly obtained interval is denoted by I'_{CLR}. This method is very effective because its process is not time-consuming compared with the error back-propagation algorithm and therefore the fast compensation can be achieved by its real-time operation. In the following, a simple approach based on this process is presented.

If the observed CLR $\hat{y}_s$ exceeds the estimated value in the exchange operation, the value of α_0 is replaced by that of $\tilde{Q}(\hat{y}_s)$. Namely, I'_{CLR} is given by

$$I'_{CLR} = \tilde{Q}_{\alpha'_0}, \qquad \alpha'_0 = \tilde{Q}(\hat{y}_s).$$

It makes the interval I_{CLR} wider so as to include the value $\tilde{Q}(\hat{y}_s)$.

The degree of widening the interval depends on the shape of $\tilde{Q}(y)$ at each value of x. In this way, the CLR estimation can be performed for guaranteeing the allowed CLR at the next CAC stage. This compensation mechanism takes advantages of the adopted inference method. It can be provided because the inference method adopted here can estimate the possibility distributions of CLR in convex forms. After the above-mentioned compensation, observed CLR data continue to be collected and then learning is conducted in the next learning period using these data. In this way, α'_0 is returned to a higher value, namely the original value α_0 before compensation, by tuning the parameters of membership functions in fuzzy rules.

As has been mentioned above, the proposed method provides the upper bound estimation of CLR and self-compensation of estimation errors at the same time. This is the advantage of estimating the possibility distribution of CLR by using the inference method based on a weighted mean of fuzzy sets.

4.3.4 Simulation Studies on Upper Bound Estimation for Cell Loss Ratio

In the simulation, one transmission-rate class characterized by PCR and SCR was considered for simplicity. The transmission rate class is set for SCR at 98 kb/s and PCR at 6.3 Mb/s. The traffic sources can select this class as long as their traffic rate is shaped in this rate pattern, which means a wide variety of traffic patterns could be multiplexed in one transmission rate class. Namely, the dispersed CLR data can be observed because a wide variety of traffic patterns may exist even in one transmission rate class as mentioned above. In order to show the ability of upper bound estimation effectively in the simulation, widely dispersed CLR data are needed. Considering this point and simplicity for simulation, two different types of traffic sources are intentionally multiplexed in the transmission rate class. One of the traffic sources is on–off traffic in which the rate in on-period is 6.3 Mb/s, the average length of on-period is 1.433 cell time, and its average rate is 98 kb/s. The other traffic source is constant bit rate (CBR) traffic at the rate of 98 kb/s. The ratio of connection numbers of these two traffic sources was decided by uniformly random numbers. These two traffic sources have quite different rate patterns and therefore the dispersed CLR data may be observed. As long as the declared traffic parameters are satisfied, such a wide variety of traffic could arrive in practice and their characteristics other than declared parameters cannot be known in advance. This requires the upper bound estimation of CLR in order to guarantee the allowed CLR.

For background traffic, two different types of traffic sources are also multiplexed in the same way as the foreground traffic. One of the traffic sources is on–off traffic in which the rate in on-period is 51.84 Mb/s, the average length of on-period is 1.433 cell time, and its average rate is 810 kb/s. The other traffic source is CBR traffic at the rate of 810 kb/s. The ratio of connection numbers of these two traffic sources was decided by uniformly random numbers and the total number of the connections is eighty.

The ATM switch was assumed to have a single buffer with 32-cell length and output rate was 155.52 Mb/s. The cells arriving from the traffic sources mentioned above were multiplexed into this buffer. The number of fuzzy rules was seven, and the membership functions for fuzzy rules and inference consequences were defined with Eqs.(4.1), (4.2), and (4.3). The membership functions $P_j(x)$ were initially placed at equally spaced positions so as to cover the input space for the simulation, as exemplified in Figure 4.7. The membership functions $Q_j(y)$ were initially placed randomly in the range of the observed data and their widths were also randomly set within the range, both of which are to be tuned by the learning algorithm. The data points in the number of connections are decimated to reduce the time required for this simulation. The simulation was conducted by using SPARC server 1000 (CPU: Super SPARC; 60 MHz, RAM: 128 Mbytes).

Figure 4.5 shows the simulation results. The observed CLR data, shown by the symbols "+," were widely dispersed. The solid line shows the upper bound of CLR which is obtained by learning from these observed CLR data. From this figure, it can be found that the upper bound of CLR is well

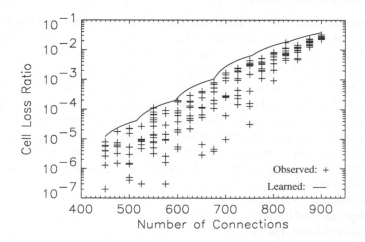

FIGURE 4.5
Upper bound of observed cell loss ratio estimated by fuzzy inference.

extracted. No observed CLR data exceed this estimated upper bound. Although it is difficult to compare the CLR estimation by the proposed method to that by the other methods, we calculated the CLR by the nonparametric method, for reference, that can estimate the upper bound of CLR. As a result, the nonparametric method estimated about 10^{-1} in the range of the number of connections shown in Figure 4.5. This calculation was performed in the condition that all traffic sources are variable-bit-rate processes because ATM switches have no way of knowing the ratio of connection numbers of the two traffic sources mentioned above. Although the nonparametric approach can guarantee the allowed CLR, an excessively high CLR is estimated, at values of 10^{-1}, which results in lower multiplexing gain. At the allowed CLR of 10^{-3}, the CLR estimation method proposed in this chapter judged that about 670 connections can be accepted whereas the nonparametric method judged that only 100 connections can be accepted. At the allowed CLR of 10^{-4}, the proposed method judged that about 540 connections can be accepted whereas the nonparametric method judged that no connection can be accepted. In the next section, a CAC algorithm is proposed which takes advantage of the proposed method for CLR estimation.

4.4   Fuzzy Connection Admission Control

In this section, a CAC scheme is proposed, applying the CLR estimation method described in the previous section. Figure 4.6 shows the overall scheme for the CAC discussed here. The variable x_i denotes the number of connections in the i-th transmission rate class. When a call request comes and the class C_i is selected, the value of x_i is incremented. Then, the vector $(x_1, x_2, \ldots, x_k)$ is input to the fuzzy inference engine. The upper bound of CLR is estimated by this fuzzy inference engine. If this estimated CLR exceeds the allowed CLR, the call is rejected. Otherwise, the call is accepted.

This section, first, clarifies the problem in the observation-based CAC. Then, CAC with successive generation of fuzzy rules is proposed in order to solve the problem. Simulation results are also provided to show its feasibility.

4.4.1   Connection Admission Control with Successive Generation of Fuzzy Rules

In the previous section, it was assumed that the observed CLR data exist in all ranges for the number of connections. In performing CAC, however, CLR data do not exist at a number of connections before accepting calls. This generally causes a problem in the application of the observation-based CLR estimation to CAC. The problem is explained below in more detail, using the CLR estimation method proposed here for convenience, although it is a basic problem for CAC with observation-based CLR estimation.[18]

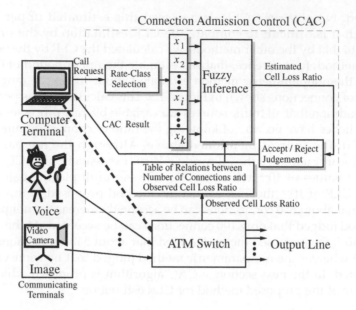

FIGURE 4.6
Connection admission control by using fuzzy inference.

In applying the proposed CLR-estimation method to CAC, the parameters for fuzzy rules should be initialized so as to perform CAC guaranteeing the allowed CLR even in the initial stage. Then, the CAC is performed by the estimated CLR obtained by the initial values of these parameters in the fuzzy rules and a number of CLR data are observed. After enough data are observed, the fuzzy rules are adjusted by the learning algorithm with these observed data. In this process, however, the area with no observed data exists because the estimation is conducted by the initial parameters and calls are accepted only in the area where the allowed CLR can be guaranteed. As a result, the upper bound of CLR is learned only in the area where calls have been accepted by the CLR estimated by the initial values. Therefore, even if the upper bound of CLR is effectively estimated in this area, higher multiplexing gain cannot be attained because the estimation cannot be conducted outside the area. Figure 4.7 illustrates this situation.

In order to solve this problem, this chapter proposes the following method which is based on the generation of fuzzy rules from adjacent ones.[18, 21] In the area with no observed CLR data, fuzzy rules are generated with adjacent fuzzy rules in the area with observed CLR data. Namely, the parameters of the fuzzy sets in then-parts are generated by extrapolation with the parameters of the fuzzy sets in the adjacent fuzzy rules.[18] Figure 4.8 illustrates the generation of fuzzy rules. Fuzzy rules have essential parameters for estimating the possibility distribution of CLR. In other words, the parameters defining fuzzy sets in the fuzzy rules determine the estima-

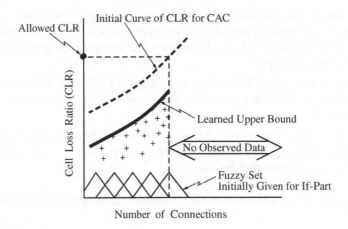

FIGURE 4.7
Area with no observed data.

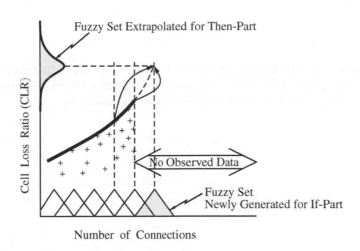

FIGURE 4.8
Successive generation of fuzzy rules.

tion curve of CLR. Therefore, the use of the parameters defining fuzzy sets in the rules is very effective in the extrapolation. The following are the steps for CAC based on the above-mentioned method.[18] The number of transmission rate classes is supposed to be one for simplicity in explanation. It is easily extended for service with a number of transmission rate classes.

Step 1: Initialize the parameters defining fuzzy sets in fuzzy rules 1, 2, . . . , n so as to estimate the upper bound of CLR guaranteeing the allowed CLR from the initial stage. Fuzzy sets for the if-parts are assigned to the area where the allowed CLR can be guaranteed by the initial estimation curve. Let $x_{\max}$ denote the maximum number of connections for which the fuzzy rules can estimate the CLR, namely the value of $x^{\circ}_{P_n}$. Perform CAC using these initialized parameters in Step 2.

Step 2: Perform CAC and store observed CLR data. When enough observed CLR data are obtained, tune the parameters for fuzzy rules 1, 2, ... , n.

Step 3: If the estimated CLR at $x_{\max}$ is larger than the allowed CLR, return to Step 2. Otherwise, generate fuzzy rule $n + 1$ by extrapolation as exemplified below.

For example, the parameters of the membership function $P_{n+1}(x)$ are given as follows:

$$x^{\circ}_{P_{n+1}} = x^{\circ}_{P_n} + \frac{\eta_{P_n}}{2} \tag{4.12}$$

$$\eta_{P_{n+1}} = \eta_{P_n} \tag{4.13}$$

The parameters of the membership function $Q_{n+1}(y)$ are generated by extrapolating those of adjacent membership functions. When the exponential extrapolation is adopted, these parameters are obtained by

$$y^{\circ}_{Q_{n+1}} = 10^c \tag{4.14}$$

$$\eta_{Q_{n+1}} = 10^w \tag{4.15}$$

where

$$c = \log_{10} y^{\circ}_{Q_n} + \frac{\log_{10} y^{\circ}_{Q_n} - \log_{10} y^{\circ}_{Q_{n-1}}}{x^{\circ}_{P_n} - x^{\circ}_{P_{n-1}}} \tag{4.16}$$

$$w = \log_{10} \eta_{Q_n} + \frac{\log_{10} \eta_{Q_n} - \log_{10} \eta_{Q_{n-1}}}{\eta_{P_n} - \eta_{P_{n-1}}} \tag{4.17}$$

According to this generation of fuzzy rules, the value of $x_{\max}$ is renewed. Then, return to Step 2, considering n in Step 2 as $n + 1$.

If the CAC has to always guarantee the allowed CLR as much as possible at the sacrifice of fast convergence to higher multiplexing gain, the extrapolation should be conducted with small steps on the axis of the number of connections. Otherwise, the larger steps on this axis can be taken and it

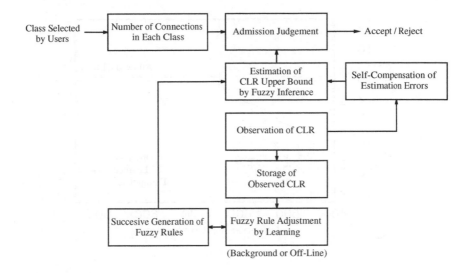

FIGURE 4.9
Overall flow of the fuzzy CAC process.

results in fast convergence to higher multiplexing gain. In this way, fuzzy rules are successively generated as long as the estimated upper bound of CLR satisfies the required value. Figure 4.9 shows the overall flow of the fuzzy CAC process proposed in this chapter.

4.4.2 Simulation Studies on Successive Generation of Fuzzy Rules

The simulation was conducted for the successive generation of fuzzy rules on the basis of the proposed method. In this simulation, the traffic sources and other conditions are the same as in the previous simulation. Let the allowed CLR be 10^{-3} in the simulation.

Figure 4.10 shows the upper bound estimation of CLRs with successive generation of fuzzy rules in the simulation. The solid lines in these figures represent the estimated upper bound by the adjusted fuzzy rules whereas the broken lines show the estimated upper bound together with the fuzzy rule generated by the extrapolation of adjacent fuzzy rules.

In Figure 4.10(a), the solid line is constructed by the four fuzzy rules learned from the observed CLR data shown with the symbols "+." In this first stage, the estimated upper bound of CLR does not exceed the allowed CLR. Thus, one fuzzy rule was generated by extrapolation. The broken line is constructed by this newly generated rule and the above-mentioned four rules. In this stage shown in Figure 4.10(a), calls are accepted according to the estimated upper bound of CLR with these five fuzzy rules, one of which was generated by extrapolation.

Then, after obtaining enough CLR data, all of the five fuzzy rules, including the generated fuzzy rule, are adjusted by the learning algorithm. The

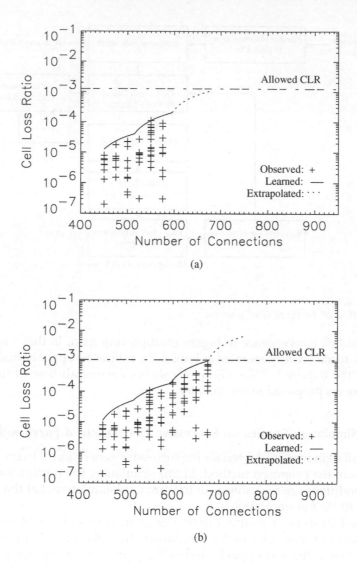

FIGURE 4.10
Simulation results for successive generation of fuzzy rules.

solid line in Figure 4.10(b) shows the estimated upper bound of CLR by these adjusted fuzzy rules. From this figure, it can be found that the upper bound of CLR is well extracted by adjusting all fuzzy rules including the newly generated fuzzy rule. In this second stage, the solid line shows that the estimated upper bound of CLR does not exceed the allowed CLR. Thus, another fuzzy rule was generated. The broken line in Figure 4.10(b) is constructed by this newly generated rule and the above-mentioned five rules. This newly obtained broken line estimates the upper bound of CLR larger than the allowed CLR. Therefore, the fuzzy rule generation is tentatively stopped and

CAC is continued with all these six rules until the fuzzy rules learned from newly observed CLR data estimate the upper bound of CLR lower than or equal to the allowed CLR.

4.5 Conclusion

In this chapter, a CAC method for use in ATM networks has been proposed, applying the fuzzy inference based on a weighted mean of fuzzy sets.[19, 20] The CLR can be estimated from observed CLR data so as to guarantee the allowed CLR, preserving higher multiplexing gain as much as possible. This can be achieved by estimating the possibility distribution of CLR and using the least upper bound of this distribution for CAC. In contrast to conventional fuzzy inference methods, the adopted inference method can control the fuzziness and specificity of its final inference consequences by adjusting those of fuzzy sets in then-parts. This property makes it possible to estimate the upper bound of CLR without excessively high value in its estimation and thus provides the higher multiplexing gain.

The fuzzy rules are adjusted automatically by the learning algorithm with the observed data. The possibility distribution of CLR is inferred by using these fuzzy rules. Taking advantage of the adopted fuzzy inference method, the learning is conducted so as to include the observed CLR data in the interval given by the α-level set of the final inference consequence. The least upper bound of this interval provides the estimated upper bound of CLR. For the learning, the energy functions were proposed which enable the inference system to avoid excessively high values in the estimation of CLR. In the energy functions, the target data are dynamically changed so that the interval is as small as possible, preserving observed CLR data in the interval. The parameters of fuzzy rules are tuned by the error back-propagation algorithm based on the energy functions. In this way, the fuzzy inference method adopted here makes possible the estimation of the upper bound of CLR without excessively high values in the estimation of CLR. The simulation results showed that the upper bound of CLR is well extracted.

The self-compensation mechanism for estimation errors is also achieved, by taking advantage of the adopted inference method which can prove convex forms of its inference consequences and the proposed energy functions mentioned above. The CLR estimation errors can be calibrated in real time when the estimated CLR is unexpectedly lower than the observed CLR. It preserves the guarantee of the allowed CLR as much as possible in operation of ATM switches.

By applying the CLR estimation method proposed here, the CAC method has been considered. The parameters for fuzzy rules are initialized so as to perform CAC guaranteeing the allowed CLR. The CLR estimation is performed beginning with these initial values for CAC. However, it results in the area with no observed CLR data because calls are accepted on the basis of the CLR's estimated only by the initial values of the parameters. Thus, no fuzzy rules can be assigned to such area because no data are observed for adjusting fuzzy rules. This means that the multiplexing gain cannot be attained at a higher level than that by the initial parameters.

In order to solve this problem, CAC with successive generation of fuzzy rules was proposed. Fuzzy rules in the area with no observed CLR data are generated by extrapolation from adjacent fuzzy rules in the area with observed CLR data. All fuzzy rules including these generated fuzzy rules estimate CLR in the area with no observed data. It makes CAC possible for the area with no observed CLR data, and CLR data can be observed in this area. These newly obtained CLR data adjust all the fuzzy rules, including the generated fuzzy rules, by the above-mentioned learning algorithm. In this way, CAC can be conducted guaranteeing the allowed CLR as much as possible with higher multiplexing gain. Simulation showed the feasibility of the proposed CAC method.

Although the CAC method proposed here has been discussed mainly in the case of one transmission rate class for simplicity, it can be easily extended for a number of transmission rate classes. Moreover, the method can be applied to other QOS criteria such as transmission delay, in the same way as the case of CLR discussed in this chapter.

References

1. H. Saito, *Teletraffic Technologies in ATM Networks*, Artech House, Boston, 1994.
2. H. Saito, New dimensioning concept for ATM networks, 7th International Teletraffic Congress, Specialist Seminar, Morristown, NJ, 1990.
3. H. Siato, Call admission control in an ATM network using upper bound of cell loss probability, *IEEE Trans. Commun.*, 40(9), 1512–1521, 1992.
4. H. Saito, Toward a future traffic dimensioning method: Non-parametric approach for cell loss rate evaluation, *IEICE Trans.* B-I, J76-B-I (3), 197–208, 1993 (in Japanese).
5. T. Yang and H. Li, Individual cell loss probabilities and background effects in ATM networks, Proc. *IEEE ICC'93*, 1373–1379, 1993.
6. K. Sohraby, On the asymptotic behavior of heterogeneous statistical multiplexer with applications, in *Proc. IEEE INFOCOM'92*, 839–847, 1992.
7. R. J. Gibbens and P. J. Hunt, Effective bandwidths for the multi-type UAS channel, *Queueing Syst.* 9, 17–27, 1991.
8. T. Murase, H. Suzuki, S. Sato, and T. Takeuchi, A call admission control scheme for ATM networks using a simple quality estimate, *IEEE J. Select. Areas Commun.*, SAC-9(9) 1461–1470, 1991.
9. H. Heffes and D. M. Lucantoni, A Markov modulated characterization of packetized voice and data traffic and related statistical multiplexer performance, *IEEE J. Select. Areas Commun.*, SAC-4(6), 856–868, 1986.

10. J. E. Neves, L. B. de Almeida, and M. J. Leitão, B-ISDN connection admission control and routing strategy with traffic prediction by neural networks, *Proc. IEEE ICC'94*, 769–773, 1994.

11. A. D. Estrella, A. Jurado, and F. Sandoval, New training pattern selection method for ATM call admission neural control, *Electron. Lett.*, 30(7), 577–579, 1994.

12. S. H. Kang and D. K. Sung, A trial multilayer perceptron neural network for ATM connection admission control, *IEICE Trans. Commun.*, E76-B(3), 258–262, 1993.

13. A. Hiramatsu, Adaptive ATM call admission control using a neural network trained with cell loss data observed from virtual output buffers, *Technical Report of IEICE*, SSE93-43, 1993 (in Japanese).

14. A. Hiramatsu, Integration of ATM call admission control and link capacity control by distributed neural networks, *IEEE J. Select. Areas Commun.*, 9(7), 1131–1138, 1991.

15. A. Hiramatsu, ATM communications network control by neural networks, *IEEE Trans. Neural Networks*, 1(1) 122–130, 1990.

16. H. Saito and K. Shiomoto, Dynamic call admission control in ATM networks, *IEEE J. Select. Areas Commun.*, 9(7), 982–989, 1991.

17. V. Paxson and S. Floyd, Wide area traffic: The failure of Poisson modeling, *IEEE/ACM Trans. Networking*, 3(3), 226–244, 1995.

18. K. Uehara, and K. Hirota, Fuzzy connection admission control for ATM networks based on possibility distribution of cell loss ratio, *IEEE J. Select. Areas Commun.*, 15(2), 179–190, 1997.

19. K. Uehara, Fuzzy inference based on a weighted average of fuzzy sets and its learning algorithm for fuzzy exemplars, in *Proceedings of the International Joint Conference of the Fourth IEEE International Conference on Fuzzy Systems and the Second International Fuzzy Engineering Symposium, FUZZ-IEEE/IFES'95* (Yokohama, Japan), IV, 2253–2260, March 1995.

20. K. Uehara and K. Hirota. Parallel fuzzy inference based on α-level sets and generalized means, *Int. J. Inf. Sci.*, 100(1–4), 165–206, 1997.

21. L. T. Koczy and K. Hirota, Approximate reasoning by linear rule interpolation and general approximation, *Int. J. Approx. Reason.*, 9(3), 197–225, 1993.

22. C. C. Lee, Fuzzy logic in control systems: fuzzy logic controller—parts I and II, *IEEE Trans. Syst. Man Cybern.*, 20(2), 404–418, 419–435, 1990.

23. D. E. Rumelhart, G. E. Hinton, and R. J. Williams, Learning representations by back-propagating errors, *Nature*, 323(6088), 533–536, 1986.

5

Congestion Control

Andreas Pitsillides and Ahmet Sekercioglu

CONTENTS

0-8493-1075-X/01/$0.00+$.50
© 2001 by CRC Press LLC

ABSTRACT   Network congestion control remains a critical issue and a high priority, especially given the growing size, demand, and speed (bandwidth) of the increasingly integrated services networks. Designing effective congestion control strategies for these networks is known to be difficult because of the complexity of the structure of the networks, nature of the services supported, and the variety of the dynamic parameters involved. In addition to these, the uncertainties involved in identification of the network parameters lead to the difficulty of obtaining realistic, cost effective, analytical models of these networks. This renders the application of classical, control system design methods (which rely on availability of these models) very hard, and possibly not cost effective. Consequently, a number of researchers are looking at alternative nonanalytical control system design and modeling schemes that have the ability to cope with these difficulties in order to devise effective, robust congestion control techniques as an alternative (or supplement) to traditional control approaches. These schemes employ artificial neural networks, fuzzy systems, and design methods based on evolutionary computation (collectively known as Computational Intelligence). In this chapter we first discuss the difficulty of the congestion-control problem and review control approaches currently in use, before we motivate the utility of Computational Intelligence based control. Then, through a number of examples, we illustrate congestion control methods based on fuzzy control, artificial neural networks, and evolutionary computation. Finally, some concluding remarks and suggestions are given for further work.

5.1   Introduction

It is generally accepted that the problem of network congestion control remains a critical issue and a high priority, especially given the growing size, demand, and speed (bandwidth) of the increasingly integrated services networks.* One could argue that network congestion is a problem unlikely to disappear in the near future. Furthermore, congestion may become unmanageable unless effective, robust, and efficient methods of congestion control are developed. This assertion is based on the fact that despite the vast research efforts spanning a few decades, and the large number of different control schemes proposed, there are still no universally acceptable congestion control solutions. Current solutions in existing networks are increasingly becoming ineffective, and it is generally accepted that these solutions cannot easily scale up—even with various proposed "fixes." In this chapter we first define congestion, what causes congestion, how it is felt, how fast it is sensed,

---

* Integrated services communication networks include high speed packet switching networks: ATM, current and future TCP/IP Internet, frame relay, etc.

and where. We then review current approaches on congestion control in the worlds of Internet and ATM. A structured approach, toward designing one of the most (if not the most) complex control systems that man ever made, is then advocated. Guided by the success of control theory in other man-made complex and large-scale systems, we assert that a control theoretic point of view is necessary. We propose that Computational Intelligence should have an essential role to play in designing this challenging control system. Nowadays, Computational Intelligence research is very active and consequently its applications are appearing in some end user products. Finally, we present several illustrative examples, based on documented studies, of successful application of Computational Intelligence in controlling congestion, and conclude with some suggestions and open questions.

5.2 Congestion Control

5.2.1 Preliminaries

According to the International Telecommunication Union (ITU) definition (ITU-T: Rec. I371[1])

> In B-ISDN,* *congestion* is defined as a state of network elements (e.g. switches, concentrators, cross-connects and transmission links) in which the network is not able to meet the negotiated network performance objectives for the already established connections and/or for the new connection requests.

A similar definition can be posed for packet-switching networks. For example, Reference 2 defines congestion as a network state in which performance degrades due to the saturation of network resources, such as communication links, processor cycles, and memory buffers. **Congestion control** refers to the set of actions taken by the network to minimize the intensity, spread, and duration of congestion. It can be said that it is that aspect of a networking protocol that defines how the network deals with congestion. Despite the many years of research efforts, the problem of network-congestion control remains a critical issue and a high priority, especially given the growing size, demand, and speed (bandwidth) of the networks. Network congestion is becoming a real threat to the growth of existing packet-switched networks, and of the future deployment of integrated services communication networks. It is a problem that cannot be ignored. In order to understand the nature of the

---

* Broadband-ISDN (Integrated Services Digital Network) is the standards-based (ITU-T) multiservice and multimedia network. Asynchronous Transfer Mode (ATM) was selected as the transport mode for Broadband-ISDN.

problem of congestion, and before we can discuss any approach toward solving the congestion-control problem, we next attempt to understand what causes congestion, how it is felt, how fast it is sensed, and where.

Congestion is caused by saturation of network resources (communication links, buffers, network switches, etc . . . .). For example, if a communication link delivers packets to a queue at a higher rate than the service rate of the queue, then the size of the queue will grow. If the queue space is finite, then, in addition to the delay experienced by the packets until service, losses will also occur. Observe that congestion is not a static resource shortage problem, but rather a dynamic resource allocation problem. Networks need to serve all users' requests, which may be unpredictable and bursty in their behavior (starting time, bit rate, and duration). However, network resources are finite, and must be managed for sharing among the competing users. Congestion will occur if the resources are not managed effectively. The optimal control of networks of queues is a well known, much studied, and notoriously difficult problem, even for the simplest of cases. For example, Papathemitriou and Tsitsiklis[3] show that several versions of the problem of optimally controlling a simple network of queues with simple arrival and service distributions and multiple-customer classes is complete for exponential time (i.e., provably intractable).

The effect of network congestion is degradation in the network performance. The user experiences long delays in the delivery of messages, perhaps with heavy losses caused by buffer overflows. Thus there is degradation in the quality of the delivered service with the need for retransmissions of packets (for services intolerant to loss). In the event of retransmissions, there is a drop in the throughput—a wastage of system resources—and leads to a collapse of network throughput when a substantial part of the carried traffic is due to retransmissions (in that state not much useful traffic is carried). In the region of congestion, queue lengths, hence queuing delays, grow at a rapid pace—much faster than when the network is not heavily loaded. This is well illustrated in the case of a single uncontrolled queue with infinite waiting space, featuring a single stream of packet arrivals with exponential packet length distribution and exponential service rate (M/M/1 queue). It can be shown analytically that the normalized average time through the queue is equal to $1/(1-\rho)$, where ρ is the normalized load (traffic intensity, utilization)[4] (Figure 5.1a) for a plot of offered normalized load vs. the normalized delay. Figure 5.1b shows the effect of excessive loading on the network throughput for three cases: no control, ideally controlled, and practically controlled.

In the case of ideal control, the throughput increases linearly until saturation of resources, where it flattens off and remains constant, irrespective of the increase of loading beyond the capacity of the system.

Obviously this type of control is impossible in practice. Hence for the practically controlled case, we observe some loss of throughput, as there is some communication overhead associated with the controls, possibly some inaccuracy of feedback state information and some time delay in its delivery. Finally, for the uncontrolled case, congestion collapse may occur, whereby, as the network is increasingly overloaded, the network throughput collapses;

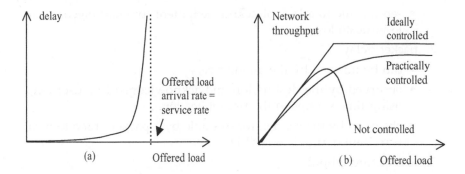

FIGURE 5.1
Network delay and throughput versus offered load.

i.e., very little useful network traffic is carried—due to retransmissions or deadlock situations. Note that a network service provider aims to control the network in such a way so as to maximize the network throughput. Examples of network controls include admission and regulation of traffic flow into the network. These controls should be achieved without causing network congestion, or at least exhibit no degradation of performance beyond acceptable levels to the user, i.e., no degradation of quality of service in terms of loss and delay, and no appreciable reduction in throughput.

Congestion is a complex process to define. It is felt by a degradation of performance. One can identify that increased loss and delays coupled with a drop in throughput (due to retransmissions) are a good indicator of congestion.

A "good"* congestion control system should be preventive, if possible. Otherwise it should react quickly and minimize the spread of congestion and its duration. A good engineering practice will be to design the system in such a way as to avoid congestion. But taken to the extreme (i.e., to guarantee zero loss and zero queuing delay), this would not be economical. For example, assuring zero waiting at a buffer implies increasing the service rate or the number of servers—at its limit to infinity. A good compromise would be to allow for some deterioration of performance, but never allow it to become intolerable (congested). The challenge is to keep the intolerance at limits acceptable to the users. Note the fuzziness present in defining when congestion is actually experienced. Before we review existing congestion controls, let us have a look at how we can measure or sense congestion and where. This will provide us with a better understanding of the many different approaches.

Congestion can be sensed (or predicted) by:

1. packet loss sensed by
   - the queue as an overflow,
   - destination (through sequence numbers) and acknowledged to a user,

---

* See later discussion for what constitutes good control.

- sender due to a lack of acknowledgment (timeout mechanism) to indicate loss.

2. packet delay

   - can be inferred by the queue size,
   - observed by the destination and acknowledged to a user (e.g., using time stamps in the packet headers),
   - observed by the sender, for example by a packet probe to measure Round Trip Time (RTT).

3. loss of throughput

   - observed by the sender queue size (waiting time in queue).

4. other calculated or observed event though which congestion can be inferred

   - increased network queue length and its growth
   - calculated from measured data, such as queue inflow and its effect on future queue behavior.

The choice of how to measure congestion and where, apart from the other practical problems such as cost and complexity, can influence to a great degree the achievable control approach, control strategy, and control location. Here we only highlight this potential problem. For example, by sensing packet loss one expects that the observed congestion is at an advanced state (has already happened), whereas delay sensing at a node does not necessarily indicate that congestion has happened. Actually, one may expect that with delay sensing, a predictive model can be built to indicate the level of the expected state of congestion. Also, the feedback information is binary (presence or absence of congestion), and the round-trip time (and feedback delay) are significantly different. For an in-depth discussion of these issues, the effect of location on quality of control, as seen though the control horizon, as well as potential problems of control, and how these influence the design of the controls, see Reference 5.

We also identify here other potential problems of control:

1. Large scale.
2. Distributed nature.
3. Large geographic spread (at its limit it covers the globe).
4. Increasingly processing delay at nodes gets smaller, in comparison to the propagation delay in the links. Large-bandwidth delay product makes the control of congestion through feedback potentially difficult.
5. Diverse nature and behavior of carried traffic (voice, video, www, ftp, . . .).
6. Unpredictable and time-varying user behavior.
7. Lack of appropriate dynamic models for control.

8. Expectation of the need for guaranteed levels of performance to each user, which can be negotiated with the network.

This array of potential control problems has caused a lot of debate as to what are appropriate control techniques for the control of congestion, and depending on one's point of view, many different schools of thought were followed, with many published ideas and control techniques. We will not, and cannot, attempt to capture that wealth of published material here. Rather, we attempt to highlight some of the better known (to us) work that has followed a certain direction before we discuss the necessity for a structured approach toward congestion control.

5.2.2 Existing Approaches to Congestion Control

For historical reasons, and due to fundamental philosophical differences in the (earlier) approach to congestion control, we present an overview of the research for traditional TCP/IP- and ATM-based networks separately. However, some convergence between the classical TCP/IP and the ATM approach is evident (see RFC2309,[6] Internet draft-kksjf-ecn-03 and RFC2481,[7] Internet draft-salim-jhsbnns-ecn-00,[8] and ATM Forum[9]). It has become clear[6] that the existing TCP congestion avoidance mechanisms (for a description see RFC2001[10]), while necessary and powerful, are not sufficient to provide good service in all circumstances. Basically, there is a limit as to how much control can be accomplished from the edges of the network. This is discussed in length in Reference 5 where the concept of an effective control horizon is discussed. Some mechanisms are needed in the routers to complement the endpoint congestion-avoidance mechanisms.* RFC2309 strongly recommends active queue management in routers and RFC2481 suggests Explicit Congestion Notification (ECN) control for IP.[7] Internet draft-salim-jhsbnns-ecn-00[8] takes it a step further and proposes Backward ECN (BECN) and Multilevel ECN (MECN), in which the feedback signal can include information on the severity of the congestion. Note the similarity in concept with the Explicit Rate (ER)-based schemes advocated by the ATM Forum Traffic Management specification[9] for managing Available Bit Rate traffic.** ATM switches (labeled Explicit Down Switches (EDS) in Reference 9) can calculate the maximum ER that they can accept over the next control interval, so that ABR traffic into the network can be regulated, for effective use of resources. This can be contrasted to the earlier preventive (open loop)-based approaches,[11] which were suggested then as the (only) effective way to control ATM (Broadband-ISDN)-based networks.

---

* Need for gateway control was realized early; e.g., see Reference 13, where for future work the gateway side is advocated as necessary.
** Advocated by the ATM Forum, against an initial push advocating preventive open loop-based control for Broadband-ISDN.

5.2.2.1 Evolution of Congestion Controls in TCP/IP

The congestion control schemes employed by the TCP/IP protocol have been widely studied. We follow several references.[12, 10, 6, 7, 8]

The Internet protocol architecture is based on a connectionless end-to-end packet service using the IP protocol. TCP is an end-to-end transport protocol that provides reliable, in-order service. End-to-end flow control is integrated with a Go-Back-N error recovery mechanism. Network level congestion control is implemented via a reactive, closed-loop, dynamic window control scheme.[13] Jacobson developed the congestion-avoidance mechanisms that are now required in TCP implementations in 1988.

These window-based mechanisms operate in the hosts to cause TCP connections to "back off" during congestion. That is, TCP flows are "responsive" to congestion signals (i.e., dropped packets) from the network. It is primarily these TCP congestion avoidance algorithms that prevent the congestion collapse of today's Internet. The window normally increases by some amount each round-trip time, however, the window decreases (usually by a larger factor—additive-increase, multiplicative-decrease) when the sender observes congestion indications (packet loss). A fundamental aspect of TCP is that it obeys a "conservation of packets" principle, where a new segment is not sent into the network until an old segment has left. TCP implements this strategy via a self-clocking mechanism: acknowledgments received by the sender are used to trigger the transmission of new segments. This self-clocking property is the key to TCP's congestion-control strategy. If the receiver's acknowledgments arrive at the sender with the same spacing as the transmissions they acknowledge, and if the sender sends at the rate that acknowledgments are received, the sender will not overrun the bottleneck link. Along with TCP's self-clocking admission control, other elements of TCP's congestion control include the congestion-recovery algorithm (i.e., slow-start), the congestion-avoidance algorithm, and the fast retransmit/recovery algorithms, see Reference 15 for details. Note that RFC 2001[10] fully documents the four algorithms in use (as of January 1997) in the Internet: slow start, congestion avoidance, fast retransmit, and fast recovery. The TCP congestion-control algorithms are reactive and generally do not prevent congestion from occurring. The algorithms try to obtain as much bandwidth as possible from the network by continually increasing the send rate until packet loss occurs. While the congestion-control algorithms are effective in certain situations, there are other situations where TCP performance is poor. Ever-increasing demands on the Internet have led to a number of incremental changes over the last 10 years designed to improve TCP/IP performance:

1. Improved round-trip time measurement algorithm (Karn's algorithm).[14]
2. Slow-start and congestion avoidance.[13]
3. Fast retransmit, fast-recovery algorithms.[15]
4. Improved operation over high-speed, large-delay networks.[16]

Even so, there is a large amount of evidence of observed TCP behavior that collectively contributes to TCP's unpredictable performance. While the majority of TCP analysis has been simulation based, there have been several empirical studies performed to illustrate that TCP can exhibit unwanted behaviors. Examples demonstrating unwanted behavior of TCP include: cyclic behavior;[17, 18] synchronization effects and ACK compression.[12] A notable analytic evaluation of the performance of congestion algorithms[13, 19] for TCP/IP is given by Lakshman and Madhow.[20] Using simple dynamic models for the slow-start and congestion-avoidance phases of the two algorithms, they insightfully demonstrate the unwanted cyclic behavior of TCP/IP and the effect of a high-bandwidth delay product and random losses on its performance. Thus, the behavior of TCP/IP congestion controls remains a critical issue and a matter of continuous research interest in the TCP/IP world (highlighted by the frequent RFCs proposing fixes or new solutions).

The congestion-control mechanisms continue to be enhanced as TCP/IP evolves to meet new and more demanding requirements. End-to-end congestion-avoidance schemes based on windows were proposed in References 21 and 22. In Reference 21 the aim is to approach the optimal and fair points quickly, as opposed to the slow start. TCP-Vegas[22] is based loosely on Reference 21 for congestion avoidance, and offers certain significant improvements. It has been debated over the past decade as to whether window or rate control is more effective at controlling congestion. A common belief among end-to-end congestion-avoidance supporters was that, due to the high bandwidth-delay product, window-based control is inferior to rate-based control at high speeds. Proposed closed-loop rate-based congestion-control schemes in a connectionless environment include XTP (Xpress Transfer Protocol) and IBM's RTP (Rapid Transport Protocol) rate-based algorithms. RTP uses a probe packet, and XTP uses explicit or implicit feedback from either the receiver or the network to adjust their send rate.

Currently we are witnessing a shift by the Internet world toward the router/gateway congestion-control approach. Router congestion control is a form of control where the router provides either explicit or implicit feedback indication. For example, the router-based congestion-control algorithm proposed by Floyd and Jacobson,[23] aims to make the probability that a connection is notified of congestion proportional to that connection's share of the bandwidth, through an active queue-management mechanism called Random Early Detection (RED). RFC2309[6] recommends active queue management, and recommends further that RED[23] be the mechanism. Earlier, Ramakrishnan and Jain[24] proposed a scheme that uses explicit binary feedback from the gateway (a.k.a. DECbit). This scheme uses a congestion indication (CI), set by the router on the header of a packet en route to the destination. The CI is set upon sensing congestion by monitoring the queue length. Once the destination decides (based on the received CIs) that congestion is imminent, it informs the source to modify its window. Even though this algorithm is prone to oscillations, exhibits bias against bursty traffic, and is not fair, it has stimulated a lot of interest in the early discussion about ABR control in the ATM Forum (Forward and Backward Explicit Congestion Notification, FECN and BECN) and for Frame Relay control. In the Internet world, the initial suggestions to introduce a methodology for adding Explicit Congestion Notification (ECN) to IP router-congestion control are outlined in Reference 25, and later in the Internet draft-kksjf-ecn-03 and RFC 2481.[7] Internet draft-salim-jhsbnns-ecn-00[8] proposes an alternative approach to the ECN mechanism proposed in Reference 7. It proposes a Backward-ECN (BECN) which uses the existing IP-signalling mechanism, the Internet Control Messaging Protocol (ICMP) Source Quench message (ISQ), to reduce the reaction time to a congestion in the network. In addition, the ISQ message can include information on the severity of the congestion allowing the end host to react accordingly so as to make maximal use of the resources while maintaining network equilibrium (which they refer to as Multilevel ECN).

A clear trend is observed: to progressively move the controls inside the network, closer to where it can be sensed, and for the feedback information to become richer (shift from binary to a more explicit value of congestion). These enable better controls (flexibility and effectiveness); see later discussion on the congestion control framework and in Reference 5. For the flexible and effective control of congestion in a TCP/IP environment, new control structures and approaches are necessary. According to Braden et al.,[6] it is imperative that work in developing congestion control be energetically pursued to ensure the future stability of the Internet. In this respect, a control framework and appropriate control techniques (e.g., Computational Intelligence) will be a necessary aid. Worth noting is the fact that not much work on the use of Computational Intelligence in controlling congestion in the Internet has been reported.

Furthermore, it should be pointed out that the congestion-control problem in the Internet is exacerbated, as the Internet is increasingly transformed into an integrated services high-speed network. See, for example, the intserv and diffserv proposed architectures.[26, 27] For Integrated Services (intserv),[26] not many congestion-control algorithms have appeared in the open literature. Its architecture is expected to provide a mechanism for protecting individual flows from congestion, and introduces its own queue management and scheduling algorithms. In Reference 26 it is speculated whether a virtual circuit model should be adopted as proposed in ATM and ST-II protocol (i.e., abandon IP). Debate is still at an early stage, but again the approach to congestion control should be based on a congestion-control framework and appropriate control techniques. The same comments can be applied for differentiated services in the Internet (diffserv).[27, 28] Note that some recent works apply feedback control theory for differentiated services.[29, 30]

5.2.2.2   Congestion Control in ATM Networks and Its Evolution for the Available Bit Rate Service

Congestion control in ATM-based networks has been extensively researched. This is evidenced by the large body of published papers. See, for example, the proceedings of INFOCOM, GLOBECOM, ICC, and ITC to name but a few; the journals devoting whole issues to ATM control; and the large number of books published spanning almost two decades. Even with this large body of published works, there are still substantial unresolved problems of control in ATM-based high-speed networks. The complexity and immensity of the task was recognized early. See, for example, the guest editorial comments in Reference 31 (they state that: "The international telecommunications community fully appreciates the complexity of the issue and, to cope with this problem, proposed a large variety of congestion control techniques. Many researchers believe that there is no silver bullet, and that control of high-speed packet networks can be obtained by executing several concurrent mechanisms . . . . "); and[32] the guest editorial comments of the JSAC special issue on congestion issues in B-ISDN for a brief discussion of some of the control difficulties (they state that: "the dynamic, heterogeneous, time-varying network environment, with different service requirements is a significant factor in the design of controls"; and that "the design of the entire system and the interaction of the various components is often more important than the optimization of individual components"). Almost ten years on, we note that the same comments are still applicable.

Initially, for Broadband-ISDN*[33] there was a push for preventive control[11] (more correct to say open-loop-type controls). This was motivated by the large bandwidth-delay product, which was seen[11] as an inhibitor to the effective application of reactive (more correct to say closed-loop or feedback) control. This view was influenced by a predominant (not often stated) view that controls must reside at the edges of the network, thus making the total delay around the feedback loop (in comparison to the bandwidth) so high as to render feedback-based control ineffective. Note that many researchers, even at an early stage, did not adopt that view.[34] Progressively, there was a shift from that view (see parallel with TCP/IP-congestion-control debate on router control) in that feedback is essential for effective (and efficient) control, and finally that controls inside the network should not be precluded, at least to supplement preventive controls. This was formalized by the ATM Forum.[9]

The initial view for preventive control was reflected in the general structure of the control framework described in the ITU recommendation I.371.[1] It consists of Connection Admission Control (CAC),[35, 36, 37] and Usage Parameter Control (UPC).[38] When a new call request is made, the user is required to inform the network about the traffic characteristics (the "contract", which contains traffic descriptors such as mean bit rate, peak bit rate, and possibly burstiness of traffic and others) and the desired QoS of the connection. It is then the responsibility of the CAC to decide whether to accept or reject the new connection. It is accepted only if the requested QoS can be satisfied without affecting the QoS of existing connections. Once a connection is accepted, the UPC polices the traffic characteristics to ensure that they do not exceed those specified at connection establishment. Such a

---

* Currently, a lack of research activity is observed for Broadband-ISDN.

preventive control framework, however, is not adequate to achieve the objectives of traffic control. Studies reveal that, for very bursty traffic such as LAN-to-LAN interconnection whose peak rate is comparable to the link speed, the achievable utilization is very small unless detailed knowledge on the traffic characteristics is available.[39] Such knowledge may include burst length, burst- and silence-length distributions, a description of any correlation between successive bursts and silences, or even higher-order statistics. In most cases, these traffic descriptors are unknown at connection establishment. Even if they are known, it is inconceivable that they can be accurately policed. Note that the effectiveness of policing units has been questioned for what is seemingly a straightforward task of controlling, or policing, the peak rate.[40] Furthermore, the variability of traffic inside the network becomes independent of the variability of traffic at the network edge (e.g., Desimone[41] has shown this to be true for moderate- to high-network utilization). Thus, cell clustering within the network may cause congestion, which cannot be prevented by network edge-preventive controls. These arguments suggest that additional controls are necessary to handle congestion. Other notable preventive-control schemes were proposed, such as (open-loop) rate-based flow control,[42] and the Fast Reservation Control Protocol (FRP)[43] which handles burst-level congestion by controlling the admission of bursts, but similar arguments can be applied to justify the need for additional reactive controls.

A number of researchers advocated the need for closed-loop controls early. Many feedback-based control schemes (a large proportion was derived using intuition) were proposed for ATM. In an ATM network, depending on the nature of the traffic sources, the closed-loop congestion-control issue can be approached in two ways:

- For delay-tolerant traffic, which is basically comprised of TCP/IP traffic, switches can send feedback signals to the sources leading them to reduce the rate at which they release cells to the network. Then, excess traffic is queued at the source and consequently delayed.

- On the other hand, since delay tolerance of video/voice traffic is very low, congestion is controlled by sending coding-rate signals to these types of sources.[44] In the presence of congestion, the sources can vary their coding rate, and so reduce the frequency of cells generated by using this feedback information. Lower coding rate inevitably reduces the image/sound quality at the receiver but network utilization is maintained at higher levels by minimizing the cell losses due to congestion.

Several feedback-based control schemes have been proposed for delay-tolerant traffic, including: end-to-end window-based flow control,[45] end-to-end binary feedback,[24] network edge-rate control,[46] end-to-end ECN-based (forward or backward) flow-rate control,[47, 48] EPRCA (Enhanced Proportional Rate Control Algorithm),[49] ERICA,[50] Predictive Adaptive control,[51, 52] Fuzzy Backward Congestion Notification,[53] Fuzzy Explicit Rate Marking (FERM),[54] hop-by-hop rate-based,[55] and credit-based control.[56] In contrast, the number of proposed congestion-control schemes for delay-sensitive traffic is much less: Fuzzy congestion control,[57] and Neural-based congestion control,[58] Fuzzy-based rate control for MPEG video.[44]

In the summer of 1993, realizing that the commercial success of ATM will depend highly on the performance of "legacy" (i.e., TCP/IP, Ethernet, token-ring LAN) applications connected to an ATM backbone, the ATM community concentrated its efforts on a mechanism to allocate bandwidth dynamically within an ATM network, while simultaneously preventing data loss.[59] This effort culminated in the introduction of a service category by the ATM Forum, called available bit rate (ABR), in order to allocate bandwidth dynamically within an ATM network, while simultaneously minimizing the cell losses. A feedback-control framework has been selected to achieve these aims.[9] The proposed framework allows downstream nodes to periodically send information to the traffic sources relating to maximum cell rates that they can handle. The cell rate information is carried by a stream of resource management (RM) cells generated by the traffic sources and relayed back to the sources by the destination end systems, or the ATM switches. During their round-trip, while these cells pass through the switching nodes, the cell rate information contents of these cells are dynamically updated by the intermediate systems. The actions of the source, destination, and intermediate switches are well defined by the ATM Forum. The calculation of the rate is not part of the standard. As a prototype of this mechanism, the ATM Forum had developed a set of algorithms for ABR sources, destinations, and switches.[9] These algorithms had been shown to be robust in a variety of scenarios that had been simulated. However, since these schemes are designed with significant nonlinearities (e.g., two-phase—slow start and congestion avoidance—dynamic windows, binary feedback, additive-increase multiplicative-decrease flow control) based mostly on intuition, analysis of the closed-loop behavior is difficult if at all possible, even for single control-loop networks. The interaction of additional nonlinear feedback loops can produce unexpected and erratic behavior,[60] and empirical evidence demonstrates poor performance and cyclic behavior of the controlled network.[12] This is exacerbated as the link speed increases to satisfy demand (hence the bandwidth-delay product* increases), and also as the demand on the network for better quality of service increases. For example, for WAN networks a multifractal behavior has been observed,[61] and it is suggested that this behavior—cascade effect—may be related to existing network controls.[62]

---

* The bandwidth-delay product indicates the relative delay in the feedback path (mainly constant due to the finite propagation delay) in relation to the time dynamics of the queues in the system.

5.2.2.3   Shifts in Control Approach

In evolutionary sense, for TCP/IP and ATM, we see a progressive shift of controls from the edges of the network (initially open loop then edge binary feedback based) to inside the network. The feedback signal has also shifted from implicit to explicit, from pure binary to multivalued and explicit.

There is a greater need now to take a step back and design the control system using a structured approach. In order to arrive at (designed) flexible and effective control structure we should combine a bottoms-up approach with a top-down approach, as well as to design the controls together with the network system (not as an afterthought).[5] In the next section we follow Reference 5 and advocate the benefits of using a control-theoretic congestion-control framework.

5.2.3   A Control-Theoretic Congestion-Control Framework

A *system* may be broadly defined as an aggregation of objects united by some form of interaction or interdependence. When one or more aspects of the system change with time, it is generally referred to as a *dynamical* system. A *control system* can be qualitatively described as a system that has the ability to direct, alter, or improve its own behavior, or that of another system, and maintain some quantities of interest more or less accurately around a **prescribed value*** (**reference point**), or even control toward a **prescribed task.**** In an abstract sense it is possible to consider every physical object as a control system, as everything alters its environment passively or actively. For example, an on-off switch (e.g., to control the flow of packets into a network) is a man-made control system.

An *open-loop* control system is one in which the control action is independent of the output, whereas a *closed-loop* (also known as *feedback****) control system is one in which the control action is somehow dependent on the output. (Note that feedback is said to exist whenever a closed sequence of cause-and-effect relationships exist among the variables of the system. The essence of the feedback concept consists of measurement, comparison, and correction.)

---

* Note that selection of an appropriate prescribed or reference value or task can be a very important issue. For example, for large systems, it may be selected by considering global optimization objectives, and can be used as an instrument to reduce interactions and provide coordination.
** Observe that task-oriented control is closer toward full automation; setpoint control can be seen as limited to regulation: L. A. Zadeh, *Fuzzy Control vs Conventional Control Debate*, Lotfi A. Zadeh, Michael Athans, EUFIT, Aachen, Germany, September 13–16, 1999.
*** Feedback, one of the most fundamental processes existing in nature, is present in almost all dynamic systems, including those within man, among men, and between men and machines. However, use of feedback concepts has not been widespread, even though recognition that this theory is directly applicable to formulating and solving problems in many fields is becoming widespread, but its use is limited because of its heavy orientation toward technological applications [Note that this was stated in 1967 by J. Di Stefano, A. Stubberad, and I. Williams, "Feedback and control systems", Schaum's outline series, McGraw Hill, 1967].

An example of an open loop is a source feeding its packets into the network based on the timer value to control congestion. The quality of service (loss and delay) delivered by the network depends on the timer setting. Knowledge of this relationship, together with the state of the network and its environment, is essential to ensure that the delivered quality of service meets the desired. Of course any changes from the assumed environmental conditions, such as other users of the network, would make the controls ineffective. To make the flow of packets a closed-loop system, a continuous measure of network congestion is required.

If a measure of the congestion is available (or could be inferred from other on-line measurements), then in principle 'good' control can be exercised at all times, irrespective of the environmental conditions at the time of control. (Such environmental conditions include the other sources feeding into the network, their number and their behavior.) This is achieved by measuring the congestion (system output), comparing it with the desired value (the reference), and using the error between the measured and the desired to take corrective action, based on some control strategy. Control theory aims to address the design of control strategies with known attributes of control quality. Note that the human user in this case is not part of the control system—the system is *self-regulating*.

The ability of a closed-loop control system to maintain some quantities of interest, more or less accurately around a setpoint (or task) is a very powerful concept often ignored (seldom highlighted) in the literature of congestion control. Controlling against a setpoint means that one can have (known, predictable) control over the state of the system (assuming that the manipulated variable, e.g., the flow, can influence the system state—i.e., cause and effect must be proven). Selection of an appropriate setpoint value can be a very important task. For example, for large systems, it may be selected by considering global optimization objectives. Local states can be driven in such a way, as to provide overall coordination toward global objectives.* Furthermore, classical control theory developed techniques for assessing the performance of the (closed-loop) controlled system.** These included as a (minimum) assessment of

1. stability and stability margin—a control system must respond in some controlled manner to applied inputs and initial conditions (generally a system is not usable if it is unstable). It is worth noting that a feedback system, if not properly designed, can cause instability, in an otherwise stable system.

2. insensitivity to modeling inaccuracies with the actual plant. Since exact matchness with the physical system is never achievable, we

---

* This approach is commonly practiced. Examples include large-scale systems, e.g., chemical plants and power plants.
** Note that the performance metrics discussed here are not directly applicable to nonlinear systems. See discussion in Section 5.3.1.

require the controlled system to be reasonably insensitive to the parameters of the mathematical model used in the design. In addition the characteristics of the system may change with time (e.g., different connection mix in a network), so again we are interested in the sensitivity of the closed-loop system to parameter changes in the system. Feedback, in other words, can offer a high degree of robustness in terms of model inaccuracies (see discussion on Robustness, in Reference 63, page 123). This idea has been successfully tested in countless closed-loop systems (millions), especially in the process industry where plants are complex, nonlinear with many (even thousands) of control loops around the plant, mostly using a general purpose local controller—the ubiquitous PID regulator.*

3. ability of feedback-control system to handle unforeseen changes in the statistics of the external inputs and disturbances or noise that may be considered as changes in the environment in which the system operates. Disturbances are present in all systems (e.g., changed statistics in the flow of packets in a buffer under control). The disturbance rejection characteristics of the system are important. Note that open-loop systems cannot offer any disturbance rejection.

4. steady-state accuracy. A feedback control system can eliminate steady-state error.

5. transient response (i.e., system behavior until it reaches steady state). A feedback-control system can improve the transient behavior of the system.

In the case of an overall congestion-control system above, attributes of a good control are not enough (these were derived with one controlled variable (loop) in mind, and so they are applicable for the single congestion-control loop). Supplementary control-performance attributes that one should take into account include:

6. the concept of fairness,

7. the complexity and cost of implementation of the overall congestion-control strategy, and

8. interoperability.

---

* To alleviate the cost of modeling every controlled loop, a general purpose controller, the PID (PID, Proportional Integral Derivative, is a 2nd order controller), is often used. The PID parameters are "tuned" on-line, based on well publicized procedures, aiming to optimize certain controlled-system attributes, such as rise time and settling time. Process plant loops are seldom a good match to a 2nd order model. Furthermore, it is worth nothing that the setpoint of the PID is often set on-line by a supervisory control system to meet global objectives.

A lot more work is needed to define a formal framework for performance characterization of the various control strategies, but the above attributes, we think are important.*

Making use of control theoretic concepts we believe has potential benefits,[60] including:

1. Simpler and/or more effective algorithms with more predictable properties.

2. Better understanding of the performance of the controlled system (including their dynamic behavior), see discussion above.

3. Better understanding of existing nonlinear algorithms, including the need for any fixes ("jacketing software").

4. Better analysis techniques for large systems of interacting algorithms.

Traditionally, most control problems** arise from design of engineering systems (e.g., power plants, chemical plants, space shuttle, communication networks, etc . . . .). Such problems are typically complex, large-scale and fuzzy. They could also span large geographic areas. Control systems theory typically deals with small-scale, well-defined problems. A major difficulty in control system design is to reconcile the large-scale, fuzzy, real problems with the simple, well-defined problems that control theory can typically handle.[63] It is however in this area that a control system designer can effectively use creativity and ingenuity. This must be based on good understanding of the fundamental control theory (which can be sophisticated and complex), as well as a deep understanding of the system under control (not necessarily in the form of an accurate mathematical model). It is useful to have some perspective of the design process and a feel for the role of the theory in the design process. A good control system may have to satisfy a large number of specifications, and there are often many equally good solutions to a design problem. Many compromises are often necessary; for example, cost of control versus control performance. What theory can contribute to the design process is to give insight and understanding. In particular, theory can often

---

* An analytic performance characterization will be very difficult, so we propose the design of a formal Common Simulative Framework (CSF). All proposed algorithms can then be tested with regard to known control-performance attributes (e.g., robustness, efficiency, transient behavior, complexity of controls, scalability, interoperability, etc.). A fairer comparison between any proposed congestion-control solutions can then be made. (See 802.14 Modelling: advantages of a common simulative framework, IEEE Working Group, January 1990.) Note that a network simulator is in common use for the Internet (UCB/LBNL/VINT Network Simulator-ns (version 2), http://www-mash.cs.berkeley.edu/ns/). It can be expanded to include the controlled system-performance objectives and their measurement for a set of sample network configurations and connection mix(es).

** Nature is abundant with examples of self-regulating systems. Application of formal control-systems theory has provided a clearer understanding of the underlying principles and working of these systems.

pinpoint fundamental limitations on control performance. If idealized design problems can be described, which can be solved theoretically, these can often give good insight into suitable structures and algorithms. It is useful to note that control problems can be widely different in nature. They can range from design of a simple control loop in a given system to design of an integrated control system for the complete system. The relation between the design of the system (in our case the network), and the design of the control system to control it (the network management and control system) is often ignored, but it is one that can make the design of controls more effective. By designing the system and its control system together an additional degree of freedom is introduced, whereby the network designers can use it to design better trade-offs. Control systems are often introduced into given systems as an afterthought to simplify or improve their operation. If designed into the system from the beginning, the control of strong interactions in the system can be more effective. If proper controllability of the system is not designed from the beginning in the system then effective control will be difficult, if not impossible—see discussion in Reference 52, where the concept of network controllability is discussed. An example of the problems of introducing controls as an afterthought can be found in the control of congestion in the Internet. A piecemeal approach of solving one problem at a time has been adopted, with well-documented problems of control, and associated fixes.[16] For example, it is now well accepted that use of network devices, such as routers/gateways/ATM switches in the control system can make the problem of congestion control simpler and more effective.[7] Network devices should be designed to enable effective implementation of the adopted control strategy, and the control strategy should be designed with the possible limitations of network devices in mind—an iterative approach is therefore necessary to find the best compromise solution.

5.2.4 Modeling

The importance of mathematical models in every aspect of the physical, biological, and social sciences is well known. Starting with a phenomenological model structure that characterizes the cause and effect links of the observed phenomenon, the parameters of the model are tuned so that the behavior of the model approximates the observed behavior. Alternatively, a general mathematical model such as a differential or a difference equation can be used to represent the input-output behavior of the given process, and the parameters of the model can be determined to minimize the error between the process and model outputs in some sense. It should be noted that no mathematical model of a physical system is exact and there is no such thing as the model of the system. Generally, increasing the accuracy of the model also increases the complexity of the mathematical model, as well as the cost of deriving the model,* but exactness cannot be achieved. We generally strive to develop a model at an appropriate level of abstractness, which is adequate for the problem at hand without making the model overly complex (and costly). Simple and manageable models are required (complicated or intractable models with an abundance of parameters are not likely to be used in practice). The model should be parsimonious and able to capture the essential dynamic behavior in the simplest way. An important question is how good should the model be. Intuitively, one may say that a good model is one that maximizes the benefits that the model offers to the behavior of the designed controlled system (such as steady-state accuracy, disturbance rejection, robustness, fast transient response, etc.—see earlier discussion). It is important to highlight that what is important is not the actual model itself, but rather the improvement the model offers in the behavior of the control system, which was designed using that model.

To design a control system it is necessary to have a model between the input and the output of the system. As an example of the difficulty of deriving such a (mathematical) model let us take the development of a model of traffic behavior.[5] One can identify several factors that affect the model, and some are time varying. For example:

1. The diverse user (human) behavior (which can be time varying and different for different humans, even for the same interactive services) will affect the way traffic is generated.

2. The inherent fuzziness, for example, in the definition of the contract between the user and the network and its policing, and in the controls (declared objectives of controls and observed behavior of the system). Examples of fuzzy attributes include the quality of service to the user (requested and measured), definition and policing of the declared user-traffic parameters, and the definition and measurement of congestion, congestion onset and congestion collapse.

3. Data generation, organization, and retrieval (long-range dependence has been shown for both the source generation, as well as the storage of data).[64]

4. Traffic aggregation (the aggregation process is a very complex one—many studies suggest that self-similarity seems to be preserved under a variety of network operations, and this holds over a wide range of network conditions).

5. Network controls (there is speculation that fractal features in network traffic remain even after network controls).

6. Network evolution (again self similarity appears robust to network changes, e.g., upgrades).

---

* It has been noted that the development of models for control-system analysis and design involves 80 to 90% of the total effort required.

Computational intelligence to handle the complexity and fuzziness present in the network system surely has an essential role to play here. We should exploit the tolerance for imprecision and uncertainty to achieve tractability, robustness, and low cost.[65]

5.2.5 Role of Computational Intelligence

A network system is a large distributed complex system, with difficult, often highly nonlinear, time-varying and chaotic behavior. There is an inherent fuzziness in the definition of the controls (declared objectives and observed behavior). Dynamic or static modelling of such a system for (open or closed-loop) control is extremely complex. Measurements on the state of the network are incomplete, often relatively poor and time delayed. Its sheer numerical size and geographic spread are mind-boggling, for example, with customers (active services) in the 10s of millions, network elements in the 100s of millions, and global coverage.

Therefore, in designing the network-control system, a structured approach is necessary. The traditional techniques of traffic engineering, queuing analysis, decision theory, etc. should be supplemented with a variety of novel control techniques, including (nonlinear) dynamic systems, computational intelligence and intelligent control (adaptive control, learning models, neural networks, fuzzy systems, evolutionary/genetic algorithms), and artificial intelligence.

Computational Intelligence (CI)[66, 67, 68] is an area of fundamental and applied research involving numerical information processing (in contrast to the symbolic information processing techniques of Artificial Intelligence (AI)). Nowadays, CI research is very active and consequently its applications are appearing in some end-user products. The definition of CI can be given indirectly by observing the exhibited properties of a system that employs CI components:[67]

> A system is **computationally intelligent** when it: deals only with numerical (low-level) data, has a pattern recognition component, and does not use knowledge in the AI sense; and additionally, when it (begins to) exhibit
>
> - computational adaptivity;
> - computational fault tolerance;
> - speed approaching human-like turnaround;
> - error rates that approximate human performance.
>
> The major building blocks of CI are artificial neural networks, fuzzy logic, and evolutionary computation.

While these techniques are not a panacea (and it is very important to view them as supplementing proven traditional techniques), we are beginning to see a lot of interest not only from the academic research community,[69] but also from telecommunication companies.[70]

It is worth pointing out that almost all published studies on congestion control using CI have concentrated on ATM networks, as opposed to TCP/IP. This can probably be attributed to the experienced difficulty in obtaining any useful dynamic models for congestion control in ATM network. ATM was conceived and designed to deliver a variety of traffic services (voice, image, data, etc. . . .) with a certain guaranteed level of QoS (i.e., controlled levels of congestion). This complexity made the use of CI techniques in the research on ATM network congestion control inevitable. As the popularity and pressure to deliver other media through the Internet increases, we expect to see more research in the application of CI techniques to TCP/IP network congestion control. This is further facilitated by the progressive shift in TCP/IP network congestion control culture from the delivery of data to integrated services, from the location of controls from the outside of the network to inside,[7] and from simplistic to progressively more sophisticated and responsive congestion sensors.[8]

In the rest of the chapter, we shall illustrate through a number of selected examples the power of the Computational Intelligence techniques that show that effective congestion control is possible.

5.3 Computational Intelligence Techniques for Effective Congestion Control

5.3.1 Fuzzy Logic Applications

A Fuzzy Logic Controller (FLC) defines a nonlinear-control law by employing a set of fuzzy-if-then rules (fuzzy sets for short). The if-part describes the fuzzy inputs and the then-part of a fuzzy rule specifies a control action (law) applicable within the fuzzy region from the if-part. Two basic approaches to design FLCs are commonly used: heuristic-based design and model-based design.

Heuristic-based FLC design may be viewed as an alternative, nonconventional way of designing feedback controllers where it is convenient and effective to build a control algorithm without relying on formal models of the controlled system and control theoretic tools (e.g., see Reference 71). For example, obtaining a formal (mathematical) model may prove infeasible for control-system design (e.g., cost of deriving a model may be prohibitive, resultant model may be too complex for control system design, linearization of the nonlinear model may result in poor controlled system behavior, etc. . . .). The control algorithm is encapsulated as a set of commonsense rules. FLCs have been applied successfully to the task of controlling systems for which analytical models are not easily obtainable or the model itself, if available, is too complex and highly nonlinear. Even though this approach is simple and appealing, a major drawback is the lack of any formal verification of the controlled-system properties (stability, performance, robustness), and the lack of any systematic way to design the control algorithm with prescribed specifications on the controlled-system performance. However, as elegantly pointed out by Mamdani,[72] overstressing the necessity of mathematically derived performance evaluations may be counterproductive and contrary to normal industry approach (e.g., prototype testing may suffice for accepting the controlled-system performance). Also, the prescribed specifications on the controlled-system performance can be embedded into the controller design as a set of fuzzy if-then rules (e.g., rise-time rules, damping rules and steady-state rules). Furthermore, for nonlinear systems, there are no systematic specifications of the desired controlled-system behavior, as these are not obvious at all because the response of a nonlinear system to one input vector does not reflect its response to another input vector (initial condition dependent). A consequence of this is that specifying the desired behavior one needs to employ some qualitative specifications of performance, including stability (which takes a different interpretation for nonlinear systems; currently there is no universally accepted definition among the experts), accuracy and response speed and robustness.[73] Another serious drawback in applying heuristic-based FLC (perhaps more important than earlier-cited criticism) is the one commonly suggested in the literature, difficulty for Multiple-Input Multiple-Output systems.

On the other hand, model-based fuzzy control deals with the design of the set of fuzzy rules given a conventional, linear or nonlinear, open-loop model of the system under control. Heuristics and the specification of the controlled-system behavior can be incorporated in the design procedure. The idea is to exploit the best of each of the traditional and fuzzy approaches. One may expect that designs that draw on the power of control theory are likely to be more powerful than the simpler heuristic approaches. Appropriate design may allow a formal verification of the stability performance and robustness of the controlled system,[73] which are expected to be better than both the fuzzy and the classical approach on their own. However, the whole design process is often very complex, and reliant on the availability of a conventional model.

A choice between the two methods (as well as between any of the other formal control design methods) is not an easy one. It depends on many factors, such as cost of developing the control system, area of application, tolerance of failure, effectiveness, and so on. Later on in this chapter, we adopt the heuristic-based design approach to illustrate the FLC design process for FERM. Our experience, based on extensive simulation has shown the resultant FLCs to be stable, robust, and effective. This is in addition to the relative easiness of developing the FLC.

In recent years, a handful of research papers have been published on the investigation of solutions to congestion-control issues in ATM networks. Given the complexity of ATM networks, rich variety of traffic sources that operate on them, and difficulty of obtaining formal models for in-depth analysis, it is not surprising to see that FLCs are favored by the researchers involved in ATM network development. As discussed in the earlier sections, purely reactive congestion-control techniques will not be effective in ATM-based multimedia and multiservice networks. Therefore, the researchers who applied the computational intelligence methods to congestion-control problems mostly looked at predictive congestion-control schemes. In general, the schemes observe the short-term behavior of a link to estimate the future of cell arrivals in order to predict the onset of congestion and take proactive measures to prevent its occurrence. In the following paragraphs, an overview of the recent research efforts is presented.

Liu and Douligeris[74] have proposed a combination system consisting of a leaky bucket and a fuzzy-logic cell-rate controller. Their system is designed for video/voice sources that negotiate their peak and mean cell rates during the call set up phase. The leaky bucket module is responsible for the compliance of the sources to the negotiated traffic parameters. The authors stipulate that, for nonconforming sources, instead of simply discarding the excess cells at the switch, it is possible to send back signals to the sources to reduce their cell generation rates and so, minimize the cell losses in order to maximize the resource utilization. The second module, which consists of two fuzzy-logic controllers, is responsible for the generation of these rate-control signals. It attempts to predict the near-future cell-discarding behavior of the switching nodes based on the short-term observation of the cell arrivals. This prediction is then fed back to traffic shapers in the sources to regulate the data generation rate to minimize cell losses in the switching nodes.

The leaky bucket is realized as a counter. It is incremented whenever a cell is transmitted to the network, and decremented for every time interval T passed until it reaches 0. Whenever the counter is bigger than a threshold value S_{th}, the incoming cells are discarded until the counter value drops below the threshold. If the counter value is 0, the leaky bucket allows the source to transmit a burst of θ cells maximum. Liu and Douligeris have found that, if the network traffic gets burstier, selecting the optimum values for T, S_{th}, and θ becomes a difficult task. In order to reduce the sensitivity of the system to these parameters, and consequently to minimize the number of cells discarded, the fuzzy system has been included into the scheme to regulate the peak cell rate of the sources.

Jensen[75] has proposed a fuzzy system for controlling the transmission rate of sources to protect links against overload in the case of connections exceeding their negotiated traffic parameters. The fuzzy system consists of three FLCs connected in a cascade formation. The scheme operates as follows: at the call admission stage a service-dependent priority is assigned to each connection. This priority is kept as a fixed value for the whole lifetime duration of the connection. Also, in the switching node, a certain buffer capacity is allocated to the connection. The fuzzy system generates the cell-service rate-control signals for each buffer. Inputs the fuzzy system are:

(a) allocated priority level,

(b) difference between the effective bandwidth at which the source is transmitting the cells and the declared bandwidth negotiated during the call set-up stage,

(c) current buffer occupancy level, and

(d) bandwidth utilization at the output link of the switching node.

The variables (a) and (b) above are the inputs of the first FLC of the cascade formation. The output generated by the first FLC and variable (c) form the inputs of the second FLC. The third FLC receives the output of the second FLC and variable (d) as its inputs and generates the cell service-rate signal to be used by the server of a particular buffer. The reason behind using a three-step fuzzy-control mechanism, each receiving two input variables instead of a single FLC having four input variables, is to keep the number of linguistic rules of the rule base in a reasonable level. FLCs suffer from a problem called curse of dimensionality: the number of linguistic rules rises exponentially when the number of input variables increases linearly. One of the solutions to this problem is the one adopted by Jensen and to use a cascade connection of FLCs each has a limited number of input variables. For a discussion of other solutions[76] can be referred.

Cheng and Chang[77, 78] have opted for a system which combines connection admission control (CAC) and congestion-control mechanisms. The congestion-control mechanism sends back coding rate-control signals to video and audio sources, and congestion-control signals to data sources to adjust the cell transmission rate of the sources, and subsequently the traffic density at the switches. The system contains seven modules, three of them are FLCs:

1. Fuzzy Congestion Controller accepts three inputs: queue length, queue-length change rate, and overall cell-loss probability for all traffic using the same queue. The input signals of the Fuzzy Congestion Controller are generated by the Performance Measures Estimator module. The Fuzzy Congestion Controller generates a control action. A negative value generated denotes a certain degree of congestion, a new call has little chance of being accepted. A negative value also initiates selective discarding for video and audio sources, and transmission rate reduction for data sources. A positive output value indicates that the system is free of congestion to a certain degree, new calls have a good chance of entering the network, and existing connections can be restored to their original rates. Coding Rate Manager and Transmission Rate Manager modules are responsible for sending the control signals to the respective traffic sources.

2. Fuzzy Bandwidth Predictor estimates the equivalent capacity of the call based on the advertised traffic parameters peak bit rate, average bit rate, and peak bit rate duration. The estimated capacity is used by the Network Resource Estimator module to calculate the total capacity in use.

3. Fuzzy Admission Controller is responsible for generation of accept/reject signals for audio/video call requests. The input variables for this module are total capacity in use, cell loss probability (generated by the Performance Measures Estimator module), and control action signal generated by the Fuzzy Congestion Controller. Based on these input variables it generates an accept/reject signal, which will be used to grant or deny the incoming call requests.

An interesting approach adopted by Cheng and Chang is the utilization of genetic algorithms to generate the linguistic rules of the three FLCs mentioned above.

Pitsillides et al.,[53, 54] and Qiu[79] have proposed congestion control schemes which operate under similar principles. The schemes, by measuring the queue length and queue growth rates at the output buffer of a switch, attempt to estimate the future behavior of the queue, and send explicit rate-control signals to the traffic sources to avoid or alleviate congestion. The explicit rate-control signals are calculated periodically by fuzzy inference engines located in the switches, and sent to the traffic sources in resource management (RM) cells.

The scheme of Pitsillides et al. is used in Fuzzy Explicit Rate Marking (FERM) algorithm. They have analyzed its performance in detail regarding fairness, responsiveness, resource utilization, and cell loss in LAN and WAN environments. The scheme has been further refined (FERM2) and as an adaptive scheme which has self-tuning capabilities (A-FERM).[80] A detailed overview of FERM2 is presented in later sections as an illustrative example.

The linguistic rules, which determine the actions to be taken by the FLCs, can sometimes pose challenges to the designers. Traditionally, the rules encapsulate the expert's experience or belief about the necessary control actions taken. It is possible that the expert's knowledge is not available, or not easily obtainable. In this case, if operational data is in hand, linguistic rules may be extracted from the data by using clustering methods. Cheng and Chang[77, 78] have used genetic algorithms to obtain linguistic rules from the operational data. Another challenge is, usually the rules are static: they do not change during the operation of the system. Naturally, this can lead to sub-optimal control actions to be taken if system dynamics change in time. The solution to this problem is to use adaptive methods to modify a set of parameters that are used to define the linguistic rules in real-time.

Takagi and Sugeno[81, 82] have proposed a method for adaptive tuning of linguistic rules of a FLC. In their method, they approximate the system under control by an open-loop fuzzy model (model based FLC design), given in terms of fuzzy rules. The controlled variable, which determines the output action defined in the linguistic rules, is chosen as a polynomial expression of some state variables whose coefficients are modified by adaptive techniques. This method can be used for controlling very complex systems and has been successfully demonstrated by Sugeno to control the flight of a helicopter. For an illustrative example of this approach see References 83 and 84. Hu, Petr, and Braun[85] have used this method to design an adaptive fuzzy congestion-control scheme. In their approach, the level of network congestion is monitored through the queue length at the output buffer of the switch, with the control target being set at a desired queue length. The FLC uses this information as its input variables:

(a) Length of the queue at the output buffer of the switch (normalized);

(b) Queue length change rate;

(c) Data traffic transmission demand. (The demand is calculated as the ratio of current rate of data traffic to the allowed rate. If the ratio is significantly less than 1, the data traffic sources do not have as much traffic to send as what the network allows.)

(d) Number of discarded data cells (normalized).

Then, the FLC calculates the allowed cell rate for the data sources. At the same time the parameters of the polynomial functions which constitute consequent parts of the individual rules are tuned using gradient descent method. The adaptation objective is chosen as the minimization of the difference between the queue length and the desired queue length.

Not many schemes were proposed for the control of real-time video traffic. A notable example is Tsang et al.[44] who propose a fuzzy logic-based scheme for real-time MPEG video to avoid long delay or excessive load at the user interface in an ATM network. They control the input and output rates of a shaper whose role is to smooth the MPEG output traffic rate. This they do at the expense of variable picture quality, but in a controlled way (by allowing a small output variation, similar to an open loop which aims for constant picture quality at the expense of variable bit rate). They use two fuzzy logic-control systems operating in two different time scales. The first fuzzy system controls the intra-Group-of-Picture traffic, in order to ensure compliance of the coded video output stream with predefined sustainable cell rate and burst tolerance parameters to avoid cell dropping by the leaky bucket. The second fuzzy system operates in the inter-Group-of-Picture time scale, to change the quantization parameter of the coder (hence the coding rate), according to information about the network-congestion level. They are able to show that the rate fluctuation of the video is reduced, as compared to the open-loop (constant-quality) scheme, without a substantial drop in picture quality. Thus the proposed scheme reduces burstiness, therefore preventing congestion from occurring.

5.3.1.1   An Illustrative Example: FERM2 Congestion-Control Algorithm

In this section the operation of FERM2 explicit rate congestion-control scheme is summarized. FERM2 is very similar to FERM, which is documented in Reference 54 , and can be considered as a further refinement of the original scheme. The main difference between the two schemes is in the former one the desired queue length is implicit; in the later one it is set by a higher-level control module to provide more dynamic resource utilization across the switches on a particular virtual connection. Figure 5.2 shows the block diagram FERM2. Overall operation of the scheme is compliant with the ATM Forum Traffic Management Specification, Version 4. The scheme uses the following three parameters as stipulated in the specification:

Parameter	Definition
PCR	Peak cell rate
ICR	Initial cell rate
MCR	Minimum cell rate

Whenever a new ABR connection is established, the values of these parameters are negotiated between the traffic source and the network.

Cell rates of data sources are adjusted by Explicit Rate (ER) information carried by Resource Management (RM) cells. RM cells are periodically generated by traffic sources, transmitted towards the destination-end systems, and initial ER information is set by the ICR. The destination-end systems bounce the RM cells back to the sources. During the return path, when a RM cell passes through an ATM switch, its ER value is examined and possibly modified. A data source, upon receiving a RM cell, adjusts its cell rate based on the value contained in the RM cell's ER field. If the ER field contains a rate bigger than PCR, the cell rate is set to PCR. Similarly, the cell rate is set to MCR for the ER values less than MCR.

The scheme, in the calculation of the ER, monitors both the current queue length and its growth rate. The queue length captures the current state of the output buffer of the switch, and the rate of change of the queue length provides some form of prediction for the near-future buffer behavior. Thus, the scheme could be expected to be more effective than schemes using feedback based on the queue-length threshold, queue length, or the rate of change of the queue length alone. The scheme provides the ER to all the active VCs at all the time so that congestion and undesired resulting behavior can be avoided. The scheme does not need to keep the state of current VC connections sharing the same semistatic VP at the switch.

Periodical ER calculations are performed by the Fuzzy Congestion Controllers (FCCs) located in each ATM switch. The structure and operation of FCCs are outlined in the next section.

5.3.1.2 *Fuzzy Congestion Controller*

Fuzzy Congestion Controller (FCC) is a fuzzy logic controller (FLC). Designing a FLC involves selection of suitable mathematical representations for t-norm, s-norm, defuzzification operators, fuzzy implication functions, and shapes of membership functions among a rich set of candidates. Particular selection of these operators and functions alter the nonlinear input-output relationship, or in other words, the behavior of a FLC. But, research has shown that same effects can be achieved by proper modification of the rule base.[86] Therefore, in practical applications, usually computationally lighter and well-studied operators and functions are selected, and desired behavior of FLC is obtained by altering the linguistic rules.

For the implementation of the FCC, the authors have chosen the most widely used and computationally lighter methods, which are

- singleton fuzzification
- t-norm algebraic product for the mathematical representation of the connective "and"
- Larsen's product rule of implication
- sup-product compositional rule of inference
- weighted mean of maximums defuzzification.

As can be observed from the control surface of the FCC (Figure 5.3), it is a nonlinear controller. For a certain queue length, it calculates different flow rate limits depending on the rate at which queue length varies.

At the end of the each filter period of N_{fp} cell-service times (control interval), two numerical values showing the average length of the ABR queue and the difference of the ABR queue length from the previous control interval (i.e., queue growth rate) are calculated and fed to FCC. Based on this data and the linguistic information stored in the rule base, FCC computes the Flow Rate Correction ($-1 <$ FRC < 1) and an Explicit Rate

$$\text{ER}_{next} \leftarrow \min(\text{LinkCellRate}, \max(0, \text{ER}_{current} + (\text{FRC} \times \text{LinkCellRate})))$$

for the sources feeding the ATM switch. If, within the current control interval, the ATM switch receives an RM cell traveling to the upstream nodes, it examines the ER field of the cell and if this rate is greater than the calculated flow rate, it modifies the ER field with the computed value and retransmits the RM cell.

5.3.1.3 Rule Base Design Process

The selection of rule base is based on the designer's experience and beliefs on how the system should behave. Design of a rule base is twofold: First, the linguistic rules (surface structure) are set; afterwards, membership functions of the linguistic values (deep structure) are determined.

The trade-off involving the design of the rule base is to have a set of minimum number of linguistic rules representing the control surface with sufficient accuracy to achieve an acceptable performance. Recently, in the fuzzy control literature, some formal techniques for obtaining a rule base by using Artificial Neural Networks or Genetic Algorithms have appeared. Nevertheless, the conventional trial and error approach under the guidance of some design rules of thumb[87] can be referred or a discussion of these have been used in this study.

Usually, to define the linguistic rules of a fuzzy variable, Gaussian-like, triangular- or trapezoidal-shaped membership functions are used. Selection of Gaussian-like membership functions leads to smoother control surfaces.

Then, the rule base is fine-tuned by observing the progress of simulation, such as cell loss occurrences and demand versus throughput curves. The tuning can be done with different objectives in mind. For example, any gain in throughput must be traded off by a possible increase in the delay experienced at the terminal queues. However, since the tuning of the fuzzy rules is intuitive, and can be related in simple linguistic terms with user's experience, it should be a straightforward matter to achieve an appropriate balance between a tolerable end-to-end delay, and the increase in throughput. Alternatively an adaptive fuzzy logic-control method can be used which can tune the parameters of the fuzzy logic controller on line, using measurements from the system. The tuning objective can be based on a desired optimization criterion, for example, a trade-off between maximization of throughput with minimization of end-to-end delay experienced by the users. The set of linguistic rules shown below in Table 5.1 define the control surface of the FCC.

TABLE 5.1

Set of Linguistic Rules Defining the Control Surface of the FCC

---

if ABR queue length is too short and queue is decreasing fast then increase flow rate sharply
if ABR queue length is too short and queue is decreasing slowly then increase flow rate moderately
if ABR queue length is too short and queue length is not changing then increase flow rate moderately
if ABR queue length is too short and queue is increasing slowly then decrease flow rate moderately
if ABR queue length is too short and queue is increasing fast then decrease flow rate moderately

if ABR queue length is acceptable and queue is decreasing fast then increase flow rate moderately
if ABR queue length is acceptable and queue is decreasing slowly then increase flow rate moderately
if ABR queue length is acceptable and queue length is not changing then do not change flow rate
if ABR queue length is acceptable and queue is increasing slowly then decrease flow rate moderately
if ABR queue length is acceptable and queue is increasing fast then decrease flow rate moderately

if ABR queue length is too high and queue is decreasing fast then do not change flow rate
if ABR queue length is too high and queue is decreasing slowly then do not change flow rate
if ABR queue length is too high and queue length is not changing then decrease flow rate moderately
if ABR queue length is too high and queue length is increasing slowly then decrease flow rate sharply
if ABR queue length is too high and queue length is increasing fast then decrease flow rate sharply

---

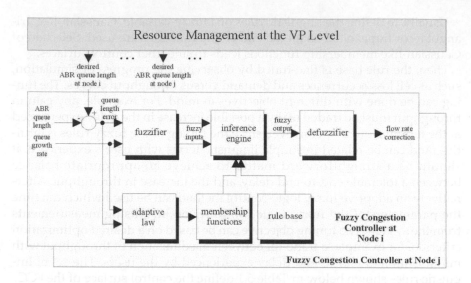

FIGURE 5.2
Block diagram of the Fuzzy Congestion Controller of the FERM2 scheme.

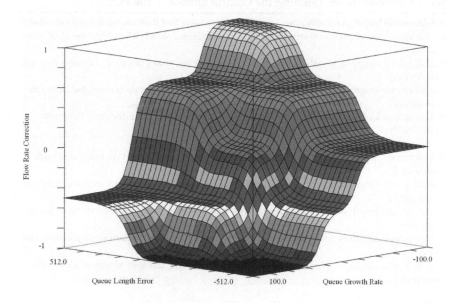

FIGURE 5.3
Control surface of the Fuzzy Congestion Controller. The control surface is shaped by the rule base and the linguistic values of the linguistic variables. By observing the progress of simulation, and modifying the rules and definitions of the linguistic values, FCC can be tuned to achieve better server utilization and lower cell loss coupled with minimal end-to-end cell delay.

5.3.1.4 The Thinking Behind the Selected Rule Base

An inspection of either the linguistic rules of Table 5.1 or the resulting control surface of Figure 5.3 hints at some of the designer's beliefs about how the system should be controlled.

The rules of Table 5.1 are more aggressive about decreasing flow rate sharply than increasing it sharply. There is only one rule that results in increasing flow rate sharply, whereas two rules result in decreasing flow rate sharply. This is shown in the surface of Figure 5.2 by a much bigger region of maximum flow rate reduction relative to maximum flow rate increase.

For intermediate queue lengths (acceptable queue length), the rules are somewhat restless. Attention is paid to the rate of change of queue length, and moderate changes of flow rate are invoked unless the queue length is almost constant. This corresponds to the small flat region in the center of the surface in Figure 5.2.

These rules reflect the particular views and experiences of the designer, and are easy to relate to human reasoning processes.

The authors have done extensive simulations on a representative ATM network (Figure 5.4) and have compared the performance of FERM against enhanced proportional rate control algorithm (EPRCA). The results of this study have been reported in Reference 54. FERM2 yields yet better throughput results than FERM in overloaded networks (Figure 5. 5 and Figure 5.6).

The following plots show the time evolution of the Explicit Rate, as calculated by FCC, for the case of a LAN (Figure 5.7) and a WAN network (Figure 5.9). The other two figures show the time evolution of the queue length for both LAN (Figure 5.8) and WAN (Figure 5.10), with the reference point set at 400 cell places. Please note the expected deterioration in performance of the controlled network for high bandwidth delay products (WAN), as opposed to the excellent controlled system performance for the case of very small propagation delay (LAN). Nevertheless, even for the WAN case, the network system is well controlled and the network losses and retransmissions are limited. Note that the distances between switches are set at 1500 km, with a maximum end-to-end round trip delay around 30 ms @ 6000 km; compare with the 2.6 ms time it takes to fill or empty a buffer of 1000 cells @ 155 Mb/s.

In the simulations, the values of the negotiated parameters are set as follows: PCR = 149.76 Mb/s, ICR = PCR and MCR = 2 Mb/s. The control interval N_{fp} is set as 32 cell-service periods.

5.3.1.5 Real-Time Implementation Issues of FERM2 in an ATM Switch

Even though fuzzy logic has demonstrated its strengths in control applications of industrial machinery and consumer appliances, its integration into high-speed communication networks presents a number of challenging issues. Today's networks are very fast: the links operate in the order of Mb/s, and they will soon be transmitting data at the rates of Gb/s and Tb/s. An

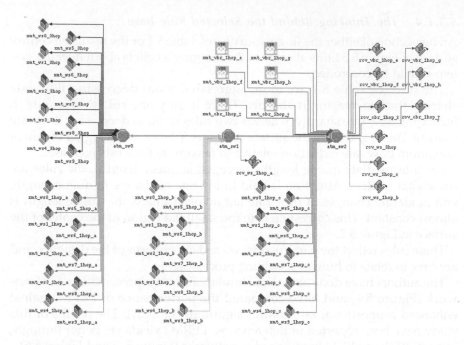

FIGURE 5.4
ATM network model used for performance analysis of FERM and FERM2 algorithm. Same network configuration has been used for the simulation of ATM WAN backbone and ATM LAN backbone except that the distances between switches have been assumed to be 1500 km and 10 km for WAN and LAN simulations, respectively. All traffic (except 1hop (b) traffic) leaving ATM switch 2 travels to a fourth ATM switch and distributed. Since no cell buffering occurs at this switch, it has not been included into the simulation model. The speed of all links have been considered as 155 Mb/s.

ATM switch connected to 155 Mb/s links has very little scope for time-consuming fuzzy inferences at the VP level. There is considerable progress on design and implementation of dedicated hardware for fuzzy inference operations, but additional cost of integrating fuzzy processors into networking equipment would never be cost effective with using today's technologies or without invention of a totally new approach. For relatively simple FLCs such as FCC of FERM2, simple table lookup methods can easily be employed. As mentioned above, the linguistic rules of the FCC basically defines a non-linear control surface and the rules themselves play the role of an interface as an aid to describe the shape of the control surface. The control surface can then be encoded as a lookup table, so that when input variables are read, they are used to determine an output value by executing just a few processor operations.

In FERM2, the lookup table of flow rate correction is stored as a two-dimensional matrix of data and a particular flow rate correction value is accessed by using the values of input variables queue length and queue growth rate as the indices.

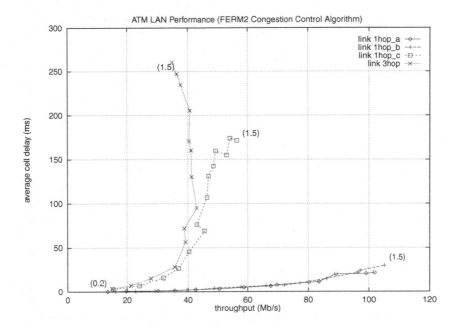

FIGURE 5.5

Plot of average end-to-end ABR cell delay vs. useful throughput of simulated ATM LAN under FERM2 congestion control. The graph has been produced by varying the offered link loads generated by the ABR traffic sources from 20% to 150% of the link capacities.

5.3.2 Artificial Neural Networks Applications

ANNs can be used to devise techniques that adapt to new network situations, changing traffic patterns because of their function estimation, and extrapolation abilities. These abilities have been exploited by a handful of researchers to design congestion and admission control techniques especially for high-speed, ATM-based multimedia networks. Most of the studies fall in the Connection Admission Control (CAC) category. In the following paragraphs a few representative works related to ANN-based congestion-control research are summarized.

Tarraf, Habib and Saadawi[88–91] have extensively investigated how ANNs can be used to solve many of the problems encountered in the development of a coherent traffic-control strategy in ATM networks. In Reference 91 they present an ANN-based congestion controller for ATM video/voice multiplexers. The congestion controller monitors the number of cells in the multiplexer buffer to predict the potential congestion problems. It then generates a rate-control signal to be fed back to the sources in order to alter the arrival rate of cells. During the periods of buffer overload, the control signal reduces the arrival rate by decreasing the coding rate at the video/voice source. When the overload period ends, the coding rate is returned back to its previous level. The ANN generates coding-rate sig-

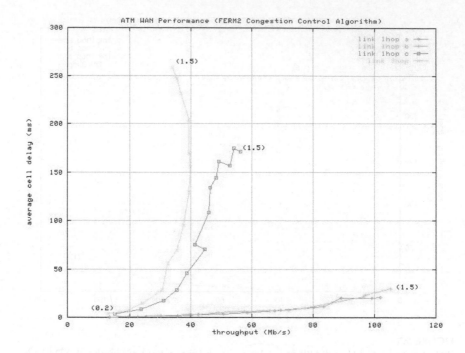

FIGURE 5.6

Plot of average end-to-end ABR cell delay vs. useful throughput of simulated ATM WAN under FERM2 congestion control. The graph has been produced by varying the offered link loads generated by the ABR traffic sources from 20% to 150% of the link capacities

nals and so attempts to maximize the overall performance of the system through a cost function. The cost function combines two system performance measures:

1. Minimization of the input multiplexer buffer overflow periods (in order to minimize cell-loss rate).

2. Maximization of the coding-rate levels at the input sources (in order to maintain the quality of the video/voice traffic). The congestion-control algorithm has self-tuning capability by using reinforcement learning technique.[92]

Chen and Leslie[93] have proposed a general adaptive congestion-control mechanism. The ANN-based controller monitors two parameters: the arrival rate of the traffic, and a QoS measure such as the cell-loss ratio or delay. Both parameters are processed as time-dependent averages, before presented to the ANN, in order to capture the dynamics of the traffic. The ANN then generates a control signal which attempts to maximize the arrival rate while maintaining the QoS. The learning is performed by using an adaptive

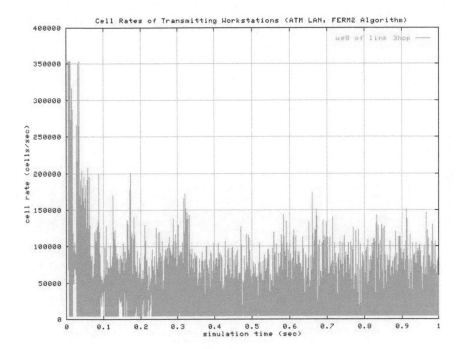

FIGURE 5.7
Time evolution of the Explicit Rate for the case of the LAN; calculated by the FCC.

backpropagation algorithm. The adaptive backpropagation algorithm has been chosen to overcome the problem of slow rate of learning usually experienced by the classical backpropagation algorithm. The adaptive backpropagation algorithm, in order to accelerate the learning process, changes the learning rate as learning proceeds.

Liu and Douligeris have also done an extensive study on applications of ANNs to ATM congestion control issues for data[94] and video/voice[95] traffic. In Reference 94, they present three different ANN models used as static and adaptive feedback-congestion controllers for data traffic and compare their performance. In their approach, ANNs are used to predict the possible cell losses in the near future. Based on these cell loss predictions, a feedback cell containing explicit rate information is sent to the data sources to regulate their transmission rates. The three schemes differ in the type of information processed by the ANNs. In the first approach, the current queue length in the buffer of the ATM switch and the cell arrival patterns in the past few periods are used to predict the amount by which sources need to reduce their rates. In the second mechanism, the cell arrival patterns are processed using the standard normal deviate (SND) model before being fed into the ANN. In the third mechanism, cell-arrival patterns are processed by a moving average data-smoothing technique.

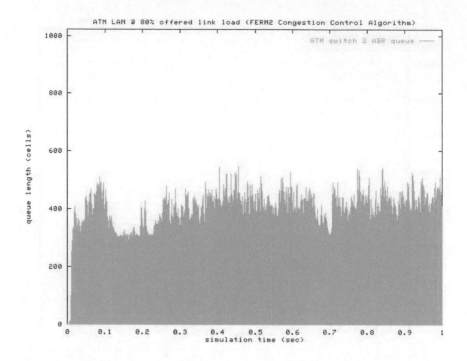

FIGURE 5.8
Time evolution of the queue length for the case of a LAN. Note that the reference value is set at 500 cell places.

5.3.2.1 *An Illustrative Example: Rate Regulation with Feedback Controller*

As an illustrative example, we present the rate-based regulation scheme proposed by Liu and Douligeris.[95] In the study, they propose an ANN-based rate-based feedback-control scheme for audio/video sources. In the scheme, a leaky bucket (LB) mechanism is used to perform cell discarding when the traffic violates a predefined threshold. They argue that selection of the optimum threshold value and depletion rate for the LB is very difficult. To overcome this difficulty, they propose an ANN model which monitors the status of the LB and predicts the amount of possible cell discarding in the LB in the near future. When possible cell discarding is detected, the coding rate of the source is regulated to a certain amount by sending a feedback signal to the traffic sources. They selected a three-layer feedforward ANN with an error back-propagation learning algorithm (Figure 5.11).

They use two real MPEG traces. The training data is only a very small percent of one of the data sets that goes through the network. They show that their model can be generalized and can be applied to different traces without the need to retrain the network. Through simulation, they show that their

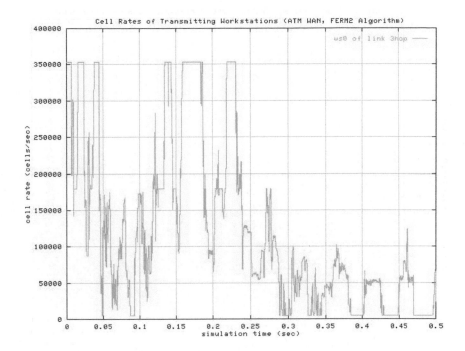

FIGURE 5.9
Time evolution of the Explicit Rate for the case of the WAN; calculated by the FCC.

mechanism outperforms the static threshold approaches in both cell loss rate (3 to 5 times), transmission delays, and channel utilization.

5.3.3   Evolutionary Computation Applications

Evolutionary computation (genetic algorithms and genetic programming) is a powerful method of system design which relies on developing systems that demonstrate self-organization and adaptation in a similar, though simplified, manner to the way in which biological systems work. The work of Holland[96] initiated so-called genetic algorithms. A genetic algorithm is a procedure maintaining a population of structures that are candidate solutions. The fitness of these structures in the current population is evaluated and on this basis a new population of candidate solutions is formed. This process of generating a new population involves the use of genetic operators such as reproduction, crossover, and mutation. Generally, this new population is fitter. This process is repeated many times, and as a result the solution will emerge. Using similar inspiration from genetic and natural selection, John Koza proposed a system that evolves computer Lisp programs. This method, called genetic programming, is extensively described in his two books.[97, 98] Genetic programming

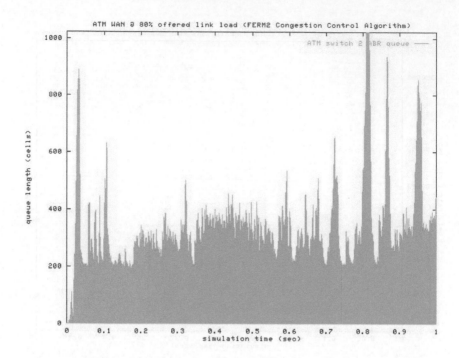

FIGURE 5.10
Time evolution of the queue length for the case of a WAN. Note that the reference value is
set at 500 cell places.

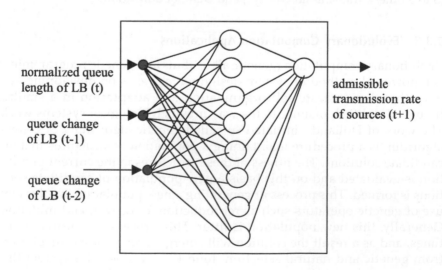

FIGURE 5.11
Proposed ANN for rate-based feedback control.

starts with an initial population of hundreds or thousands of randomly generated computer programs. Then, by using the Darwinian principle of survival and reproduction of the fittest, new offspring populations are created. The reproduction operations involve selection of programs (individuals) from the current population in proportion to their fitness and copy of them to the next population. The best individuals survive, and finally optimal or near-optimal programs are obtained. Evolutionary computation can be used in engineering applications for synthesis of hardware or software modules.

It appears that this framework has not been investigated thoroughly for solving congestion control issues in telecommunication networks, most probably because of its very computationally intensive nature (there are, however, many examples of its use for network optimization). The strength of evolutionary computation-based techniques for time-series prediction has been extensively reported in the literature. This can be exploited for designing predictive congestion controllers similar to the ANN-based ones. For example, observation of traffic patterns and resulting sequences at a switching mode can be used to design a module that is capable of estimating the future behavior of the traffic and so generate rate signals.

5.3.3.1   *An Illustrative Example: Queue Prediction Using Genetic Programming Techniques*

Jagielski and Sekercioglu[99] have demonstrated this capability. They have presented a scheme for prediction of queue dynamic behavior in an ATM switch. The scheme is based on estimator functions that are generated automatically by using genetic programming techniques in the form of C language procedures. These estimator functions can be used in rate-based control schemes for effective control of network congestion. According to the results of their simulations, very accurate forecasting can be achieved due to the nonlinear data relationship-capturing capability of genetic programming techniques.

To generate a suitable estimator function, a set of training data representing the temporal queue behavior must be obtained. The data is acquired by sampling and recording the length of the ABR queue at an ATM switch every regular time interval (they have selected the interval as 32 cell service periods) for a duration of 1 s under offered ABR traffic load of link capacity (Figure 5.12).

Then, they have used a genetic programming system developed in-house with this data set to evolve an estimator function. An example of evolved estimator functions is shown below:

```
float estimate(float a, float b, float c)
{
float estimated_value;
estimated_value = (IFG( ( DIV((c), (((( IFG( (888.788), (595.93), (c))) +
        (DIV((c), (a)) ))>(164.152))+(a)))),
        (min(((a)*(c)), (((c)+(( DOWN(a,b,c)) >
```

$$(((b)*(\min((c), (DIV((c), (a)))))) <$$
$$(DIV((c), (a)))))) * (\min((b), (DIV((c), (b)))))))),$$
$$(((DOWN(a,b,c)) > (b)) < (DIV((c), (a)))));$$
      return estimated_value;
    }

The estimator function, which is a C procedure, predicts the value of the queue length $Q(t + 1)$ at time $t + 1$, taking into consideration three known preceding values $Q(t)$, $Q(t - 1)$, $Q(t - 2)$. IFG, DIV, DOWN are macro expressions, and the constants are generated randomly during the initialization stage.

After generating the candidates for the estimator function, they have selected the one that has the best prediction potential (given above). To test its generalization capability, they have run a series of simulations under varying traffic conditions. In these simulations they have sampled the ABR queue length at the end of every 32 cell-service period, and recorded the predicted value of the queue length for the next sampling time as evaluated by the estimator function. The estimator function used the current and past two values of the queue length $Q(t - 2)$, $Q(t - 1)$ and $Q(t)$ to predict the queue length $Q(t + 1)$ one step ahead. The authors have shown that, on average, error of the one step ahead estimation of possible queue length remains less than six cells. Figure 5.13 and Figure 5.14 illustrate the dynamic behavior of the estimator function for the case of running the ATM LAN at 40% output link

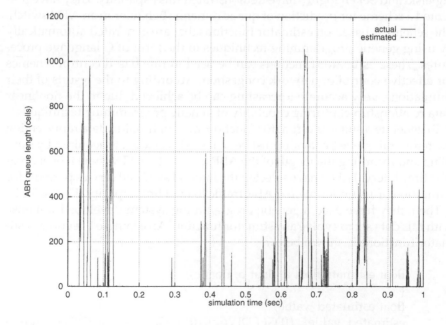

FIGURE 5.12

The data set used for generating the estimation functions. It is obtained by running the simulated ATM LAN with a traffic load of 60% of the link capacity.

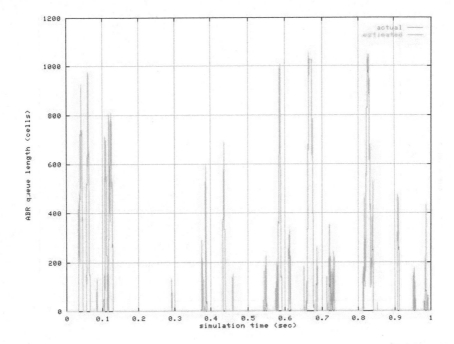

FIGURE 5.13
Graph of actual and estimated ABR queue length at the ATM switch for an offered link load of 40% of link capacity.

capacity. Figure 5.14 shows the actual and estimated length of the queue in more detail between 0.718 an 0.734 s of the simulation time. As can be observed, the estimator function can predict the queue length quite accurately in most of the situations except in the cases where sudden changes in queue length are experienced. Even in these situations of sudden changes, estimation error remains within ±30 cells limit.

5.4 Conclusions

In the "fight" against congestion, despite the research efforts spanning a few decades and the large number of different schemes proposed, there are no universally acceptable solutions (a control strategy, a control system, or a package of control solutions). Congestion control remains a critical issue and a high priority, especially given the growing size, demand, and speed (bandwidth) of the increasingly integrated services network.

In this chapter we have reviewed existing literature on IP- and ATM-congestion control. We have presented illustrative examples of using CI to control congestion using Fuzzy Logic, Neural, and Evolutionary approaches.

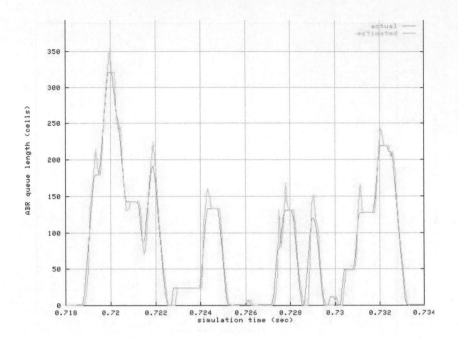

FIGURE 5.14
A detail section of the graph in Figure 5.13.

These and the literature we review on CI methods applied to ATM networks show that CI can be effective in the control of congestion. There is no doubt that we will see more and more use of these techniques, including their use in the IP world. We also expect that, as in other commercial products, CI techniques will finally make it into real products in this area, and we expect with tremendous success.

Of course, many challenges to the control of congestion remain unresolved (after all, the network control system is one of the most, if not the most, complex control system that man ever made—it is complex as well as large-scale). The challenges include:

- To get agreement on a structured approach to congestion control for the network. Control-theoretic concepts and techniques have an essential role to play.
- To add further credibility to CI-based control systems for congestion control, in addition to the empirical evidence of the effectiveness of the approach, formal verification of controlled system stability, performance, and robustness for CI-based systems should be explored. Also, a more systematic approach to design Congestion Control with known controlled system performance should be sought. Note though, that for nonlinear systems there are no systematic specifications of the desired controlled system behavior. For FCC some pre-

liminary results in that direction can be found in References 73, 80, and 100. The complexity versus benefit of these schemes for congestion control should be explored further. As elegantly pointed out by Mamdani,[72] overstressing the necessity of mathematically derived performance evaluations may be counterproductive and contrary to normal industry approach. Also, the prescribed specifications on the controlled system performance can be embedded into the fuzzy controller design as a set of fuzzy if-then rules (e.g., rise-time rules, damping rules, and steady-state rules).

- To define desirable controlled-system features. In addition to network metrics, such as throughput and QoS provision, the desirable features will include fairness toward all sources, effectiveness of the congestion-control system, rise and settling time of controlled system following a disturbance, robustness of the control system, regulation, efficiency, implementation complexity, ease of tuning, scalability, internetworking with other schemes, and policing of connections. These should be incorporated into the control-system design brief.

- To engineer the network system with the network-control system together in order to add another degree of flexibility.

  - It is commonly acknowledged that for the effective control of congestion a combination of controls, such as input-rate control, window control, connection-admission control, resource-reservation control, and bandwidth allocation, will be necessary. These controls can reside at the edge of the network, within the network, or in a combination. Due to the complexity of the problem and lack of established and accepted methodologies, combined overall congestion-control schemes that take into account interactions between the different control objectives are not currently available.

  - Different input-output pairs and feedback sensors, together with control structures should be investigated. The idea is to match the input-output pairs to ensure that there is structural flexibility in the system for a control design to be effective. For example, if the propagation delay exceeds a certain value, the controls may not be effective—so the feedback-delay path should be reduced, or different input-output pairs should be chosen so that feedback and system dynamics are matched. Another consideration for the feedback sensor is its accuracy, taking into consideration the complexity of implementation in fast switches.

- To deploy these control systems in the large-scale, geographically distributed network system. Theoretical advances in handling large-scale complex systems are required, including decomposition and organization of controls (possibly hierarchical, multilayer, multi-level, decentralized co-ordinated, and decentralized overlapping), and the selection and tuning of a possibly large fuzzy rule base.

- To globally optimize the overall network objectives.
- To develop a framework for the evaluation of the performance of the controlled systems for different control solutions. The framework will possibly have to be simulative (a Common Simulative Framework, CSF). The CSF will have to define and include a number of predefined scenarios of test loads, test networks, and controlled system performance indices (these will include the indices discussed in Section 5.2.3).

In conclusion, there is a real challenge in the control of congestion in communication networks, especially the ones supporting video, voice, and data applications simultaneously. Computational Intelligence techniques are expected to play a central role, especially in the large scale, geographically distributed network systems. Hybrids are also expected to supplement these techniques and prove useful, especially in optimizing the overall network objectives.

Problems

1. Consider an idealized FIFO single server M/M/1 queue, described by a dynamic fluid flow state model:[101]

$$\dot{x}(t) = -\frac{x(t)}{1 + x(t)}C + \lambda(t) + \lambda^b(t),$$

where $x(t)$ is the ensemble average of the buffer state (number of packets), $C(t)$ is the ensemble average of the server rate (packets/s), $\lambda(t)$ is the ensemble average of the controlled arrival rate (packets/s) and $\lambda^b(t)$ is the ensemble average of the interfering (background) traffic, initially set at 0.

Design a fuzzy control system to control the rate of traffic (packets) into the queue.

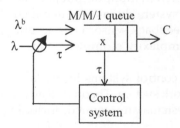

The objective is to regulate the controlled-traffic rate, irrespective of any variations in the rate of the background traffic, so as to maintain the buffer state equal to the reference (say 500 packets for a physical buffer of 1000). Consider three different propagation time delays τ, and investigate the controlled-system performance for two different sets of fuzzy rules (3 and 5, so that you may interpret the results). Tune the fuzzy rules and set the linguistic values manually, by observing the performance of the controlled system. You can obtain such measures of controlled-system performance as stability, rise time, and if the behavior is cyclic, measure of period and amplitude of oscillation. Note that the fluid flow-based model is used to gain insight into the system dynamics. Its closeness to an event-based simulation model of an ATM switch has recently been demonstrated.[102] Critically discuss the results, noting how the propagation delay influences the controlled-system behavior. Choose a value of time delay for which the performance of the controlled system is not acceptable, and investigate whether by increasing the number of rules the performance improves. Observe that the 3 fuzzy rule-based system may be likened to the additive-increase, multiplicative-decrease control law. Comment. Note that you may use any simulation, or mathematical tool to implement and simulate the system. A sample MATLAB-based simulation system can be found in http://www.cs.ucy.ac.cy/~cspitsil/CIbook/problem.html

2. Repeat 1, but replace the fuzzy controller with a neural-based one.
3. Repeat 1, but replace the fuzzy controller with a combination of fuzzy and neural. The fuzzy controller regulates the flow (as in 1), and the neural network tunes the fuzzy rules of the controller.
4. Repeat 1, but replace the fuzzy controller with a combination of fuzzy and evolutionary. The fuzzy controller regulates the flow (as in 1), and the evolutionary algorithm tunes (optimizes) the fuzzy rules of the controller.
5. Compare 1, 2, 3, and 4 in terms of performance and cost of implementation (e.g., complexity of control law).
6. Repeat 1 to 5, but now include an interfering background traffic:

   a. sinusoidally varying with an average value of half the server rate and a peak of one quarter,
   b. with reasonable correlation (can be simulated using an AR model, as in Reference 103)
   c. self-similar traffic
   d. traces of real traffic.

7. Repeat 1 to 6, replacing the model by a network simulator, e.g., ns2 and OPNET.[102]

References

1. ITU-T Recommendation 1.371 (previously CCITT Recommendation), *Traffic Control and Congestion Control in B-ISDN*, March 1993.
2. C-Q Yang and A. V. S. Reddy, A taxonomy for congestion control algorithms in packet switching networks, *IEEE Networks*, July/August 1995.
3. C. H. Papadimitriou and J. N. Tsitsiklis, The complexity of Optimal Queueing network control, to appear in *Mathematics of Operations Research*. (Available to download at http://lids.mit.edu/~jnt/exprev.ps)
4. M. Schwartz, *Telecommunication Networks: Protocols, Modeling, Analysis,* Addison-Wesley, Reading, MA, 1988.
5. A. Pitsillides and A. Sekercioglu, Intelligent Congestion Control Framework for Integrated Services Communication Networks, Techn. Rep. TR-99-1, Department of Computer Science, University of Cyprus, Nicosia, Cyprus, January 1999.
6. R. Braden et al., Recommendations on queue management and congestion avoidance in the Internet, RFC2309, April 1998.
7. K. K. Ramakrishnan and S. Floyd, A proposal to add explicit congestion notification (ECN) to IP, Internet draft-kksjf-ecn-03.txt, October 1998, RFC2481, January 1999.
8. H. Salim, A proposal for Backward ECN for the Internet Protocol (IPv4/IPv6), Internet draft-salim-jhsbnns-ecn-00.txt, June 1998.
9. ATM Forum, Traffic Management Specification Version 4.0, Tech. Rep. AF-TM-0056.000, April 1996.
10. W. Stevens, TCP Slow Start, Congestion Avoidance, Fast Retransmit, and Fast Recovery Algorithms, *RFC 2001*, January 1997.
11. G. M. Woodruff et al., A congestion control framework for high speed integrated packetized transport, *GLOBECOM'88*.
12. J. Martin and A. Nilsson, The evolution of congestion control in TCP/IP: from reactive windows to preventive flow control, CACC Tech. Rep. TR-97/11, North Carolina State University, August 1997. (available on line http://www2.ncsu.edu/eos/info/ece_info/www/ccsp/tech_reports/faculty/nilsson.html)
13. V. Jacobson, Congestion avoidance and control, ACM SIGCOMM88, 1988.
14. P. Karn and C. Partridge, Improving round-trip time estimates in reliable transport protocol, ACM SIGCOMM87, October 1987.
15. W. Stevens, TCP/IP illustrated, Vol. 1, *The Protocols*, Addison-Wesley, Reading, MA, 1994.
16. V. Jacobson, R. Braden, and D. Borman, TCP extensions for high performance, *RFC 1323*, May 1992.
17. L. Zhang, A new architecture for packet switching network protocols, Ph.D. dissertation, Massachusetts Institute of Technology, Labratory Computer Sciences, Cambridge, MA, 1989.
18. S. Shenker, L. Zhang, and D.D. Clark, Some observation on the dynamics of a congestion control algorithm, *Computer Communications Review,* 30–39, October 1990.
19. V. Jacobson, Modified TCP congestion avoidance algorithm, message to end2end-interest mailing list, April 1990.
20. T. V. Lakshman and U. Madhow, The performance of TCP/IP for networks with high bandwidth delay products and random loss, *IEEE/ACM Trans. Networking*, 5, 336–350, June 1997.
21. Z. Wang and J. Crowcroft, A new congestion control scheme: slow start and search (Tri-S), *ACM Computer Communications Review,* October 1994.

22. L. Bracmo, S. O.'Malley, and L. Peterson, TCP Vegas: new techniques for congestion detection and avoidance, *ACM SIGCOMM94*, 1994.

23. S. Floyd and V. Jacobson, Random early detection gateways for congestion avoidance, *IEEE/ACM Trans. Networking*, August 1993.

24. K. K. Ramakrishnan and R. Jain, A binary feedback scheme for congestion avoidance in computer networks, *ACM Trans. Comp. Syst.*, 8, 2, 158–181, 1990. (Also see K.K. Ramakrishnan, and R. Jain, A binary feedback scheme for congestion avoidance in computer networks with a connectionless network layer, *ACM SIGCOMM*, 1988.)

25. S. Floyd, TCP and explicit congestion notification, *ACM Comp. Commun. Rev.*, 24, 5, pp. 10–23, October 1994.

26. R. Braden, D. Clark, and S. Shenker, Integrated services in the internet architecture: an overview, *RFC 1633*, July 1994.

27. D. Black, S. Blake, M. Carlson, E. Davies, Z. Wang, and W. Weiss, An architecture for differentiated services, Internet Draft, May 1998.

28. Y. Bernet, et al., A framework for differentiated services, Document: <draft-ietf-diffserv-framework-02.txt>, Internet Draft, February 1999.

29. S. Kalyanaraman, S. Arora, K. Wanglee, G. Guarriello, and D. Harrison, *A one-bit feedback enhanced differentiated services architecture*, Internet Draft, Internet Engineering Task Force, Apr. 1998. Work in progress.

30. C. Hungkei and A. Leon-Garcia, A feedback control extension to differentiated services, http://www.comm.utoronto.ca/~keith/ietf-id/draft-chow-diffserv-fbctrl-00.ps.

31. M. Decina and V. Trecordi, Traffic management and congestion control in ATM networks, Guest Editorial, *IEEE Network*, 6, 5, September 1992.

32. K. Sohraby, L. Fratta, I. Gopal, and A. A. Lazar, Congestion control in high-speed packet switched networks, Guest Editorial, *IEEE J. Selected Areas in Commun.* 9, 7, September 1991.

33. CCITT: Recommendation I.211, B-ISDN Broadband Aspects of ISDN, Geneva, Switzerland, 1991.

34. A. Pitsillides, Control structures and techniques for Broadband-ISDN communication systems, Ph.D. dissertation, Swinburne University of Technology, Melbourne, Australia, 1993.

35. G. Gallassi, G. Rigolio, and L. Fratta, ATM: Bandwidth assignment and bandwidth enforcement policies, IEEE GLOBECOM'89, 49.6, 1788–1793, 1989.

36. H. Suzuki, T. Murase, S. Sato, and T. Takeuchi, A simple and burst-variation independent measure of service quality for ATM traffic control, 7th ITC Specialist Seminar, 13.1, New Jersey, October 1990.

37. R. Guerin, H. Ahmadi, and M. Naghshineh, Equivalent capacity and its application to bandwidth allocation in high-speed networks, *IEEE J. Selected Areas Commun.*, 9, 7, September 1991.

38. E. Rathgeb, Modelling and performance comparison of policing mechanisms for ATM networks, *IEEE J. Selected Areas Commun.*, April 1991.

39. J.W. Roberts, Traffic control in the B-ISDN, *Comp. Networks ISDN Syst.*, 25, 1991.

40. F. Guillemin, P. Boyer, A. Dupuis, and L. Romoeuf, Peak rate enforcement in ATM, IEEE INFOCOM'92, 753–758, Florence, Italy, May 1992.

41. A. DeSimone, Generating burstiness in networks: a simulation study of correlation effects in networks of queues, *Comp. Commun. Rev.*, 21, January 1991.

42. G. Ramamurthy and R. S. Dighe, A multidimensional framework for congestion control in B-ISDN, *IEEE J. Selected Areas Commun.*, December 1991.

43. P. Boyer, A congestion control for ATM, 7th ITC Seminar, Morristown, NJ, 1990.
44. D.H.K. Tsang, B. Bensaou, and S.T.C. Lam, Fuzzy based rate control for real-time MPEG video, *IEEE Trans. Fuzzy Syst.*, accepted for publication, 1998.
45. D. Mitra and J.B. Seery, Dynamic adaptive windows for high speed networks with multiple paths and propagation delays, INFOCOM'91.
46. N. Yin and M.G. Hluchyj, A dynamic rate control mechanism for source coded traffic in a fast packet network, *IEEE J. Selected Areas Commun.*, September 1991.
47. O. Aboul-Magd and H. Gilbert, Incorporating congestion feedback in B-ISDN traffic management strategy, ISS'92 Intern. Switching Symp., Osaka, Japan, October 1992.
48. P. Newman, Backward explicit congestion notification for ATM local area networks, GLOBECOM'93, Houston, December 1993.
49. L. Roberts, Enhanced Proportional Rate Control Algorithm (EPRCA), Tech. Rep. AF-TM 94-0735R1, August 1994.
50. R. Jain, S. Kalyanaraman, R. Goyal, S. Fahmy, and R. Viswanathan, ERICA switch algorithm; a complete description, ATM Forum, AF/96-1172, August 1996.
51. A. Pitsillides and J. Lambert, Adaptive connection admission and flow control: quality of service with high utilisation, 13th Conf. Comp. Commun., Toronto, Ontario, Canada, June 1994, 1083–1091.
52. A. Pitsillides and J. Lambert, Adaptive congestion control in ATM based networks: quality of service with high utilisation, *J. Comp. Commun.*, 20, 1997, 1239–1258.
53. A. Pitsillides, A. Sekercioglu, and G. Ramamurthy, Fuzzy backward congestion notification (FBCN) congestion control in asynchronous transfer mode (ATM) networks, Global Telecommun. Conf., Singapore, November 13–17, 1995, 280–285.
54. A. Pitsillides, A. Sekercioglu, and G. Ramamurthy, Effective control of traffic flow in ATM networks using fuzzy explicit rate marking (FERM), *IEEE J. Selected Areas Commun.*, 15, issue 2, 209–225, February 1997.
55. P. P. Mishra and H. Kanakia, A hop by hop rate-based congestion control scheme, ACM SIGCOM, Baltimore, Maryland, 1992.
56. H. T. Kung, The FCVC (Flow-Controlled Virtual-Channels) proposal for ATM networks, Intern. Conf. Network Protocols, San Francisco, October 1993.
57. C. Douligeris and G. Develekos, A fuzzy logic approach to congestion control in ATM networks, IEEE Int. Conf. Commun., ICC '95, 1969–1973, 1995.
58. A.A. Tarraf, I.W. Habib, and T.N. Saadawi, Congestion control mechanism for ATM networks using neural networks, IEEE Int. Conf. Commun., 206–210, 1995.
59. K.W. Fendick, Evolution of controls for the Available Bit Rate, *IEEE Commun. Mag.*, 35–39, November 1996.
60. C.E. Rohrs, R.A. Berry, and S.J. O'Halek, A control engineer's look at ATM congestion avoidance, IEEE Global Telecommun. Conf., Singapore, 1995.
61. A. Feldmann, A.C. Gilbert, and W. Willinger, Data networks as cascades: investigating the multifractal nature of the Internet WAN traffic, SIGCOMM 98, Vancouver, 1998.
62. A. Feldmann, P. Huang, A.C. Gilbert, and W. Willinger, Dynamics of IP traffic: A study of the role of variability and the impact of control, To appear at ACM/SIGCOMM 1999.
63. K. J. Astrom and B. Wittenmark, *Computer Controlled Systems: Theory and Design*, Prentice Hall International (2nd ed.), 1990.
64. P. Pruthi and A. Popescu, Effect of controls on self similar traffic, 5th IFIP ATM Workshop, Bradford, UK, July 1997.

65. L.A. Zadeh, *Soft Computing and Fuzzy Logic*, IEEE Software, November 1994.
66. J.C. Bezdek, On the relationship between neural networks, pattern recognition and intelligence, *Int. J. Approximate Reasoning*, 6:85–107, 1992.
67. J.C. Bezdek, What is Computational Intelligence?, in Computational Intelligence: Imitating Life, edited by J. M. Zurada, R. J. Marks II and C. J. Robinson, IEEE Press, 1–12, 1994.
68. W. Pedrycz, *Computational Intelligence: An Introduction*, CRC Press, Boca Raton, FL, 1998.
69. Special issue on Computational Intelligence, IEEE Journal on Selected Areas in Communications (JSAC), vol. 15, issue 2, February 1997.
70. B. Azvine (chairman), ERUDIT Technical committee D on Traffic and Telecommunications: Application of soft computing techniques to the telecommunication domain, Aachen, Germany, September 1997.
71. E. H. Mamdani and S. Assilian, An experiment in linguistic synthesis with a fuzzy logic controller, International journal of Man-Machine studies, 7, 1–13, 1975.
72. E. H. Mamdani, Twenty years of fuzzy control: Experiences gained and lessons learned, IE³ International conference on fuzzy systems, San Franscisco, 339–344, 1993.
73. R. Palm, D. Driankov, and H. Hellendoorn, *Model Based Fuzzy Control*, Springer-Verlag, 1996.
74. Y. C. Liu and C. Douligeris, Static vs. adaptive feedback congestion controller for ATM networks, IEEE Global Telecommunication Conference, Singapore, 1995.
75. D. Jensen, B-ISDN network management by a fuzzy logic controller, IEEE Global Telecommunications Conference, 799–804, 1994.
76. B. Kosko. *Fuzzy Engineering*. Prentice Hall, 1997.
77. C. Chang and R. Cheng, Traffic control in an ATM network using fuzzy set theory, IEEE INFOCOM'94 Conference, Toronto, Canada, pp. 1200–1207, June 1994.
78. R-G. Cheng and C-J. Chang, Design of a fuzzy traffic controller for ATM networks, *IEEE/ACM Trans. Networking*, 4(3), 460–469, June 1996.
79. B. Qiu. A predictive fuzzy logic congestion avoidance scheme, *IEEE Global Telecommunications Conference*, 2, 967–971, 1997.
80. A. Sekercioglu, Fuzzy Logic Control techniques for Asynchronous Transfer mode (ATM) based multimedia networks, Ph.D. thesis, Swinburne University of Technology, Melbourne, Australia, Submitted. 1999.
81. T. Takagi and M. Sugeno, Fuzzy identification of systems and its application to modelling and control, *IEEE Transactions on Systems, Man, and Cybernetics*, 15(1), 116–132, 1985.
82. M. Sugeno, *Industrial Applications of Fuzzy Control*, North-Holland, 1985.
83. L. A. Zadeh, Fuzzy Logic, *IEEE Computer*, pp. 83–93, April 1988.
84. M. Sugeno and M. Nishida. Fuzzy Control of a Model Car. *Fuzzy Sets and Systems*, 10, 105–113, 1985.
85. Q. Hu, D. W. Petr, and C. Braun, Self-tuning fuzzy traffic rate control for ATM networks, IEEE International Conference on Communications, ICC'96, Dallas, Texas, USA, 424–428, 1996.
86. R. Jager, Fuzzy Logic in Control, Ph.D. thesis, Technische Universiteit Delft, 1995.
87. D. Driankov, H. Hellendoorn, and M. Reinfrank, *An Introduction to Fuzzy Control*, Springer-Verlag, 1993.

88. A. A. Tarraf and I. W. Habib, A novel neural network traffic enforcement mechanism for ATM networks, *IEEE J. Selected Areas in Communications*, 12(6), 1088–1096, August 1994.
89. A. A. Tarraf, I. W. Habib, and T. N. Saadawi, Intelligent traffic control for ATM broadband networks, *IEEE Communications Magazine*, 33(10), 76–82, October 1995.
90. A. A. Tarraf, I. W. Habib, and T. N. Saadawi, Congestion control mechanism for ATM networks using neural networks, 1995 IEEE International Conference on Communications ICC '95, 206–210, 1995.
91. A. A. Tarraf, I. W. Habib, and T. N. Saadawi, Reinforcement learning-based neural network congestion controller, Military Communications Conference MILCOM'95, 2, 668–672. 1995.
92. A. G. Barto, Reinforcement Learning and Adaptive Critic Methods, in *Handbook of Intelligent Control: Neural, Fuzzy, and Adaptive Approaches*, D.A. White and D.A. Sofge, Eds., 469–491, Van Nostrand Reinhold, New York, 1992.
93. X. Chen and I. M. Leslie, A neural network approach towards adaptive congestion control in Broadband ATM networks, IEEE Global Telecommunications Conference GLOBECOM'91, 115–119, 1991.
94. Y. C. Liu and C. Douligeris, Static vs. adaptive feedback congestion controller for ATM networks, IEEE Global Telecommunications Conference GLOBECOM'95, Singapore, 1995.
95. Y. C. Liu and C. Douligeris, Rate Regulation with Feedback Controller in ATM Networks-A Neural Network Approach, *IEEE J. Special Areas in Communications*, 15, (2), 200–208, February 1997.
96. J. H. Holland, *Adaptation in Natural and Artificial Systems*, The University of Michigan Press, 1975.
97. J. R. Koza, *Genetic Programming: On the Programming of Computers by Means of Natural Selection*, MIT Press, 1992.
98. J. R. Koza, *Genetic Programming II: Automatic Discovery of Reusable Programs*, MIT Press, 1994.
99. R. Jagielski, A. Sekercioglu, Dynamic forecast of boundaries of queue in ATM networks, ATNAC'96 Australian Telecommunication Networks Applications Conference, Melbourne, Australia, December 1996.
100. A. Sekercioglu, A. Pitsillides, and G. Egan, An adaptive fuzzy logic control system based on adapting the relative rule weights, ANZIIS, Brisbane, Queensland, Australia, Nov 30–Dec 2, 204–208, 1994.
101. A. Pitsillides, P. Ioannou, and D. Tipper, Integrated control of connection admission, flow rate, and bandwidth for ATM based networks, IEEE INFOCOM'96, 15th Conference on Computer Communications, San Francisco, USA, March 24–28, 785–793, 1996.
102. A. Pitsillides, P. Ioannou, and L. Rossides, Comparison of a fluid flow based model with OPNET simulated ATM switch model, Internal report, Department of Computer Science, University of Cyprus, June 1999.
103. B. Maglaris, D. Anastassiou, P. Sen, G. Karlsson, and J. Robbins, Performance models of statistical multiplexing in packet video communications, *IEEE Trans. Commun.*, 36, (7), 834–844, July 1988.

6

Fuzzy Queue Function for ERICA ATM Switch Controller

Yaron Klein and Abraham Kandel

CONTENTS

0-8493-1075-X/01/$0.00+$.50
© 2001 by CRC Press LLC

ABSTRACT In this paper we propose a modification to the popular ERICA algorithm for controlling Available Bit Rate (ABR) traffic in Asynchronous Transfer Mode (ATM) networks. The ERICA algorithm monitors the load of each link and advises the sources of the links about the rate at which to transmit. Designed to achieve efficiency, fairness, and control of queuing delays, the ERICA algorithm has proven to be a simple, efficient scheme. We show that we can achieve better results with a similar complexity by using a fuzzy rule base to determine some ERICA parameters.

KEY WORDS: *ATM, ABR, ERICA Switch Algorithm, Fuzzy Logic.*

6.1 Introduction

ATM (Asynchronous Transfer Mode) is the most promising transfer technology for implementing B-ISDN. It supports applications with distinct QoS requirements, such as delay, jitter, and cell loss, and distinct demands, such as bandwidth or throughput. Thus, ATM enables multimedia communication in a networked environment. However, in order to support old applications or applications that do not choose to use performance-guaranteed services, ATM networks also provide best-effort services, such as LAN emulation and IP-over-ATM. To provide these services for a wide variety of applications, the ATM Forum Standards Group has defined a family of service categories including CBR (Constant Bit Rate) service, rt-VBR (real-time Variable Bit Rate), nrt-VBR (nonreal-time VBR), UBR (Unspecified Bit Rate) services, and ABR (Available Bit Rate) service.[1]

The ABR services category provides economical data transport for a variety of applications.[9] Any nontime-critical application running over an end-system capable of varying its transmission rate can exploit the ABR service. This category provides economical support to applications that possess vague requirements for throughput and delay and that require a low Cell Loss Ratio (CLR). Examples include LAN interconnection/internetworking services that are driving the business service market for ATM. These are typically run over router-based protocol stacks like TCP/IP that can easily vary their emission rate as required by the ABR rate-control policy. Using ABR for these services will likely result in an increased end-to-end performance. Another application environment suitable for ABR is LAN Emulation. Other application examples are critical data transfer (e.g., defense information, banking services), super computer applications, and data communications services, such as remote-procedure call, distributed file services, and computer process swapping/paging. This rich set of applications has led to the recent introduction of several feedback flow-control schemes for ABR services.[10, 13, 16–18, 20, 21, 23]

6.1.1   ABR Flow Control

We first briefly introduce the basic operation of the closed-loop rate-based control mechanism.[1] The rate at which an ABR source is allowed to schedule cells for transmission is denoted by ACR (Allowed Cell Rate). The ACR is initially set to the Initial Cell Rate (ICR) and is always bounded from below by the Minimum Cell Rate (MCR) and above by the Peak Cell Rate (PCR). Transmission of data cells is preceded by sending an ABR Resource Management (RM) cell. The source will continue to send RM cells, typically after (Nrm-1) user cells have been transmitted, more frequently when its ACR is low. The source rate is controlled by the return of these RM cells, either by the destination or by a virtual destination.

The source places the rate at which it is allowed to transmit cells (its ACR) in the Current Cell Rate (CCR) field of the RM cell and the rate at which it wishes to transmit cells (usually the PCR) in the Explicit Rate (ER) field. The RM cell travels forward through the network circuit, thus providing the switches in its path with the information they need to allocate bandwidth among different ABR connections. Switches may also reduce the value of the explicit rate field ER or set the Congestion Indication bit CI to 1. Switches supporting only the Explicit Forward Congestion Indication (EFCI) mechanism (in which an indicator in the header of each data cell is set under congestion) will ignore the content of the RM cell. Switches may also generate a controlled number of ABR RM cells on the return path, in addition to those originally supplied by the source. Switch-generated RM cells must have the Backward Notification (BN) bit set to 1 and either the CI bit or the No Increase (NI) bit set to 1.

When the cell arrives at the destination, the destination should change the direction bit in the RM cell and return the RM cell to the source. If the destination is congested and cannot support the rate in the ER field, the destination should then reduce ER to whatever rate it can support. If, when returning an RM cell, the destination observes a set EFCI since the last RM cell was returned, then it should set the RM cell's CI bit to indicate congestion.

As the RM cell travels backward through the network, each switch may examine the cell and determine if it can support the rate ER for this connection. If the ER is too high, the switch should reduce it to a rate that it can support. No switch should increase the ER, since information from switches previously encountered by the RM cell would then be lost. A switch should try to modify the ER only for those connections that are a bottleneck since this promotes a fair allocation of bandwidth. Also, switches should modify the ER content of the RM cells traveling on either their forward or backward journeys, but not on both (Figure 6.1).

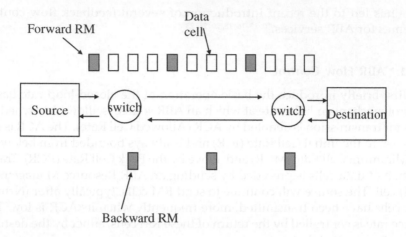

FIGURE 6.1
ABR traffic model.

When the source receives a backward RM cell, it modifies its ACR using additive increase and multiplicative decrease. The new ACR is computed as follows, using the values of the CI and ER fields in the RM cell:

$$ACR = \max(\min(ACR + RIF \cdot PCR, ER), MCR), \quad \text{if } CI = 0,$$

$$ACR = \max(\min(ACR \cdot (1 - RDF), ER), MCR), \quad \text{if } CI = 1,$$

where RIF is the rate increase factor and RDF is the rate decrease factor.

6.1.2 EFCI Scheme

ATM switches may be classified as EFCI switches or ER switches, depending on their congestion monitoring. In the EFCI scheme,[24] when congestion occurs, the switch sets the EFCI bit to one in the header of each passing data cell (EFCI = 1). If a cell with EFCI = 1 has been received, the DES marks the CI bit (CI = 1) to indicate congestion in each backward RM cell. In most cases, the queue length is used to decide whether congestion occurs or not. As the queue length exceeds a threshold, denoted by Q_t, congestion is claimed. When the queue length falls below the threshold, congestion is relieved.

6.1.3 Explicit Rate Feedback Schemes

In ER schemes, the switch computes the fair share of bandwidth that can be allocated to a VC, determines the load, and determines the actual explicit rate. When each RM cell passes, the switch sets its ER field to the determined explicit rate. Note that no switch is allowed to increase the ER field. Thus, a source receives the minimum allowed cell rate of all the switches along the circuit path. Examples of ER switch mechanisms are the EPRCA, ERICA, CAPC, Charny Max-Min, and Tsang Max-Min schemes.[3, 6, 8, 12, 22] The methods are summarized and compared in Reference 25.

6.1.3.1 Enhanced Proportional Control Algorithm (EPRCA)[22]

Each switch maintains a mean allowed cell rate (MACR) using a running, exponentially weighted average. When a switch receives a forward RM cell during the congestion period, MACR is updated as

$$MACR = (1 - \alpha) \cdot MACR + \alpha \cdot CCR$$

where α is the exponential averaging factor, generally set to be 1/16, and CCR is the current cell rate of the VC recorded in the RM cell. The fair bandwidth share is computed as a fraction of the MACR:

$$\text{Fair share} = DPF \cdot MACR,$$

where DPF is a switch Down Pressure Factor set close to but below 1. When a switch receives a backward RM cell, it reduces the ER field to the fair share if its queue length is larger than Q_t.

6.1.3.2 Explicit Rate Indication for Congestion Avoidance (ERICA)[12]

The ERICA uses a load factor z to indicate the overload or underload state of a switch. The load factor is defined as

$$z = \frac{\text{Input Rate}}{\text{Target Rate}}$$

The input rate is measured over a fixed time window, and the target rate is usually set slightly below the link bandwidth. Since the goal of this algorithm is to keep the load factor close to one, the sources must update their transmission rate by a factor inversely proportional to the load factor z. The VC share and fair share are as follows:

$$\text{VC Share} = \frac{CCR}{z}$$

$$\text{Fair share} = \frac{\text{Target Rate}}{\text{Number of active connections}}$$

A switch updates the ER field in each backward RM cell to be the maximum value of the fair share and VC share.

6.1.3.3   Congestion Avoidance Using Proportional Control (CAPC)[3]

As in the ERICA scheme, the switches set a target utilization slightly below 1 and compute the load factor. The main difference lies in the way the fair share is computed, and this depends on whether $z < 1$ or $z > 1$. Thus, we have

Fair share = Fair share $\cdot$ min(ERU, $1 + (1 - z) \cdot$ Rup),   if $z < 1$, and

Fair share = Fair share $\cdot$ max(ERF, $1 - (z - 1) \cdot$ Rdn),   if $z > 1$,

where Rup is a slope parameter between 0.025 and 0.1, and Rdn is between 0.2 and 0.8. ERU and ERF determine the maximum allowed increase and minimum allowed decrease, respectively. Usually ERU is set to 1.5, and ERF is set to 0.5. When a returning RM cell arrives at a switch, the ER field is updated to be the fair share.

6.1.3.4   Charny Max-Min Scheme[6]

In this scheme, the fair share is computed using an iterative procedure. Initially, the fair share is set to the link bandwidth divided by the number of active VCs. Some VCs cannot achieve the fair share at a switch because of the constraints imposed by other switches along their paths. For this switch, these VCs are called "constrained VCs". The switch can determine whether a VC is constrained or not by comparing the fair share with the CCR field in a forward RM cell on that VC. If the CCR field is less than the fair share, the VC is a constrained VC. Otherwise, it is an unconstrained VC.

For high throughput, the available bandwidth that the constrained VCs cannot use should be utilized by the unconstrained VCs. Hence the fair share is computed as follows:

$$\text{Fair Share} = \frac{\text{Link Bandwidth} - \sum \text{Bandwidth of constrained VCs}}{\text{Number of VCs} - \text{Number of constrained VCs}}$$

As a forward RM cell traverses the network, each switch determines whether the VC is constrained or not, recomputes the fair share, and reduces the ER field and CCR field of the RM cell to its fair share. The ER field and CCR field of a backward RM cell may be further reduced down to the most current fair share on the forward path.

6.1.3.5   Tsang Max-Min Scheme[8]

This scheme is similar to the Charny Max-Min method, except for three differences:

1. The switch does not update the CCR field of the RM cell.

2. The switch determines the VC state based on the ER field, rather than the CCR field, of the RM cell.

3. The switch determines the VC state and computes the fair share on forward and backward RM cells, not just the forward RM cell.

6.2 Fuzzy Logic in ATM

6.2.1 ATM Traffic Management Using Fuzzy Logic

The growing success of Fuzzy Logic in various fields of application, such as control, decision support, knowledge base systems, database and information retrieval, and pattern recognition is due to its inherent capacity to formalize control algorithms that can tolerate imprecision and uncertainty, emulating the cognitive process that human beings use every day. Fuzzy logic systems have been widely applied to control nonlinear, time-varying, and ill-defined systems where they can provide simple and effective solutions. The motivation for using Fuzzy Logic for ATM traffic management and congestion control arose from an analysis of the limitations of traditional mechanisms.

6.2.2 Related Work

In this section we review the use of fuzzy theory in ATM. Considerable work had already been done in the area of ATM traffic management using Fuzzy Logic. Our brief survey indicates the solutions that have been investigated so far, emphasizing their inconclusive or insufficient points.

In Reference 11 two fuzzy traffic control schemes for ATM networks are presented. The first scheme is a fuzzy CAC system, and the second scheme is fuzzy UPC that replaces more common UPC functions, such as the leaky bucket. Both sources take as input three ATM traffic parameters: Maximum Burst Size (MBS), Sustainable Cell Rate (SCR), and Peak Cell Rate (PCR).

Another fuzzy CAC solution is presented in Reference 7. The traffic controller designed in that paper incorporates both CAC and congestion control, and it is based on fuzzy set theory. In the fuzzy admission controller, the declared traffic parameters of Peak Bit Rate (PBR), Average Bit Rate (ABR), and Peak Bit Rate Duration (PBRD), the extent of network congestion and the cell loss probability are employed in making call-acceptance decisions.

The Fuzzy Logic-based UPC function for ATM networks or a source traffic policing mechanism ensures that each source conforms to its negotiated parameters. This time-window-based control mechanism is introduced in Reference 4. The number of cells that can be accepted per window is dynamically updated in accordance with the source's compliance with the negotiated average cell rate. Finally, a traffic controller for an ATM switch is proposed in Reference 15. This work describes the structure of a leaky bucket based on fuzzy control.

6.3   The Fuzzy Based Controller

6.3.1   Design Goals

In this section, we enumerate the design goals of the ABR switch algorithm. The goals describe the desired "steady state" operation, and priorities are used for a graceful degradation of the goals under transient conditions. Under such conditions, lower priority goals are traded off to achieve higher priority goals. Stability and robustness are implicit goals for this algorithm. The goals of the ABR switch algorithm are as follows:

- Utilization: Maximize link utilization $\rho(t)$. Since a link could be shared by several classes, and the switch algorithm controls only the ABR class, the goal is actually the maximization of ABR utilization.

- Queuing Delay: Link or ABR utilization is only a partial indicator of "efficiency." For example, $\rho(t)$ or z could be unity while a huge queue backlog exists, creating a bottleneck. An important goal is to achieve a target queuing delay at the bottleneck. The combination of the delay and utilization goals is often called "congestion avoidance."

- Fairness: The next priority is to allocate rates in a fair manner. One commonly used criterion for describing fairness is the max-min allocation.[5] Intuitively, fairness calls for maximizing the allocation of the minimum rate source, i.e., to give each contending source a "maximum possible equal share" of the bandwidth. In a configuration with n contending sources, where the ith source is allocated a bandwidth x_i, the allocation vector $\{x_1, \ldots, x_n\}$ is said to be feasible if all link load levels are less than or equal to 100%. If we order the elements of the vector in ascending order, then the max-min allocation is defined as the vector that is lexicographically the largest among the set of vectors. The ATM Forum specified several fairness definitions for cases in which ABR VCs have nonzero Minimum Cell Rates (MCRs).[2]

6.3.2   The Fuzzy Controller Architecture

The design of a controller for the ABR flow control in an ATM switch should address the following issues:

- The switch should work in a hard time-constrained environment. The computation of the explicit rate should be done in real time, that is, each arriving RM cell should be set with minimum delay.

- The ERICA algorithm is an efficient, robust mechanism, but its choice of parameters is not optimal.
- The choice of parameters requires a learning and adaptation mechanism.

While taking into account these points we propose the following principles for the improved fuzzy based controller:

- The controller is based on the ERICA switch algorithm, because it is efficient, stable, robust, and simple.
- The parameters of the ERICA switch are determined by a simple fuzzy rule base.
- The rule base parameters are updated by an off-line algorithm for monitoring the switch's ABR traffic and performance.

Taking these points into consideration, we propose the following architecture (see Figure 6.2). Arriving cells are queued in the link buffer. The ERICA controller monitors the load and computes the feedback ER according to the prescribed inference mechanism. The algorithm uses the queue function $f(Q)$ given by the fuzzy rule base. The fuzzy rule base monitors the queue load and produces the appropriate $f(Q)$ value.

These actions are performed in real time, i.e., for each arriving cell. This time constraint limits the rule base complexity to no more than three or four fuzzy rules. In order to get the optimal rule base (i.e., the membership functions parameters), we use an off-line predictor. Since a good prediction based on analysis of traffic and switch performance requires a more complex algorithm, this module is implemented off-line. That is, the predictor will update the rule base parameters periodically at relatively large intervals.

6.3.3 The Queue Function

In Section 6.1.3.2, we note that the Target ABR Capacity is a fraction of the Total ABR Capacity. Moreover, this fraction is a function of the queuing delay, $f(Q)$. For instance,

$$\text{Target ABR Capacity} = f(Q) \times \text{Theta Total ABR Capacity}$$

The function $f(Q)$, called the "queue control function" allows only a select fraction of the available capacity to be allocated to the sources. The remaining capacity is used to drain the current queue. The original ERICA philosophy was that correct rate assignments depend more upon the aggregate input rate than on the queue length (see Figure 6.3). However, three facts about queuing delays are important to consider in feedback calculation: a) nonzero steady-state queues imply 100% utilization, b) a system with very long queues is not

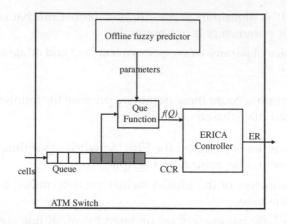

FIGURE 6.2
The switch architecture.

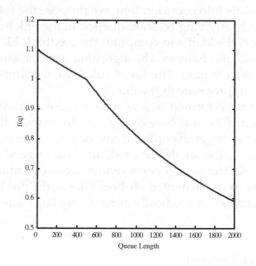

FIGURE 6.3
The ERICA queue function.

operating efficiently, and c) a service providing controlled-queuing delay may help support applications with delay requirements (e.g., variable quality video).

A simple queue control function is the constant function, i.e., a fixed parameter. This function (called *Target Utilization-U*) has been used for the OSU scheme[14] and for earlier versions of ERICA. This function is adequate in representative LAN (small round trip) and low error/variance WAN (large round trip) cases. The drawbacks of the constant function are

- it restricts the system utilization to a maximum of U in the steady state,

- the system cannot achieve a queuing delay target, and
- it does not provide compensation when measurement and feedback are affected by errors.

The alternative is to have $f(Q)$ vary depending upon the queuing delay. A number of such functions can be designed. One such function is the following:

$$f(Q) = \frac{a \cdot Q_0}{(a - 1) \cdot Q + Q_0} \quad \text{for } Q > Q_0$$

and

$$f(Q) = \frac{b \cdot Q_0}{(b - 1) \cdot Q + Q_0} \quad \text{for } 0 \leq Q \leq Q_0$$

$f(Q)$ is a number between 1 and 0 in the range Q_0 to infinity, and between b and 1 in the range 0 to Q_0. Both curves intersect at Q_0, where the value is 1. Observe that these are rectangular hyperbolic functions that assume a value 1 at Q_0. This function is lower bounded by the queue drain limit factor (QDLF):

$$f(Q) = \max\left(QDLF, \frac{a \cdot Q_0}{(a - 1) \cdot Q + Q_0}\right) \quad \text{for } Q > Q_0$$

6.3.4 The Fuzzy Queue Function

The queue function reflects the queue delay versus the utilization of the switch. The greater the queue delay, the greater the utilization, and vice versa. The target capacity depends on our definition of the queue function. The basic rules of thumb we use are

- With low queue delay, increase the target capacity.
- With high queue delay, reduce the target capacity.

Variations of these may occur.

While the ERICA algorithm uses the algorithm described in Section 6.3.2,[12] the use of hyperbolic functions is complicated and may not reflect the goals of the control mechanism. Furthermore, the parameters in the queue function are dependent on the current queue delay, while the effect will take place in the future in other conditions.

In Reference 19 a method of calculating the ER with a fuzzy controller using queue-load prediction is presented. However, using sophisticated algorithms in a time-constrained system such as the ATM switch is all but impossible.

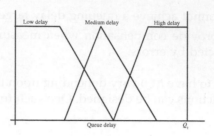

FIGURE 6.4
Queue delay membership function.

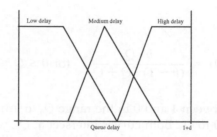

FIGURE 6.5
Queue function membership function.

The fuzzy queue function is based on the rules of thumb in Section 6.3.2. Two linguistic variables are defined: Queue Delay and Queue Function. Membership functions for these variables are shown in Figure 6.4 and Figure 6.5 in which the exact parameters are subject to fine tuning by the fuzzy predictor. In designing the rule base, we take into consideration the key ideas of simplicity regarding the rule base size and the rules of thumb in Section 6.3.2. We propose the following rule base:

- *if queue delay is low, then target function is high.*
- *if queue delay is medium, then target function is medium.*
- *if queue delay is high, then target function is low.*

The rule base has three rules enabling us to compile in the time constraints required by an ATM real time switch.

6.3.5 The Fuzzy Predictor

The fuzzy predictor is a fuzzy rule base to monitor the switch parameters: load, queue delay, fairness, and switch performance. The controller processes these parameters in an interval of time and computes the ideal parameters of the membership functions in the fuzzy queue function (see Sections 6.3.2 and 6.3.4).

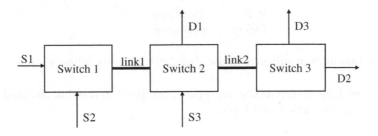

FIGURE 6.6
Simulation model.

The predictor's output is the parameters of the membership functions of the queue load and the queue function. The parameters include the definitions of the high-, medium-, and low-queue level, and the definition of high-, medium-, and low-queue function. The input to the predictor is the input rates of the sources, the load of the links, the queue length, the fairness, and the utilization at the switch.

The queue predictor is structured as a regular feedback controller with added expert knowledge. The rules for the queue predictor are as follows:

- *if queue load is more than Queue Size, then reduce Queue Delay.*
- *if queue load is less than Queue Size, then increase Queue Delay.*
- *if utilization is less than Utilization Size, then increase Queue Function.*
- *if utilization is more than Utilization Size, then reduce Queue Function.*

where *Queue Delay* is a membership parameter (low, medium, or high) as in Figure 6.4, and *Queue Function* is a membership parameter (low, medium, or high) as in Figure 6.5. *Utilization* is the average utilization, and *queue load* is the average queue size (over the current interval). *Utilization Size* is the desired utilization, and *Queue Size* is the desired queue length.

6.4 Results

In our simulation, we use the following configuration:

The traffic model for S1, S2, and S3 are taken from monitoring TCP traffic from three computers in Tel-Aviv University over the period of a day. The parameters taken for the simulation are as follows:

Link 1	20 Mbps
Link 2	10 Mbps
MCR	1 Mbps
PCR	20 Mbps

ICR   15 Mbps
Delay   600 μsec

The parameters monitored in the simulations are as follows:

Utilization: Calculated as the average throughput in the link divided by the link capacity for Link 1 and Link 2.

Maximum Queue Length: Calculated as the maximum value of the output queue of each switch; output of Switch 1 entering Link 1 and output of Switch 2 entering Link 2.

Fairness: The ratio of allocation of bandwidth to the fair share of bandwidth, calculated as

$$F = \max(1 - \max(|x_i - 1|, 0))$$

where x_i is the ratio of the actual throughput to the fair throughput for source i.

6.4.1   ERICA Results

We first run the ERICA algorithm with fixed sizes of $f(Q)$, where we measure the maximum queue size of each switch, the utilization of each switch, and the fairness of each source channel. The results are shown in Table 6.1 and Figure 6.7.

Applying the fuzzy mechanism to set $f(Q)$ without the fuzzy predictor, we have obtained the results in Table 6.2.

Using the fuzzy predictor to set the membership functions' values, we get the results in Table 6.3.

TABLE 6.1

ERICA Switch Results

$f(Q)$	Max queue 1	Max queue 2	Utilization S1	Utilization S2	Fairness
0.900000	0	9	0.878496	0.881743	0.466265
0.920000	1	10	0.900000	0.881813	0.452938
0.940000	1	11	0.900052	0.881873	0.485754
0.960000	3	13	0.938795	0.902650	0.436679
0.980000	3	14	0.950047	0.903060	0.445605
1.000000	5	23	0.972106	0.947657	0.432828
1.020000	62	15	0.999998	0.948078	0.408781
1.040000	2210	14	1.000000	0.950392	0.438833
1.060000	55349	1348	1.000000	0.999265	0.453717
1.080000	60139	1627	1.000000	0.999853	0.481096

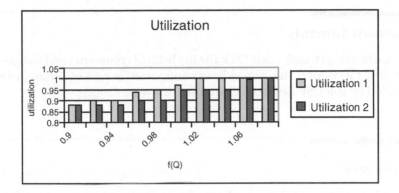

FIGURE 6.7
ERICA utilization results.

TABLE 6.2

Fuzzy Based ERICA Results (Without Fuzzy Predictor)

$f(Q)$ low	$f(Q)$ medium	$f(Q)$ high	Max queue 1	Max queue 2	Utilization S1	Utilization S2	Fairness
0.98	1.02	1.06	20	14	1.000000	0.948488	0.408757

TABLE 6.3

Fuzzy Based ERICA Results (With Fuzzy Predictor)

Max queue 1	Max queue 2	Utilization S1	Utilization S2	Fairness
10	16	0.997950	0.946948	0.408300

We can see from the results in Tables 6.1 through 6.3 that the fuzzy-based controller is performing better than the ERICA controller without the fuzzy queue function setting. The fuzzy predictor reduces the queue delay while maintaining high utilization and good fairness.

6.5 Conclusions

In this paper we introduced a method to improve the ERICA switch in order to control ABR traffic in ATM networks. We showed that by adding a fuzzy mechanism that sets the ERICA parameters, we achieve better performance. The fuzzy predictor improved the results more, setting the internal membership functions of the fuzzy controller. This improvement is achieved without increasing complexity and satisfies all the time constraints of the switch.

Acknowledgments

We would like to thank Scott Dick for his helpful comments and suggestions regarding this paper. This research was supported in part by Grant 21-08-004 provided by the USF Research Center for Software Testing.

References

1. ATM Forum, "ATM Forum Traffic Forum Traffic Management Specification Version 4.0," ATM Forum af-tm-0056.000, April 1996.
2. "The ATM Forum Traffic Management Specification Version 4.0," ATM Forum Traffic Management AF-TM-0056.000, April 1996. Available as ftp://ftp.atmforum.com/pub/approved-specs/aftm-0056.000.ps
3. A.W. Barnhart, "Explicit Rate Performance Evaluations," ATM Forum 94-0983, October 1994.
4. V. Catania, G. Ficili, S. Palazzo, and D. Panno, "A Comparative Analysis of Fuzzy Versus Conventional Policing Mechanisms for ATM Networks," *IEEE/ACM Trans. on Networking*, 4(3), 449–458, June 1996.
5. A. Charny, D.D. Clark, and R. Jain, "Congestion Control with Explicit Rate Indication," in Proc. ICC'95, June 1995.
6. A. Charny, K.K. Ramakrishnan, and A. Lauck, "Time Scale Analysis and Scalability Issues for Explicit Rate Allocation in ATM Networks," *IEEE/ACM Trans. on Networking*, 4(4), 569–581, August 1996.
7. R.G. Cheng and C.J. Chang, "Design of a Fuzzy Traffic Controller for ATM Networks," *IEEE/ACM Trans. on Networking*, 4(3), 460–469, June 1996.
8. Danny H.K. Tsang and Wales K.F. Wong, "A New Rate-based Switch Algorithm for ABR Traffic to Achieve Max-Min Fairness with Analytical Approximation and Delay Adjustment," Proc. IEEE Infocom'96, pp. 1174–1181, 1996.
9. B. Flavio and W.F. Kerry, "The Rate-Based Flow Control Framework for the Available Bit Rate ATM service," *IEEE Network*, 9(2), 25–39, March 1995.
10. L. Gerda and O. Casals, "A Simulation Study of Switching Mechanisms for ABR Service in ATM Networks," Lloren Cerd Olga Casals. Interim Rep. UPC-DAC-1996-21, February 1996.
11. H. Hellendoorn, W. Metternich, M. Nissel, R. Seising, and C. Thomas, "Traffic Management for Broadband Networks with Fuzzy Logic-Call Admission Control and Usage Parameter Control," EUFIT 96, pp. 1579–1583, September 1996.
12. R. Jain, S. Kalyanaraman, R. Goyal, S. Fahmy, and R. Viswanathan, "The ERICA Switch Algorithm: A Complete Description," ATM Forum 96-1172, August 1996.
13. R. Jain, "Congestion Control and Traffic Management in ATM Networks: Recent Advances and A Survey," *Comp. Networks and ISDN Syst.*, 28(13), 1723–1738, October 1996.
14. R. Jain, S. Kalyanaraman, and R. Viswanathan, "The OSU Scheme for Congestion Avoidance in ATM Networks: Lessons Learnt and Extensions," 2, *Performance Evaluation J.*, 31/1–2, December 1997.

15. A. Kandel, O. Manor, Y. Klein, and S. Fluss, "ATM Traffic Management and Congestion Control Using Fuzzy Logic," *IEEE Trans. Syst., Man and Cybern.*, Part C, 29(3), 474–480, August 1999.
16. A. Kolarov and G. Ramamurhy, "Comparison of ER and EFCI Flow Control Scheme for ABR Service in Wide Area Networks," Proc. IFIP Broadband Commun., April 1995.
17. H.T. Kung and R. Morris, "Credit-Based Flow Control for ATM Networks," *IEEE Network*, 9(2), 40–48, March 1995.
18. P. Newman, "Backward Explicit Congestion Notification for ATM Local Area Networks," Proc. IEEE Globecom, pp. 719–723, December 1993.
19. S.Y. Oh and D.J. Park, "Predictive Fuzzy Explicit Rate Allocation (PFERA) for Traffic Control in ATM Networks," 1999 IEEE Intern. Syst. Conf. Proc., Seoul, Korea, August 22–25, 1999.
20. H. Ohsaki, M. Murata, H. Suzuki, C. Ikeda, and H. Miyahara, "Rate-Based Congestion Control for ATM Networks," *Comp. Commun. Rev.*, April 1995.
21. C.M. Ozveren, R. Simcoe, and G. Varghese, "Reliable and Efficient Hop-by-Hop Flow Control," *IEEE JSAC*, 13(4), 642–650, May 1995.
22. L. Roberts, "Enhanced Proportional Rate Control Algorithm," ATM Forum 94-0735R1, August 1994.
23. D. Sisalem and H. Schulzrinne, "Switch Mechanisms for the ABR Service: A Comparison Study," http://www.lirmm.fr/atm/articles.html, June 1996.
24. N. Yin and M.G. Hluchy, "On Closed-Loop Rate Control for ATM Cell Relay Networks," Proc. IEEE Infocom'94, pp. 99–108, 1994.
25. Y.-C. Lai and Y.-D. Lin, "Interoperability of EFCI and ER Switches for ABR Services in ATM Networks," 12(1), 34–42, 1998.

7

Fuzzy Rate Regulation with Feedback Controllers in ATM Networks

Christos Douligeris and Yao-Ching Liu

CONTENTS

ABSTRACT   In this chapter, we examine the use of fuzzy logic systems as rate-based feedback congestion-control mechanisms in Asynchronous Transfer Mode (ATM) networks. The information used by the fuzzy logic systems is based on the short-term observation of the network status and the system's acquired experience. A feedback value, which is between 0 and 1, is generated by the fuzzy logic systems and is sent back to all the active sources to throttle their transmission rates for every fixed time interval. Two source rate regulation mechanisms are also proposed. In the Rate Reduction (RR) mechanism, the current transmission rate of an active source is reduced by the value of the feedback signal. In the Rate Reduction Rate Increment (RRRI)

0-8493-1075-X/01/$0.00+$.50
© 2001 by CRC Press LLC

mechanism, we assume that there is an infinite buffer space at each source terminal, where cells are put with no rate regulation. When congestion is detected at the network, the source buffers transmit cells to the network at the same reduced rate as in the case of the RR mechanism. If the bandwidth of the server is plenty enough, the source buffers transmit cells at the speed of the peak cell rate. An experimental study shows that the proposed fuzzy logic control systems have lower time delay as well as lower cell-loss rate compared with traditional feedback control mechanisms. The RRRI mechanism has better performance than the RR mechanism at the cost of extra buffer space each source.

KEY WORDS: *ATM Networks, Congestion Control, Feedback Regulators, Fuzzy Control, Rate Regulation, Source Regulation, Traffic Shaping, MPEG sources.*

7.1   Introduction

The Asynchronous Transfer Mode (ATM) has been proposed by ITU-T as the transport method for the Broadband Integrated Services Digital Network (B-ISDN).[7] The ATM network provides services to sources with different traffic characteristics by statistically multiplexing "cells," fixed-length packets 53 bytes long. These services include data, video, and voice transmissions. Due to the uncertainties of broadband traffic patterns, unpredictable statistical fluctuations of traffic flows can cause congestion in the network elements (e.g., switches, concentrators, cross-connects, and transmission links).

ATM layer congestion control refers to the set of actions taken by the network to minimize the intensity, spread, and duration of congestion.[7] Feedback flow control is one of the solutions which has been extensively studied in the literature.[2, 10 12] In feedback controls, when possible traffic congestion is detected at any network element, feedback signals are sent back to all sources and each source reduces its current transmission rate by a given factor. In Reference 12 for example, when a queue in an ATM switch exceeds a certain threshold, it sends a congestion notification cell (CNC) back to the sources of the virtual channels currently submitting traffic to it. In Reference 2, instead of monitoring the instantaneous queue occupancy, a short-term arrival rate measurement in relation to thresholds and system capacity is performed to adjust the peak cell rates of the sources. However, a proper definition of the appropriate congestion thresholds must consider the characteristics of traffic sources, buffer size, and quality of service (QOS) requirements for all sources, an effort which needs complicated computations. In References 1 and 13, the authors addressed congestion-control schemes based on the assumption that the traffic follows certain statistical models. But the integration of diverse services on

ATM networks introduces poorly understood traffic parameters and user behaviors, and consequently accurate traffic models may not be able to be produced.

In this chapter, two fuzzy logic subsystems are integrated to predict possible buffer overflows in the near future. Propagation delays to send the feedback signals are explicitly considered. Based on the prediction, a feedback cell carries an explicit value to the active sources, at which value they have to regulate their source transmission rates. The first fuzzy logic subsystem generates an output which is the rate regulation factor and the second fuzzy logic subsystem performs a Kalman Filter-like function by adjusting the rate regulation factor (generated by the first fuzzy logic subsystem) based on the prediction errors in the past few cycles.

To validate the proposed algorithms, we perform several experiments by simulating and comparing the performance of the traditional feedback control mechanisms with that of the ones proposed in this chapter. In the traditional feedback congestion controller case, a CNC is generated when the ratio of occupied queue length, q/ξ (q is the current queue length and ξ is the buffer size of the shared buffer) is bigger than a threshold value $Q_{threshold}$. In this case all the active sources reduce their transmission rates to 50% of their current cell transmission rates when a CNC is received. Different $Q_{threshold}$ values are selected (since an appropriate congestion threshold is difficult to define) to compare the performance of the traditional feedback congestion controller with our mechanisms. Naturally, the smaller the $Q_{threshold}$ selected, the lower the cell loss rate observed, but higher transmission delays occur, and vice versa. The results show that fuzzy logic-based mechanisms have better performance in terms of both cell loss rate and transmission delays than the traditional feedback control approach. Two MPEG traces from UC-Berkeley (advertisement trace)[4] and Bellcore (movie trace)[5] are used as input traces. Similar comparison results are observed for both traffic traces, an indication that our model can be generalized and be applied to different traffic traces without updating the parameters of the fuzzy logic systems.

This chapter is organized as follows: Section 7.2 describes a background study of the fuzzy logic systems. Section 7.3 presents the feedback congestion-control mechanisms and the implementation of the fuzzy logic systems and Section 7.4 describes the integration of the source rate regulation mechanisms with the fuzzy logic systems. Section 7.5 presents the characteristics of the traffic traces used in this chapter and the simulation results, while Section 7.6 concludes the chapter.

7.2 Fuzzy Logic Systems

Fuzzy modeling is the method of describing the characteristics of a system using fuzzy inference rules. The method has a distinguishing feature in that it can express complex nonlinear, time-varying, ill-defined systems linguistically. This technology has been successfully applied to numerous problems, mostly in the control area,[8, 9] where the complexity of the system tends to preclude an analytic solution.

Fuzzy logic systems differ from classical math-model controllers because they are model-free systems. Fuzzy controllers do not require any mathematical model of how control outputs functionally depend on control inputs. Several applications of fuzzy logic systems in control theory have been reported, for example, in Reference 9, a Kalman-Filter controller was successfully implemented by the use of fuzzy logic systems. A fuzzy logic system has been implemented to optimize the cell service scheduling in ATM networks for multiple QOS requirements of traffic classes.[3, 11]

In the design of a fuzzy controller, one must identify the main control parameters and determine a term set that is at the right level of granularity for describing the values of each "linguistic" variable. Fuzzy logic systems store banks of fuzzy associations or common sense "rules."

Figure 7.1 shows an example of a fuzzy logic system that has two IF-THEN rules:

$$\text{IF } x \text{ is } A_1 \quad \text{and} \quad y \text{ is } B_1 \text{ THEN } z \text{ is } C_1;$$
$$\text{IF } x \text{ is } A_2 \quad \text{and} \quad y \text{ is } B_2 \text{ THEN } z \text{ is } C_2;$$

where x and y indicate possible crisp inputs to the systems. They are fuzzified by the membership functions and w_1 and w_2 are the outputs of fuzzy results

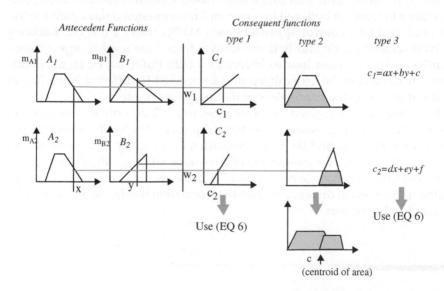

FIGURE 7.1
Commonly used membership functions and defuzzification mechanisms.

(intersections of fuzzy sets defined as above). There are many popular types of fuzzy reasoning mechanisms proposed in the literature. *Type 1*[14] uses a monotonically nondecreasing function to get the output associants or rule consequents c_i. The Weighted Average method is used to get a crispy output:

$$C = \frac{w_1 \times c_1 + w_2 \times c_2}{w_1 + w_2} \tag{7.1}$$

Type 2 uses different shapes of membership functions for the consequent part. Those rule antecedents combined with *AND* use the minimum fit value to activate consequents. The system adds the minimum-scaled output fuzzy sets and computes the fuzzy centroid of this waveform. *Type 3* was used in Reference 15 which proposed crisp functions for the consequent part. The output of each rule is a linear combination of input variables plus a constant term. The Weighted Average in Equation (7.1) is then used as a defuzzification scheme.

7.3 Fuzzy Logic-Based Feedback Congestion Controls

The architecture of our proposed control model is shown in Figure 7.2. Assume that there are N sources which are connected to the same destination server D. The server removes S_r cells from the buffer every second. A shared buffer with size ξ cells is located at the input of the server. Each source has a Source Rate Regulator (SRR) to throttle its transmission rate when a congestion state is detected in the network. Assume that all the cells (including the

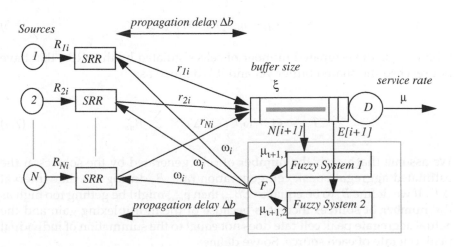

FIGURE 7.2
Simulation model for feedback traffic regulation.

feedback cells) transmitted between the sources and the destination server experience the same propagation, which is Δb seconds. A traffic source n, with n between 1 and N, declares its peak cell rate as B_{np} cells/s, and its mean cell rate as B_{nm} cells/s. Its nonregulated data generation rate beginning at time T_i is equal to R_{ni} cells/s. The aggregate cell generation rate of the N sources at time T_i is R_i and R_i is shown as:

$$R_i = \sum_{n=1}^{N} R_{ni} \tag{7.2}$$

We separate time into fixed-length intervals, ΔT where $\Delta T = T_i - T_{i-1}$. If R_{ni} is not regulated for each source, then $(R_i \times \Delta T)$ cells arrive at the destination buffer between time T_i and T_{i+1} from all sources. In our proposed feedback controller, the fuzzy logic systems sample the status of the network and generate an output ω_i between 0 and 1 for every ΔT seconds. The rate R_i is regulated to a new rate denoted as γ_i (the aggregated regulated cell transmission rate for all sources) by the SRRs if likely buffer overflow is detected in the near future. So each source rate R_{ni} can be regulated to a new value using equation $\gamma_{ni} = R_{ni} \times \omega_i$, where $\Sigma\gamma_{ni} = \gamma_i$.

7.3.1 Fuzzy Logic Subsystem I

In the first fuzzy logic subsystem, we define DT_i as the total number of cells discarded due to buffer overflows between time T_i and T_{i+1}. To monitor the changes of the queue length if R_i is not regulated by the SRRs but is fed into the server directly (since the server and SRRs are not located at the same sites and R_i is the traffic rate that we want to predict and regulate), we define:

$$\delta_i = (q_i - q_{i-1}) + DT_i + CT_i \tag{7.3}$$

where CT_i is the estimated number of cells regulated by the SRRs, that have not reached the shared buffer yet and it is defined as:

$$CT_i = (1 - \omega_{i-1}) \times \Delta T \times \sum_{n=1}^{N} B_{nm} \tag{7.4}$$

We assume that ρ_{max} is the number of cells generated by the sources at the estimated aggregate peak cell generation rate, B_p during the time interval ΔT. If we define B_p to be equal to ΣB_{np}, then ρ_{max} might be getting too high as the number of sources increases because of the multiplexing gain and the actual aggregate peak cell rate does not equal to the summation of individual peak cell rate of each source. So we define:

$$B_p = M_a + \beta\sigma \tag{7.5}$$

with

$$\beta = \sqrt{-2 \ln(\epsilon) - \ln(2\pi)} \qquad (7.6)$$

where $M_a = \Sigma_{n=1, N} B_{nm}$ is the aggregate mean cell rate and σ is the standard deviation of the aggregate cell transmission rate ($\sigma^2 = \Sigma_{i=1, N} \sigma_i^2$). Equation (7.5) was used by Guerin et al. as a reference term to estimate the "equivalent capacity" or bandwidth requirement of both individual and multiplexed connections for Connection Admission Control.[6] ϵ is the desired cell loss probability, which is assumed to be equal to 10^{-9} in this study.

From the system description, we then define δ_{max} as the maximum increment of the queue in ΔT:

$$\delta_{max} = \left[\left(1 - \frac{S_r}{B_p} \right) \times \rho_{max} \right] \qquad (7.7)$$

There are two linguistic variables in the input vector $N[i]$ and one linguistic variable $\mu_{i,1}$ as the output in this fuzzy logic system:

$$N[i] = \left[\frac{q_i}{\xi}, \frac{\delta_i}{\delta_{max}} \right], \qquad (7.8)$$

where q_i is the occupied queue length of the shared buffer at time T_i. We normalize the input variables so these fuzzy logic systems can be used in different network environments (different number of active connections, buffer size, etc.).

For each linguistic variable, there are three linguistic labels (*High, Medium, Low*) used to describe its membership functions. We select the membership functions as shown in Figure 7.3 by monitoring a single advertisement trace fed into the network and heuristically adjusting the parameters to avoid any buffer flows. There are nine fuzzy rules in this system as shown in Figure 7.4, which are the combinations of the linguistic labels of the linguistic variables.

7.3.2 Fuzzy Logic Subsystem II

The second fuzzy logic subsystem keeps track of the prediction errors of the integrated fuzzy logic subsystems in the past cycles and adjusts $\mu_{i,1}$ by using a function F. The error of a prediction error is defined as

$$\epsilon_i = \begin{cases} \dfrac{DT_i}{\delta_{max}} & \text{if } DT_i > 0 \\ 1 - \omega_{i-1} & \text{only} \quad \text{if } DT_i = 0 \text{ and } (Q - q_i) > \delta_{max} \\ 0 & \text{otherwise} \end{cases} \qquad (7.9)$$

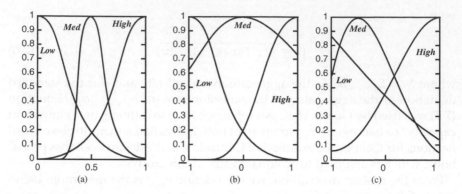

FIGURE 7.3
Membership functions for (a) q_i/ξ, (b) δ/δ_{max} and (c) $\mu_{1,1}$.

IF q_i/ξ is *High* and δ_i/δ_{max} is *High*, THEN the $\mu_{i,1}$ is *Low*.
IF q_i/ξ is *High* and δ_i/δ_{max} is *Medium*, THEN the $\mu_{i,1}$ is *Low*.
IF q_i/ξ is *High* and δ_i/δ_{max} is *Low*, THEN the $\mu_{i,1}$ is *Medium*.
IF q_i/ξ is *Medium* and δ_i/δ_{max} is *High*, THEN the $\mu_{i,1}$ is *Low*.
IF q_i/ξ is *Medium* and δ_i/δ_{max} is *Medium*, THEN the $\mu_{i,1}$ is *Medium*.
IF q_i/ξ is *Medium* and δ_i/δ_{max} is *Low*, THEN the $\mu_{i,1}$ is *High*.
IF q_i/ξ is *Low* and δ_i/δ_{max} is *High*, THEN the $\mu_{i,1}$ is *Medium*.
IF q_i/ξ is *Low* and δ_i/δ_{max} is *Medium*, THEN the $\mu_{i,1}$ is *High*.
IF q_i/ξ is *Low* and δ_i/δ_{max} is *Low*, THEN the $\mu_{i,1}$ is *High*.

FIGURE 7.4
Fuzzy rules used in fuzzy logic subsystem I.

The negative sign of ϵ_i shows that ω_{i-1} was over-estimated, which may be a cause for buffer overflow. On the other hand, the positive sign represents an under-estimation of ω_{i-1}. The first condition says a good prediction should not cause any cell loss. Any cell loss must have been caused by a previous over estimation. The second condition states that, if the length of the occupied queue is still long enough, no reduction of the transmission rate should be done. We use two linguistic variables for the input vector $E[i]$ in this fuzzy logic system. These linguistic variables represent the errors of the past two cycles: ϵ_i and ϵ_{i-1}. This fuzzy logic system generates an output $\mu_{i,2}$. The membership functions and fuzzy IF-THEN rules used in this system are shown in Figure 7.5 and Figure 7.6, respectively. Note that the membership functions for ϵ_i and ϵ_{i-1} are the same in this system.

The output of this system is used to adjust $\mu_{i,1}$ using the following function:

$$\omega = F(\mu_{i+1,1}, \mu_{i+1,2}) = \min\{(0.5 + \mu_{i+1,2}) \times \mu_{i+1,1}, 1.0\} \quad (7.10)$$

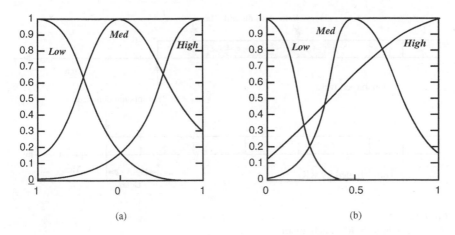

(a)

(b)

FIGURE 7.5
Membership functions for (a) ϵ_i and $\epsilon_{i,1}$, (b) $\mu_{i,2}$.

IF ϵ_i is *High* and ϵ_{i-1} is *High*, THEN the $\mu_{i,2}$ is *High*.
IF ϵ_i is *High* and ϵ_{i-1} is *Medium*, THEN the $\mu_{i,2}$ is *High*.
IF ϵ_i is *High* and ϵ_{i-1} is *Low*, THEN the $\mu_{i,2}$ is *Medium*.
IF ϵ_i is *Medium* and ϵ_{i-1} is *High*, THEN the $\mu_{i,2}$ is *High*.
IF ϵ_i is *Medium* and ϵ_{i-1} is *Medium*, THEN the $\mu_{i,2}$ is *Medium*.
IF ϵ_i is *Medium* and ϵ_{i-1} is *Low*, THEN the $\mu_{i,2}$ is *Low*.
IF ϵ_i is *Low* and ϵ_{i-1} is *High*, THEN the $\mu_{i,2}$ is *Medium*.
IF ϵ_i is *Low* and ϵ_{i-1} is *Medium*, THEN the $\mu_{i,2}$ is *Low*.
IF ϵ_i is *Low* and ϵ_{i-1} is Low, THEN the $\mu_{i,2}$ is *Low*.

FIGURE 7.6
Fuzzy rules used in fuzzy logic subsystem II.

The value of $\mu_{i,2}$ is in the range of [0, 1], when $\mu_{i,2}$ is bigger than 0.5. This means that previous predictions were too small and $\mu_{i,2}$ should be increased, and vice versa. By adding 0.5 to $\mu_{i,2}$, the value of $\mu_{i,1}$ can be increased or decreased following the previous prediction correctness.

7.4 Traffic Shaping

In this section, we present two traffic regulation mechanisms which accept our fuzzy system's crisp output to perform traffic shaping.

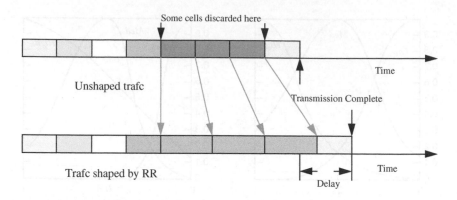

FIGURE 7.7
Rate regulation by RR mechanism.

7.4.1   Rate Reduction (RR)

In the RR mechanism, if possible buffer overflow is detected, the transmission rate of source n is reduced to $\gamma_{ni} = R_{ni} \times \omega_i$. The transmission rate is updated every ΔT seconds. The objective of this method is to avoid buffer overflows, so when the traffic is regulated, usually some delay time is needed to complete the requested transmission.

For example, assume an unshaped traffic source is to generate data at a constant rate R_{ni} cells/s for T_c seconds, $T_c = m \times \Delta T$. The source generates data at R_{ni} for only $k\Delta T$ seconds without any rate regulation, with $k < m$. It is then regulated to generation rate rn_i cells/s for the rest of the transmission ($r_{ni} \leq R_{ni}$). We are able to find that it takes $(m - k)\Delta T (R_{n_i} / r_{ni} - 1)$ seconds of delay to complete the transmission. The transmission diagram is shown in Figure 7.7. In Figure 7.7, data generation is constant in each time interval, and the fill density represents different source rates. The higher the density in a time interval the higher the source rate. In the highest density area, traffic is regulated to lower density with the penalty of longer time delays. However, the cell loss rate is lower. In this mechanism, the regulated transmission rate r_{ni} is always smaller than or equal to nonregulated transmission rate R_{ni}.

7.4.2   Rate Reduction and Rate Increment (RRRI)

In this mechanism, there is an infinite buffer space located at each source terminal. The source buffer is able to serve at rate equal to or smaller than the declared peak cell rate. All cells generated by a source pass through the buffer in a FIFO sequence. The sources put cells into their source buffers at the nonregulated rates. However, the source buffers transmit cells to the network at the regulated rates. So even if ω_i is equal to 0, the sources can still transmit cells to the source buffers.

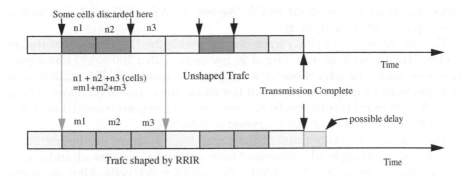

FIGURE 7.8
Rate regulation by RRRI mechanism.

The difference between RR and RRRI is that RRRI allows the transmission rate of a source n increased to B_{np} when both ω_{i-1} and ω_i are equal to 1. It means that if there are cells delayed in the source buffer and no congestion is detected in the near future (from the trend of ω_{i-1} and ω_i), we allow those cells in the source buffers to be transmitted into the network as soon as possible (by the declared peak cell rates). So in this case, r_{ni} might be greater than R_{ni}. The diagram is shown in Figure 7.8. This mechanism might still introduce some extra time delay if the transmission rates of the last few time intervals are higher than the cell service rate of the server.

7.5 Implementation and Numerical Results

This section presents the characteristics of two real-world MPEG traffic traces and the performance metric used to compare the performance between the proposed mechanisms. The simulation results of single- and multiple-sources cases are also presented.

7.5.1 MPEG Traces

Two different real world traffic traces are applied to our mechanisms. The MPEG-coded data set is a variable bit rate video trace using MPEG code from the movie "Star Wars." The movie length is approximately 2 h and contains a diverse mixture of material ranging from low complexity scenes to scenes with high action. The data set has 174,138 patterns, each pattern representing the number of bits generated in a frame time, F_t. In this trace, 24 frames are coded per second, so F_t is equal to 1/24 s. The peak bit rate of this trace is 185,267 bits/frame and mean bit rate is 15,611 bits/frame, and the standard deviation of bit rate is about 18,157. We assume ΔT is $F_t/5 = 8.3$ *ms* which is also equal to Δb for simplicity.

The advertisement MPEG trace is coded with the UC-Berkeley software coder. The stream is 10 min long at 30 frames/s with a 160×120 frame size. It is a sequence of advertisements for graphics products. This data stream was generated independently from the movie trace described above. These two models provide different burst characteristics that are essential in evaluating the performance of our proposed architecture.

The membership functions of the fuzzy logic systems are heuristically selected when a single advertisement trace is fed into the network and causes no cell loss in the case of $S_r = 1200$ cells/s and $\xi = 200$ cells. After the membership functions are selected, the parameters of both fuzzy logic subsystems are not changed while doing all the simulations presented in this chapter.

7.5.2 Performance Metric

In ATM networks, cell loss rate and transmission delay are two of the major considerations of the quality of service (QOS). In this chapter, we define the cell loss rate as the total number of cells discarded at the shared buffer divided by the total number of cells generated at the sources. For the delays, we assume that the time to complete the transmission without feedback control is T_k seconds, and that the time to complete the transmission with feedback control is T_h seconds. Delay is then defined as:

$$\text{Delay} = 100\% \times \frac{T_h - T_k}{T_k} \tag{7.11}$$

From the definition of Equation (7.11), it is obvious that there is no delay for the sources without SRRs.

Cell loss rate and transmission delay are the two metrics we will use for the comparison of the traditional schemes and the fuzzy logic based ones. We want to examine whether there is any improvement in performance through the use of fuzzy logic over a wide range of buffer sizes and transmission rates. The use of the MPEG traces as inputs to the simulator instead of artificially generated traffic puts an extra burden to the models in terms of their predictive capabilities—Markovian models can be predicted by many classical techniques.

7.5.3 Simulation Results

In this section, we run several simulations by integrating the proposed RR and RRRI mechanisms with the fuzzy logic system. In the traditional feedback control mechanism, when the queue ratio q/ξ exceeds a defined threshold, $Q_{\text{threshold}}$, a CNC is sent to all the active sources to reduce their transmission

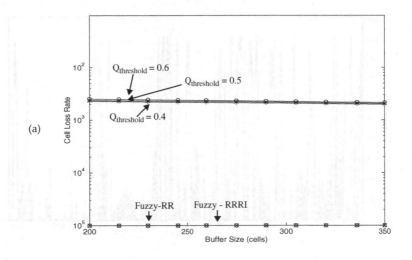

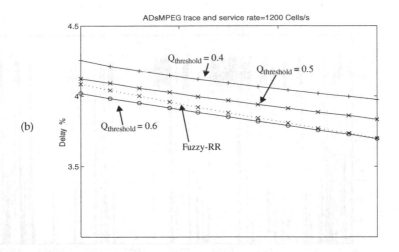

FIGURE 7.9
(a) Cell loss rate of the advertisement trace. (b) Transmission delay of the advertisement trace.

rates by a certain factor. The problem is how to define the congestion threshold and how much the source rate should be reduced. In our simulations, we select different values for the $Q_{threshold}$, namely 0.4, 0.5, and 0.6. Each source reduces its rate to 50% of its current cell generation rate when a CNC is received. The results of this model are then compared with the proposed adaptive approaches.

The inputs to the simulator model are the MPEG traffic traces mentioned in the previous sections. The fuzzy systems follow the described rules and use the aforementioned membership functions. We follow the ways buffer fill and empty and see what is the final effect on cell loss rates as well as the

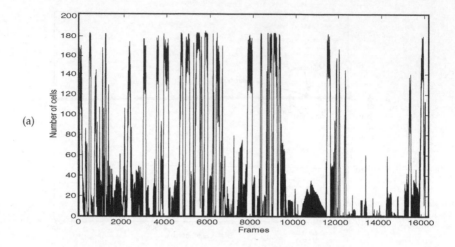

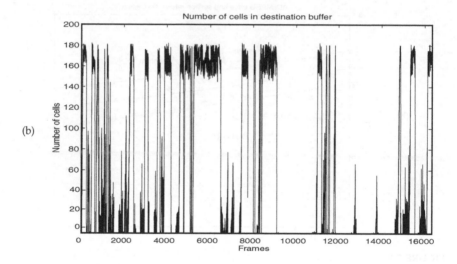

FIGURE 7.10
(a) Number of cells in the shared buffer using the RR mechanism and (b) Number of cells in the shared buffer using the RRRI mechanism.

increase in the average delay. The simulation ends when the trace ends (this is responsible for some lack of smoothness in the following figures).

In Figure 7.9a and Figure 7.9b, we select $S_r = 1200$ cells/s and change the buffer size to run the simulations when a single advertisement trace is fed into the network. Usually, S_r is a value selected between the mean cell rate and the peak cell rate to achieve reasonable cell loss rates for performance comparisons. We have experimented with different selections of S_r and similar results are observed but we do not present all the results due to the space

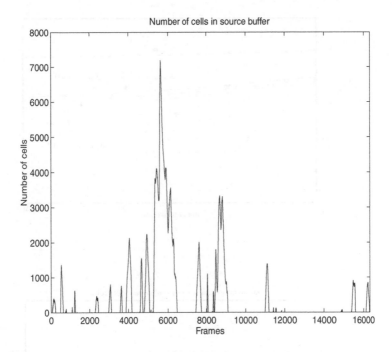

FIGURE 7.11
Number of cells in the source buffer using the RRRI mechanism.

limit. In Figure 7.9a, the cell loss rates of the fuzzy controllers using RR and RRRI mechanisms are 0. To increase the visibility of the figure, a small value of 10^{-5} is added to this simulation result. We find that the cell loss rate of the traditional approach is about 1.5×10^{-2} which is much higher than our approaches. Figure 7.9b shows the extra delay of all mechanisms, the delay of RR mechanism is between $Q_{\text{threshold}} = 0.5$ and $Q_{\text{threshold}} = 0.6$. The delay decreases faster than the traditional approach as the buffer size increases. The delay of RRRI is very close to 0—not shown in the figure.

Figure 7.10a and Figure 7.10b show the occupancy of the shared buffer of the RR and RRRI mechanisms. The RRRI mechanism has better utilization of the shared buffer which means it has better bandwidth utilization of the output link because the "waiting" cells spend less time in the server. From the above simulation results, it can be concluded that the RRRI mechanism has the best performance in both cell loss rate and delay. However, there is the trade-off of extra buffer space needed for each source. We have assumed that every source buffer has an infinite space and Figure 7.11 shows that the maximum occupied of the source buffer is about 7300 cells which means that a minimum source buffer space of 7300 cells is needed to avoid cell loss in this case.

In Figure 7.12a and Figure 7.12b, a single movie MPEG trace is fed into the network and the cell service rate is selected as 1680 cells/s. The parameters of the fuzzy logic systems are the same as those in the previous simulations.

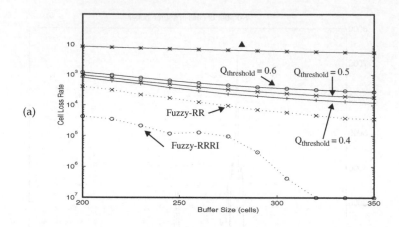

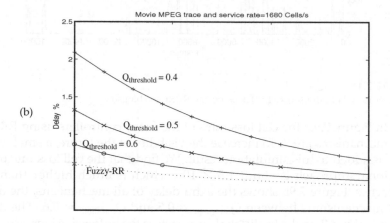

FIGURE 7.12
(a) Cell loss rate of the movie trace.   (b) Transmission delay of the movie trace.

The cell loss rate of the RR and RRRI mechanisms observed in this simulation show an average of 5 to 10 times improvement compared with the traditional mechanism. The delay time of these mechanisms is also smaller than all three selections of $Q_{threshold}$ in the traditional approach.

In Figure 7.13a and Figure 7.13b, the network is fed by the advertisement and the movie MPEG traces at the same time. Because the advertisement trace is shorter than the movie trace, the transmission of this trace is restarted again whenever the end of the trace is encountered. The simulation is ended when the transmission of the movie MPEG trace is completed. The simulation results in this case are similar to the results in Figure 12a and Figure 12b which show a more than 10 times cell loss rate improvement. However, the cell loss rates of RR and RRRI mechanisms are very close in this case.

(a)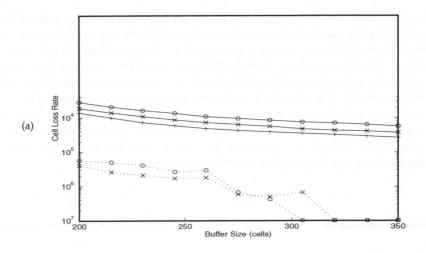

(b)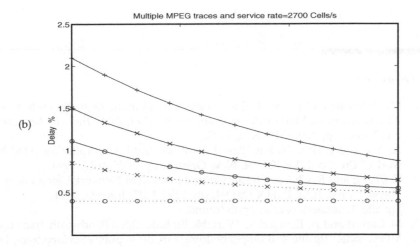

FIGURE 7.13
(a) Cell loss rate of multiple traces. (b) Transmission delay of multiple traces.

7.6 Conclusions

A fuzzy logic system which includes two subsystems has been proposed to predict possible buffer overflows in the destination-shared buffer. The prediction is based on the status of the shared buffer and no complicated statistical computations are needed. The prediction considers the propagation delay bringing the system closer to real-world applications. Two closed-loop source-shaping mechanisms which use the output of our fuzzy logic system also proposed: RR is a simple shaper whose data generation rate is directly controlled by the feedback signal; RRRI needs buffer space for each source but has the advantage of less overhead of transmission time. Simulation results show that the proposed mechanisms greatly improve the cell loss rate in different selections of cell service rates and buffer sizes. The proposed fuzzy logic system is a model-free controller that can be used in different characteristics of traffic sources and can also be applied in multiple sources without changing the parameters of the fuzzy logic systems.

Acknowledgments

This work has been partially supported by NSF under grant no. ANI-9810534.

References

1. M. Abdelaziz and I. Stavrakakis, "Some optimal traffic regulation schemes for ATM networks: a Markov decision approach," *IEEE/ACM Trans. on Networking*, vol. 2, no. 5, pp. 508–519, Nov. 1993.
2. A. Atai and J. Hui, "A Rate-Based Feedback Traffic Controller for ATM Networks," Proc. of IEEE ICC 94, New Orleans, pp. 1605–1615, 1994.
3. C. Chang and R. Cheng, "Traffic Control in an ATM Network Using Fuzzy Set Theory," Proc. IEEE INFOCOM'94, pp. 1200–1207, Toronto, 1994.
4. Ftp Site: tenet.berkeley.edu/pub/dbind.
5. M. Garrett and A. Fernandez, "Variable Bit Rate Video Bandwidth Trace Using MPEG code," ftp site = thumper.bellcore.com, dir = pub/vbr.video.trace, 1994.
6. R. Guerin, H. Ahmadi, and M. Naghshineh, "Equivalent Capacity and Its Application to Bandwidth Allocation in High-Speed Networks," *IEEE JSAC*, vol. 9, no. 7, pp. 968–981, September 1991.
7. ITU-T Recommendation I.371, "Traffic Control and Congestion Control in B-ISDN," Helsinki, 1993.
8. J. Jang, "ANFIS: Adaptive-Network-Based Fuzzy Inference System," *IEEE Trans. Syst. Man Cybern.*, vol. 23, no. 3 pp. 665–685, May 1993.
9. B. Kosko, *Neural Networks and Fuzzy Systems: a Dynamical Systems Approach to Machine Intelligence*, Englewood Cliffs, NJ, Prentice Hall, 1992.
10. Y. C. Liu and C. Douligeris, "Adaptive vs. Static Feedback Congestion Controller," Proc. IEEE GLOBECOM'95, pp. 291–295, Singapore, Nov. 1995.
11. Y. C. Liu and C. Douligeris, "Nested Threshold Cell Discarding with Dedicated Buffers and Fuzzy Scheduling," Proc. IEEE ICC'96, 429–433, Dallas, TX, 1996.

12. P. Newman, "Backward Explicit Congestion Notification for ATM Local Area Network," Proc. IEEE GLOBECOM'93, pp. 719–723, Houston, TX, 1993.
13. A. Pitsillides and J. Lambert, "Adaptive Connection Admission and Flow Control: Quality of Service with High Utilization," Proc. IEEE INFOCOM'94, pp. 1083–1091, 1994.
14. Y. Tsukamoto, "An Approach to Fuzzy Reasoning Method," in M. M. Gupta, R. K. Ragade, and R. R. Yager, editors, *Advances in Fuzzy Set Theory and Applications*, pp. 137–149. North-Holland, Amsterdam, 1979.
15. T. Takagi and M. Sugeno, "Derivation of Fuzzy Control Rules from Human Operators's Control Actions." Proc. IFAC Symp. Fuzzy Inf., Knowledge Representation and Decision Anal., pp. 55–60, July 1983.
16. L.A. Zadeh, "Fuzzy sets," *Information and Control*, vol. 8, pp. 338–353, 1965.

Problems

1. Explain in detail the parameters involved in Equation (7.3) and Equation (7.4) and their interconnections. Show the evolution in time of the parameters by assuming appropriate initial conditions and feedback messages. Study the effect of the feedback messages in this evolution, i.e., change the values of the feedback messages and see how they influence the other parameters.

2. Assume that you have a traffic that has a constant rate that is periodically doubled for half the time and then it is again reduced to the initial value. Show how the Rate Reduction (RR) and the Rate Reduction and Rate Increment (RRRI) shaping algorithms will operate. Make any necessary assumptions regarding the values of the various parameters.

3. Access the sites with the MPEG traces and verify the values given in Section 7.5.1.

4. The proposed algorithms are adaptive. Can you propose other adaptive algorithms that may be useful in such a rate regulation? What could be their advantages and their disadvantages vis-à-vis the proposed algorithms?

5. Observe Figure 7.12 and Figure 7.13 closely. What are the common characteristics of the figures? What are the main differences? Explain why this happens. Can you imagine another type of traffic with completely different performance characteristics than these?

6. Why do multiple sources present different performance characteristics than a single source by itself? Can you generalize with basic probability theorems?

7. Traffic shaping through rate regulation is a technique widely used and researched. Use the World Wide Web to create a list of

references that address this topic. Categorize, compare and evaluate the various methodologies in terms of the models used, the traffic characteristics assumed, and the topologies evaluated.

8. In this chapter we addressed traffic regulation as an ATM network problem. Is this problem valid in an Ethernet or an Internet environment? How is traffic controlled in these environments?

9. Evaluate the performance of the proposed algorithms using autoregressive Markov arrival models. Study the sensitivity of the results if we add or delete rules in the fuzzy systems.

10. Can you qualitatively explain changes in the performance of the algorithms by changing the parameters or the shapes of the membership functions? Would these changes be more profound in the single or the multiple source cases?

8

Applicability of Reinforcement Learning Algorithms to Usage Parameter Control

Antonios F. Atlasis and Athanasios V. Vasilakos

CONTENTS

ABSTRACT   Usage Parameter Control (UPC), one of the most fundamental
preventive congestion control mechanisms, monitors the traffic generated by
the sources and enforces the misbehaving ones to abide by the traffic contract
negotiated during the call set-up phase. Most of the known UPC mechanisms
do not succeed in detecting every possible violation of the traffic contract, while
their reaction time is in some cases unacceptable for real-time applications. Consequently, they cannot enforce the source to behave compliantly, while the

0-8493-1075-X/01/$0.00+$.50
© 2001 by CRC Press LLC

excess undetected traffic deteriorates the Quality of Service (QoS) requirements of the conforming users in an internodal node. In this chapter, after a brief review of the most known UPC mechanisms and a more detailed description of the Leaky Bucket, the best known of all, a UPC mechanism that uses a Reinforcement Learning Algorithm to enhance the performance of the Leaky Bucket, is proposed. A simulation study is used to show the effectiveness of this mechanism both in enforcing suitably and quickly the misbehaving sources and in guaranteeing the QoS parameters of the conforming sources. Finally, the hardware implementation concept is discussed and our conclusions are drawn.

KEY WORDS: *ATM Networks, UPC Algorithm, Policing Function, Reinforcement Learning Algorithms.*

8.1　What is Usage Parameter Control?

Call Admission Control (CAC), which is responsible for deciding whether a new call-request should be accepted or not, is not adequate to prevent congestion. This happens because some users, deliberately or not, are likely to violate the traffic contract and not to abide by the traffic parameters negotiated during the call set-up phase. Moreover, sometimes the source traffic parameters either cannot be estimated accurately in advance or cannot be estimated at all (as for example in the case of ABR traffic). Even technical malfunctions at the users' equipment can cause increased traffic rate and thus, congestion in the network. In all these cases it must be ensured that the users abide by the negotiated traffic parameters so that congestion is avoided and the *Quality of Service* (QoS) requirements of all users is guaranteed.

Usage Parameter Control (UPC) is the mechanism that has been proposed by ITU-T (Recommendation I.371)[1] and ATM Forum (Traffic Management Specification version 4.0)[2] to be applied to the *User-to-Network Interface* (UNI) in order to enforce the traffic parameters to abide by the negotiated values. When the same mechanism is applied to the *Network-to-Network Interface* (NNI) it is called *Network Parameter Control* (NPC). However, the term UPC is generally used for both cases. UPC can also be applied to *Virtual Paths* or to *Virtual Channels*. The terms *traffic policing* and *traffic enforcement* are also used in the literature.

The aim of UPC is to monitor and control the traffic so as to detect the users that violate the traffic contract as soon as possible and to protect the network resources and the QoS requirements of the compliant users from the excess traffic generated by these misbehaving users. The traffic parameters that a UPC mechanism should control are defined by the network manager and are usually the ones that CAC employs so as to decide whether a new call-request should be accepted or not. Moreover, UPC is responsible for controlling the validity of the ATM connections. This is achieved by checking the correctness of the *Virtual Path Identifiers* and the *Virtual Channel Identifiers* of the cells that pass through them. UPC should act transparently to the compliant sources without affecting their traffic. However, it should be noted that small deviations from the negotiated values should be allowed due to practical uncertainties, while the probability of enforcing the traffic when it is not required should be minimized. Finally, UPC should act in real time having small reaction time and it should be cost-effective to implement in hardware.

As it was explained above, UPC should guarantee the QoS parameters of the conforming sources. Especially as far as *Cell Loss Ratio* (CLR) is concerned, an upper bound P_{QoS} of it, which is specified by the traffic contract, should be guaranteed. On the other hand, when the source does not comply with the negotiated values, the steps that a UPC mechanism can take are cell dropping, cell tagging (marking), and traffic shaping. In the first two cases, the cell loss/marking probability should be $P_{des} > P_{QoS}$, so that the QoS parameters of the conforming sources are protected and the network resources are saved. That is because, as it has been noted in Reference 3, when a source deviates from the negotiated values, UPC should attempt to shut it off so that waste of network resources is avoided. The value of P_{des} is specified by the network manager.[4]

8.2 A Brief Review of Related Work

Several UPC mechanisms have been proposed so far in the literature. Some of the most known are the *Leaky Bucket*, the *Jumping Window*, the *Triggered Jumping Window*, the *Exponentially Weighted Moving Average* and the *Moving Window*. A review as well as a comparative performance study of them can be found in Reference 5. The Leaky Bucket, which will be described in the next section, has been shown to be the most efficient among them in terms of detection and reaction time as well as hardware implementation.

All the traditional UPC mechanisms attempt to estimate the source traffic parameters so as to monitor them and, if necessary, to enforce them. However, their estimate, especially of the long-term ones, is difficult and long observation periods are required. This results in a lot of detection and reaction time and thus, ineffective control of noncompliant sources. On the other hand, in order to achieve fast reaction, the observation periods must be decreased, which, however, results in ineffective estimate of the traffic parameters. Moreover, as it has been discussed in Reference 6, the performance of a UPC mechanism depends not only on the traffic parameters of the source, but also on the distribution function of its traffic. For all these reasons, the mechanisms mentioned above cannot detect small deviations from the negotiated values and cannot react quickly and effectively, especially as far as the mean rate, the mean burst period, and the other long-term parameters are concerned.[4-9]

The conflicts mentioned above between the need for an accurate estimate of the source traffic parameters and the requirement of fast reaction that the traditional UPC mechanisms face have led to the use of Computational Intelligence techniques. Some of the proposed mechanisms that are based on such methodologies use *Fuzzy Logic*[10-13]; a review of them can be found in Reference 14. The most significant one, which was presented in Reference 15, achieves to detect very small deviations of the traffic parameters from their negotiated values. This mechanism is described in another chapter of this book. In Reference 16 a mechanism based on *Artificial Neural Networks* (ANNs) was proposed; this mechanism controls the probability density function (pdf) of the traffic that the source produces instead of its traffic parameters. The pdf includes all the statistical properties of a source and so, its policing is more important and more effective than the policing of traffic parameters such as the Sustainable Cell Rate (SCR), the Peak Cell Rate (PCR), etc. In order to enforce the pdf effectively, the scheme proposed in Reference 16 uses two ANNs; the first one is trained to learn the pdf of the traffic in case the source is a conforming one, while the other is trained for every possible kind of traffic that the source may produce. This last case includes a conforming behavior, the cases of every possible deviation of anyone of the source traffic parameters as well as their combination. Despite the fact that this training is done off-line, it is very difficult, if not impossible, for this training to include all the possible cases of traffic that the source may produce. The number of the required neurones of the ANN increases as the number of the possible cases that the ANN should be trained for increases. For these reasons, as well as because both the size of the ANN and its training are restricted for practical reasons, the possibility of an incorrect convergence of it is not negligible.

8.3 The Leaky Bucket Mechanism

The Leaky Bucket, in its simple form, consists of a counter with size Q the value of which increases each time a new cell arrives and decreases according to a constant rate r as far as its value is positive (Figure 8.1). When the value of the counter reaches the upper limit Q, the additional incoming cells are dropped or marked (tagged).

Although the Leaky Bucket has been shown to be the most effective among the traditional UPC mechanisms,[5] it cannot control effectively long-term parameters, such as the SCR and the mean burst size.[4, 5, 7] The policing of these traffic parameters are absolutely essential since they are used almost from any CAC algorithm.

As far as the enforcement of the SCR is concerned, setting $r = cm_0$, where m_0 is the negotiated value of the SCR and c constant $(c \geq 1)$, the leak rate r

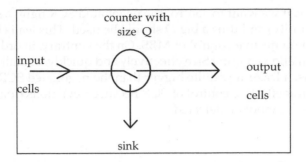

FIGURE 8.1
The Leaky Bucket mechanism.

should be set equal to m_0 ($c = 1$) so that the Leaky Bucket will police it effectively. However, it has been shown in Reference 7 that in this case a counter of an extremely big size Q should be used in order to guarantee the negotiated CLR. This results in large allowed *Maximum Burst Size* (MBS), according to the relation

$$\text{MBS} = \frac{pQ}{p - r}$$

(where p is the source Peak Cell Rate), and large reaction time (see equation 1 below). On the other hand, in order to reduce them, a small counter should be used; in this case, c and consequently r should be increased so that the Leaky Bucket will guarantee the negotiated CLR. In this case, if we define the deviation $d = m/m_0$ (where m is the real SCR), then for any value of $d < c$, the Leaky Bucket cannot detect the deviation from the negotiated value and so, it cannot police the SCR. The value of c (and consequently the value of r) is critical, because it determines how tight the control is. This is very important, because the tighter the control exerted by the policing procedure, the higher the bandwidth saved.

The average time required to fill up the counter after the instant a deviation from m_0 commences is used as a performance measure of the ability to detect and prevent long bursts. This time is given by the equation:[7]

$$T_{\text{fill-up}}(d) = \begin{cases} \dfrac{(Q - \bar{Q})n_{\text{cell}}}{(d - c)m_0}, & \text{if } d > c \\ \infty, & \text{otherwise} \end{cases} \tag{8.1}$$

where $\bar{Q}$ is the average value of the counter. As it becomes evident from the above equation, the bigger the size of the counter Q, the bigger the reaction time of the Leaky Bucket.

In summary, if we want to control the negotiated SCR tightly, small values of c (close to unity) and thus a big Q should be used. This leads to long reaction time and ineffective control of MBS. On the contrary, in order to control the Maximum allowed Burst Size effectively and quickly, a small Q should be used and thus, a leak rate r quite bigger than the negotiated SCR is required. This results in ineffective control of SCR because deviations from the negotiated value $d < c$ cannot be detected.

8.4 Enhancement of the Leaky Bucket Using a Reinforcement Learning Algorithm

Due to the problems that the Leaky Bucket encounters, a mechanism that employs a Reinforcement Learning Algorithm, called Stochastic Estimator Learning Algorithm (SELA), to enhance the performance of the Leaky Bucket was proposed in Reference 17. In order to take its decision, SELA uses as feedback a function of the distribution of the values that the counter of the Leaky Bucket takes. It was shown that this mechanism, which was suitable for policing only bursty sources, detects much smaller deviations from the negotiated values of the source traffic parameters, while it also achieves much faster reaction than the Leaky Bucket. Its effectiveness was also confirmed after examining its performance in an internodal node where, as it was shown, the proposed mechanism guarantees much better the QoS requirements of the compliant sources that are multiplexed with misbehaving ones in a single buffer.

In Reference 18 the above mechanism was improved using a different feedback function which results in even smaller reaction time. However, the most significant improvement regards its applicability not only to bursty sources, but, moreover, to any other type of sources. Its efficiency, as it was shown, results also in better guarantee of the QoS requirements of the conforming sources because it saves more bandwidth for them.

In the rest of this section, this UPC mechanism that was proposed in Reference 18 and uses a Reinforcement Learning Algorithm to enhance the Leaky Bucket will be presented, while in the next section its performance will be examined through a simulation study. However, first of all, the Reinforcement Learning Algorithm that is used, SELA, will be described.

8.4.1 The Stochastic Estimator Learning Algorithm (SELA)

Generally, a Reinforcement Learning Algorithm is a finite-state machine that interacts with a stochastic environment, trying to learn the optimal action the environment offers through a learning process. At any iteration the automaton chooses an action, according to a probability vector, using an output function. This function triggers the environment that responds with an answer (reward or penalty). The automaton takes into account this answer and jumps if necessary to a new state.

The Stochastic Estimator Learning Algorithm (SELA) used in the described methodology is a powerful and flexible ergodic Reinforcement Learning Algorithm, especially when operating in a nonstationary stochastic environment.[19] Ergodic means that it converges on the optimal action with a distribution independent of the initial state. SELA has also been used successfully in other problems, e.g., routing (see Atlasis', Saltouros et al. Chapter on "Routing") and CAC.[20] For reasons of completeness of this presentation, after some necessary definitions, a concise description of SELA algorithm is given:

$A = \{a_1, a_2, \ldots, a_n\}$ is the set of the n actions ($2 \leq n < \infty$) offered by the environment. The action selected at instant t is symbolized as $a(t) = a_k$. B is the set of the possible environment responses (feedback). The feedback of action a_i ($1 \leq i \leq n$) at instant t is symbolized as $b_i(t)$. $P = \{P_1, P_2, \ldots, P_n\}$ is the probability vector of choosing each action, i.e., P_i is the probability of choosing action a_i.

$D(t) = \{d_1(t), d_2(t), \ldots, d_n(t)\}$ is the True Estimate Vector. The True Estimate d_k of the selected action a_k is the mean reward which this action received the last W times that it was selected. It is computed as:

$$d_k(t) = \frac{\sum_{i=1}^{W} b_k(i)}{W} \tag{8.2}$$

where $\sum_{i=1}^{w} b_k(i)$ is the total reward received by the automaton during the last W times that action a_k was selected. The parameter $W \in \mathbf{N}^*$ (where $\mathbf{N}^* = \{1, 2, 3, \ldots \}$) is an integer internal automaton parameter called *learning window* and is used for ignoring old—and probably invalid—environmental responses.

$M(t) = \{m_1(t), m_2(t), \ldots, m_n(t)\}$ is the Oldness Vector; $m_i(t)$ of action a_i at any instant t is a nonnegative integer number which expresses the time passed (counted in number of iterations) from the last time that action a_i was selected.

$U(t) = \{u_1(t), u_2(t), \ldots, u_n(t)\}$ is the Stochastic Estimate Vector. The Stochastic Estimate $u_i(t)$ of action a_i is defined as:

$$u_i(t) = d_i(t) + N(0, \sigma_i^2(t)) \tag{8.3}$$

where $N(0, \sigma_i^2(t))$ symbolizes a random number selected with a normal probability distribution, with mean equal to 0 and standard deviation $\sigma_i = \min\{\sigma_{max}, am_i(t)\}$ proportional to the time passed from the last time each action was selected. Specifically "a" is an internal automaton parameter that determines how rapidly the Stochastic Estimates become independent from the True Estimates, while σ_{max} is the maximum permitted standard deviation of the Stochastic Estimates from the True Estimates.

8.4.1.1   The SELA Algorithm

INITIALIZATION: Set $P_i = 1/n$ and $d_i(t) = m_i(t) = u_i(t) = 0$, $\forall\, i \in \{1, 2, \ldots, n\}$.
STEP 1: Select an action $a(t) = a_k$ according to the probability vector.
STEP 2: Receive the feedback $b_k(t)$ of action a_k from the environment.
STEP 3: Compute the new True Estimate $d_k(t)$ of the selected action a_k according to equation (2).
STEP 4: Update the Oldness Vector by setting $m_k(t) = 0$ and $m_i(t) = m_i(t-1) + 1 \,\forall i \neq k$.
STEP 5: Compute the new Stochastic Estimate $u_i(t)\,\forall i$ according to Equation (8.3).

STEP 6: Select the optimal action a_m that has the highest Stochastic Estimate $u_m(t) = \max\{u_i(t)\}$.

STEP 7: Update the probability vector in the following way:
    For every action a_i $(i = 1, 2, \ldots m-1, m+1, \ldots, n)$ with $P_i(t) > 0$ set: $P_i(t+1) = P_i(t) - 1/N$ where N is a parameter called resolution parameter and determines the stepsize $\Delta(\Delta = 1/N)$ of the probability updating.
    For the optimal action a_m set:

$$P_m(t+1) = 1 - \sum_{i \neq m} P_i(t+1), \quad \text{with} \quad 0 \leq P_i \leq 1 \quad \text{and} \quad \sum_{i=1}^{n} P_i = 1.$$

STEP 8: Go to step 1
    A complete formal description of SELA can be found in Reference 19. It should be noted that the choice of the parameters σ_{max}, a, and N is a critical issue relative to the performance of the algorithm under various switching environments. A learning algorithm is called ϵ-*optimal*,[21] if there is an internal parameter N such that:

$$\lim_{N \to \infty} (\lim_{t \to \infty} E\{P_m(t)\}) = 1$$

where $E\{P_m(t)\}$ is the expected value of the probability of choosing the optimal action.

THEOREM 1: SELA is ϵ-*optimal* in every stochastic environment that offers symmetrically distributed noise. Let $d_1, d_2, \ldots, d_n$ be the mean rewards offered by the environment to the actions $a_1, a_2, \ldots, a_n$, respectively. If action a_m is the optimal one $(d_m = \max_i \{d_i\}$ for $i = 1, \ldots, n)$ and $P_m(t) = P[a(t) = a_m]$, then for every value $N \geq N_0$ $(N_0 > 0)$ of the resolution parameter there is instant $t_0 < \infty$ such that for every $t \geq t_0$ it holds that $E\{P_m(t)\} = 1$.

The *proof* of the theorem is given in Reference 19. The assumption of symmetrically distributed noise is not unreasonable. The noise of all known stochastic environments is symmetrically distributed about the mean rewards of the actions.

8.4.2   Application of SELA to the Leaky Bucket

In this methodology, SELA is used in order to enhance the performance of the Leaky Bucket, the best traditional UPC mechanism up to now. Specifically, SELA employs the distribution function of the values that the counter of the Leaky Bucket takes so as to detect smaller deviations from the negotiated values and to achieve faster detection and reaction time than the Leaky Bucket. With this approach, the whole behavior of a source can be policed, and not just a single traffic parameter as the traditional UPC mechanisms do.

This methodology is based on the fact that any deviation of the source traffic parameters from the negotiated values results in different distribution of the values of the Leaky Bucket counter in comparison with the distribution function that corresponds to a conforming behavior of the source. SELA, receiving as feedback in real time the values that the counter of the Leaky Bucket takes, tries to understand whether the distribution function that corresponds to the real traffic is due to a compliant behavior of the source or not. For this purpose, SELA uses two actions: Action 1 corresponds to a conforming behavior of the source, while action 2 corresponds to a nonconforming one. When one or more source traffic parameters diverge from their negotiated values, the counter of the Leaky Bucket generally takes bigger values than when the source complies with the traffic contract. Exploiting this difference, SELA tries to understand whether the distribution of the real traffic corresponds to a conforming source and consequently, whether the source misbehaves or not. In order to achieve this, the feedback of SELA is defined as follows:

8.4.3   Definition of the Feedback that SELA Receives

Let $F_1(q)$ be the steady-state distribution function of the values of the Leaky Bucket counter when the source abides by the negotiated source traffic parameters and $F_2(q)$ the steady-state distribution function that corresponds to a case that the source transmits with deviation d_{max} from the negotiated Sustainable Cell Rate (SCR) (where d_{max} is the maximum permitted deviation). In order to exploit the difference between the two distributions, a threshold value q_{thr} is used. The feedback $b_1(q)$ of Action 1 is computed as:

$$b_1(q) = F_1(q_{thr}) - F_1(q) \qquad (8.4a)$$

while the feedback $b_2(q)$ of Action 2 is computed as:

$$b_2(q) = F_2(q) - F_2(q_{thr}) \qquad (8.4b)$$

A qualitative analysis of the feedback is described below:

When $q < q_{thr}$ it holds $F_1(q_{thr}) > F_1(q)$ since every distribution function is nondecreasing. In this case $b_1(q) > 0$ and action 1 is rewarded, while, on the other hand, $F_2(q) < F_2(q_{thr})$, and so $b_2(q) < 0$ and action 2 is penalized. On the contrary, when $q > q_{thr}$ it holds $F_1(q) > F_1(q_{thr})$ and $F_2(q) > F_2(q_{thr})$; consequently, $b_2(q) > 0$ and $b_1(q) < 0$. In this case, action 2 is rewarded while action 1 is penalized. A typical example of the feedback that the two actions receive is presented in Figure 8.2. As it is explained in Reference 22, the values that the counter of the Leaky Bucket takes are exponentially distributed for any type of source. This, in combination with the fact that the values of $F_1(q)$ and $F_2(q)$ are constant for a specific $q = q_{thr}$ result in the form of the curves of Figure 8.2.

8.4.4 Theoretical Explanation of the Feedback

Let $f(q)$ be the probability density function (pdf) of the values of the Leaky Bucket counter when the source abides by the negotiated source traffic parameters and d_1 and d_2 the mean rewards that the two actions of SELA receive. These mean rewards can be computed by the following equations:

$$d_1 = \int_0^Q b_1(q)f(q)\,dq \quad \text{and} \quad d_2 = \int_0^Q b_2(q)f(q)\,dq \qquad (8.5)$$

The threshold value q_{thr} is computed so that when a source is a conforming one, action 1 is rewarded more than action 2. That is because in this case the mean reward of action 1 d_1 should be bigger than the mean reward of action 2 d_2 so that SELA will converge on action 1 (according to theorem 1). On the contrary, when one or more source traffic parameters diverge from their negotiated values, action 2 should be rewarded more than action 1. In this case, d_2 should be bigger than d_1, so that SELA converges on action 2. Typical examples of the mean rewards of the two actions vs. q_{thr} are presented in Figure 8.3, where d_i^1 and d_i^2 are the mean rewards of action i when the source is conforming and nonconforming, respectively. The curves were computed using Equation 8.5. As we can observe, in case of a compliant source:

$$\exists\, q_1 : \forall q_{thr} > q_1, d_1 > d_2$$

On the contrary, when a source misbehaves,

$$\exists q_2 > q_1 : \forall q_{thr} < q_2, d_2 > d_1$$

Consequently, if we choose a value of q_{thr} so that $q_{thr} \in (q_1, q_2)$, the mean reward of the optimal action is always bigger than the corresponding one of the nonoptimal action and thus, SELA always converges on the correct action.

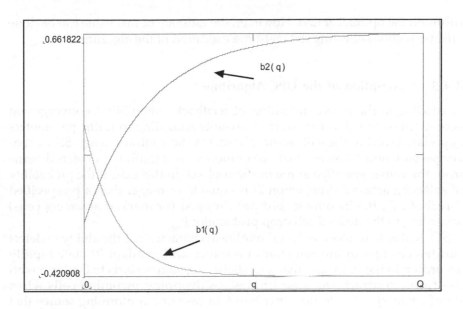

FIGURE 8.2
A typical example of the feedback that the two actions of SELA receives for the values that the counter of Leaky Bucket.

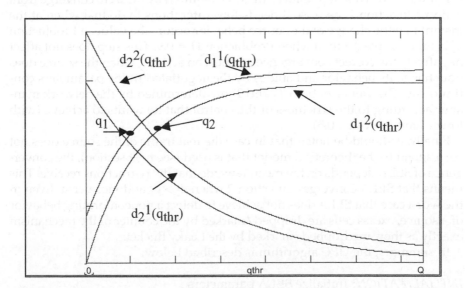

FIGURE 8.3
A typical example of the mean rewards of the two actions in case of a conforming and a nonconforming behavior of a source.

From the above comments it is concluded that the appropriate definition of the value of q_{thr} is vitally important for the correct convergence of the algorithm on the optimal action. However, the estimate of this value can be done off-line without affecting the real-time execution of the algorithm.

8.4.5 Description of the UPC Algorithm

According to the above definition of feedback, when SELA converges on action 1 it means that the source transmits according to traffic parameters negotiated during the call set-up phase. On the contrary, when SELA converges on action 2 it means that one or more source traffic parameters diverge from the values specified at the traffic contract. In this case, if the probability of selecting action 2 Prob[action 2] is equal to or bigger than a prespecified threshold T_1, the incoming cells are dropped (or marked as excess cells) according to the desired cell drop probability P_{des}.

SELA, due to its stochastic and nonlinear character, has the ability to detect sudden changes in the behavior of a source and to adapt its state rapidly according to them. In case that a misbehaving source starts to conform with the traffic contract, the algorithm stops dropping incoming cells when Prob[action 2] $< T_1$. On the other hand, in case of a conforming source that starts to misbehave suddenly, aiming at a reduction in the time required to converge on action 2, a threshold value T_2 of the probability of selecting action 1 Prob[action 1] is specified such that when Prob[action 1] $\geq T_2$, SELA is reinitialized. This step reduces the time required by SELA to converge from action 1 to action 2, because it needs fewer iterations to do that when at the instant that the change of the source behavior occurs the value of Prob[action 1] is for example 0.6 than when Prob[action 1] = 0.9. This step does not affect negatively the correct convergence of SELA on action 1 when the source does not change its behavior and abides by the negotiated traffic parameters continuously. The values of both T_1 and T_2 are determined by the network manager according to the strictness of the control that he wants to achieve (with lower limit the value 0.5).

Finally, it should be noted that in case the real traffic of the source does not correspond to the theoretical model that is used (see next section), the convergence of SELA depends on the mean rewards that the two actions receive. This means that SELA converges on action 2 when $d_2 > d_1$ and vice versa. Even in the worst case that SELA does not achieve to detect a nonconforming behavior of a source, excess cells are dropped/marked by the counter of the mechanism exactly as they are dropped/marked by the Leaky Bucket.

In summary, the UPC algorithm is described below:

INITIALIZATION: Initialize SELA parameters.

STEP 1: (In the instants that are specified in the Sections 8.4.6–8.4.7) choose an action according to the probability vector.

STEP 2: Estimate the feedback of the chosen action according to the current value of the counter q.

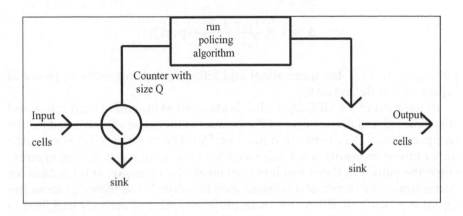

FIGURE 8.4
The proposed UPC mechanism.

STEP 3: Run steps 3–7 of SELA.

STEP 4: If Prob[of action 2] > T_1 drop (or mark) the incoming cells according to the desired cell drop (mark) probability P_{des}.

If Prob[of action 1] > T_2 re-initialize the parameters of SELA.

STEP 5: Go to step 1.
   In Figure 8.4 the described mechanism that uses SELA to enhance the Leaky Bucket is presented.

8.4.6   Applicability to Bursty Sources

The steady-state distribution of the Leaky Bucket counter in case of a bursty source can be computed by the approximation presented in Reference 8. According to this approximation, which was obtained using the two-state (ON-OFF) Markov model with exponentially distributed active and silent periods, if q is the current value of the counter ($0 \le q \le Q$) and $F(q) = Pr\{x \le q\}$ is the steady-state distribution function of its values, then it holds:

$$F(q) = \frac{1 - \rho.\exp(z.q)}{\Delta} \quad 0 \le q \le Q \tag{8.6}$$

where

$$\rho = \frac{p}{r}\frac{T_{OFF}}{T_{OFF} + T_{ON}}, \quad z = \frac{-(T_{ON} + T_{OFF})(1 - \rho)}{p - r}$$

$$\Delta = 1 - \frac{T_{OFF}}{T_{ON}} \frac{p-r}{r} \exp(zQ)$$

while T_{OFF} and T_{ON} the mean silent and active periods respectively (proof of Equation 6 in Reference 8).

An iteration of the UPC algorithm is executed at the end of each active and silent period and the values q of the counter at these instants are used for the computation of the distribution function $F(q)$. The fact that SELA is executed only at these moments is not a drawback for the accuracy of its convergence, since the values at these instants contain all the necessary information for determining the steady-state distribution function. Nevertheless, a more frequent execution of SELA (for example after each cell arrival) will lead to increased computational overhead and processing delay which may affect negatively the performance of the mechanism.

8.4.7 Applicability to Video Sources

According to the Discrete-State Continuous-Time Markov Process,[23] a video source can be approximated by M Markov (ON-OFF) bursty sources with parameters:

$$p = \frac{C(0)}{E[\lambda]} + \frac{E[\lambda]}{M}, \quad T_{ON} = \frac{1 + \frac{E^2[\lambda]}{MC(0)}}{\gamma}, \quad T_{OFF} = \frac{1}{\gamma - \frac{1}{T_{ON}}} \quad (8.7)$$

In the above equations, $E[\lambda]$ is the expected value of the bit rate and $C(\tau) = Ae^{-\gamma\tau}$ the autocovariance while A and γ are constants and τ a time variable. According to Reference 23, $M = 20$ ON-OFF sources are adequate to approximate a video source.

Generally, the Leaky Bucket can be modelled as a G/D/1/N queue with constant service rate r equal to the leak rate of the counter.[4, 22] The value of N is equal to the value of the counter Q plus the customer (cell) that is being served (transmitted). Consequently, using the Discrete-State Continuous-Time Markov Process, a Leaky Bucket mechanism with a counter of size Q and a constant leak rate r that polices a video source, can be approximated by M Markov ON-OFF sources that are statistically multiplexed in a buffer with size Q and constant service rate r. For large buffers, the distribution function of the buffer (which is equivalent to the distribution function of the Leaky Bucket counter) can be approximated by the following equation:[24]

$$F(q) \cong 1 - \beta.\exp(z_0 q) \quad (8.8)$$

where z_0 and β are constants. The way of how these parameters can be computed can be found in Reference 24. Although these computations are complicated, they are made only once and off-line without affecting the real-time operation of the UPC mechanism.

In the case of policing video sources, an iteration of the UPC algorithm is executed at the end of each video frame. The values of the counter q at these instants are used for the computation of the steady-state distribution $F(q)$ for each action using Equation (8.8). As in the case of bursty sources, the sampling of the counter and the SELA iteration at the end of each frame (instead of the end of each cell) is not a drawback for the quick detection of small deviations from the negotiated values.

8.4.8 Applicability to Other Types of Sources

As it is explained in Reference 22, the distribution function of each source that is enforced by the Leaky Bucket can be approximated by Equation (8.8). Moreover, in Reference 22 a method for computing β and z_0 is proposed. Using this method, the distribution function $F(q)$ of a source can be approximated by Equation (8.8) and, consequently, the use of the described UPC mechanism for policing any type of source is possible. The computations of β and z_0 can be made off-line without affecting the performance of the algorithm. However, as it is explained in Reference 22, in some cases the parameters β and z_0 can be estimated in real time, which results in real time estimate of the distribution function $F(q)$ too. In these cases, the proposed UPC mechanism can be applied to the case of possible renegotiation of the source traffic parameters. Finally, another advantage of the methodology described in this section is the fact that any other approximation that exists in the literature and estimates the constants β and z_0 for any type of traffic source can also be used.

8.5 Simulation Study

In Reference 18 the simulation study that was presented examined the performance of the described UPC mechanism for various types of sources, both on an access and on an internodal node. The confidence intervals of the measurements were 95% constructed with the method of independent replications. Several pilot runs were made to determine the most efficient values of the SELA parameters, as well as the value of threshold q_{thr}. These values were chosen so that in every case SELA will converge on the correct action. This is achieved choosing a value of resolution parameter N that is greater than the value of parameter N_0 (see theorem 1). The threshold values T_1 and T_2 were set to 0.9 and 0.7, respectively, while P_{des} was set to 0.1.

8.5.1   Performance Evaluation on an Access Node

At first, the effectiveness of the described UPC mechanism was examined as far as its detection capability and the reaction time that is required are concerned for a bursty (Table 8.1) and a video (Table 8.2) source. For the simulation study, the two-state (ON-OFF) Markov model and the Continuous-State Autoregressive Markov model proposed in Reference 23 were used respectively. In Figures 8.5 and 8.6 the results concerning the cell loss/marking probability vs. the deviation from the negotiated value of the SCR are presented correspondingly. As we can easily observe, the UPC mechanism that uses SELA to enhance the Leaky Bucket achieves to detect much smaller deviations than the Leaky Bucket and approximates very much the ideal behavior that a UPC mechanism should have. For example, in case of a bursty source (Figure 8.5) it controls ideally a misbehaving source for deviations $d \geq 1.05$. The fact that it does not detect deviations $d \in (1, 1.05)$ is not a drawback, since, as it was explained in Section 8.1, there should be a margin of tolerance due to practical uncertainties.

In Figures 8.7 and 8.8 the reaction time required by the examined UPC mechanisms to detect the misbehaving behavior of the sources is presented. In the case of the bursty source (Figure 8.7), the time that the Leaky Bucket requires was estimated using equation 8.1 (there is no corresponding equation for a video source). As we can observe, there is a significant difference between the performance of the two mechanisms. Thus, not only does the described mechanism detect much smaller deviations from the negotiated values, but it also detects them and reacts (dropping or marking the excess cells) much faster.

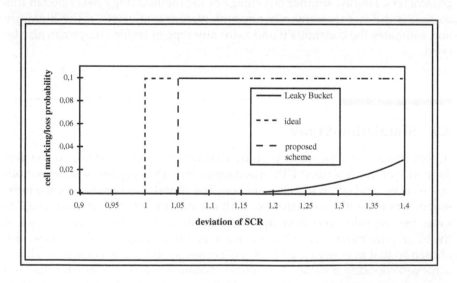

FIGURE 8.5
Cell marking/loss probability vs. deviation of SCR for a bursty source (Table 8.1).

TABLE 8.1

Traffic Parameters for a Bursty Source

Negotiated source traffic parameters	LB parameters	Traffic parameters of Action 1 and Action 2		SELA parameters
$p_0 = 10$ Mb/sec	$Q = 2500$	$p_1 = p_0$	$p_2 = p_0$	$N = 2500$
$m_0 = 1$ Mb/sec	$c = 1.465$	$m_1 = m_0$	$m_2 = 1.05$ Mb/sec	$\sigma_{max} = 10^{-5}$
$L_0 = 100$ cells		$L_1 = L_0$	$L_2 = L_0$	$a = 10^{-6}$
$T_{ON0} = 0.00424$ sec		$T_{ON1} = T_{ON0}$	$T_{ON2} = T_{ON0}$	$W = 30$
$T_{OFF0} = 0.03816$ sec		$T_{OFF1} = T_{OFF0}$	$T_{OFF2} = 0.03614$ sec	$q_{thr} = 117$
$P_{QoS} = 10^{-5}$				

TABLE 8.2

Traffic Parameters for a Video Source

Negotiated source traffic parameters	LB parameters	Traffic parameters of Action 1 and Action 2		SELA parameters
$E[\lambda] = 3.9$ Mbits/sec	$Q = 10000$	$E[\lambda]_1 = E[\lambda]$	$E[\lambda]_2 = 4.29$ Mbits/sec	$N = 2500$
$A = 0.0536$	$c = 1.502$	$A_1 = Á$	$A_2 = Á$	$\sigma_{max} = 10^{-5}$
$\gamma = 3.9$ sec^{-1}		$\gamma_1 = \gamma$	$\gamma_2 = \gamma$	$a = 10^{-6}$
$P_{QoS} = 10^{-5}$				$W = 30$
				$q_{thr} = 20$

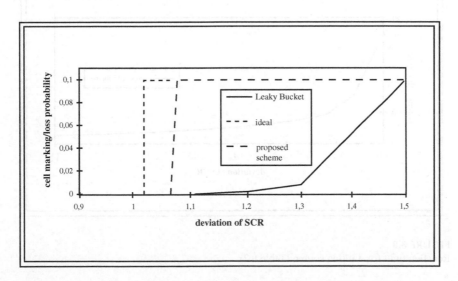

FIGURE 8.6

Cell marking/loss probability vs. deviation of SCR for a video source (Table 8.2).

One of the most difficult cases for a UPC mechanism regards the case that a conforming source starts to misbehave suddenly. In case of SELA, the reaction time that is required depends on the value of Prob[action 1] the instant

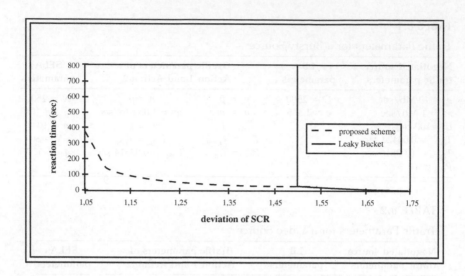

FIGURE 8.7
Reaction time for a bursty source (Table 8.1).

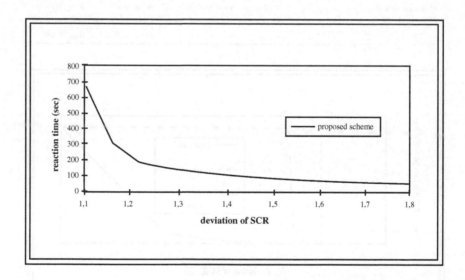

FIGURE 8.8
Reaction time for a video source (Table 8.2).

this change happens. In Figure 8.9 this time vs. Prob[action 1] is presented for a bursty source ($d = 1.35$). As we can observe, this reaction time, which is generally less than 50 sec, is not only reasonable, but it is approximately equal with the corresponding one in Figure 8.7. From Figure 8.7 we can also observe that the reaction time required by Leaky Bucket when $d = 1.35$ theoretically tends to infinity.

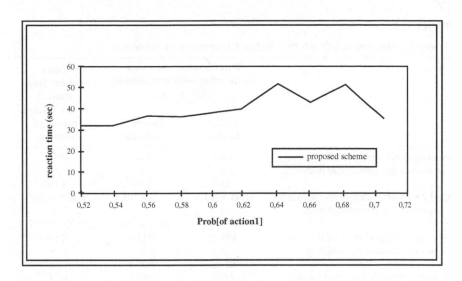

FIGURE 8.9
"Changing Rates" for a bursty source (Table 8.1) with deviation of SCR $d = 1.35$.

Similar results were obtained for the other source traffic parameters too.[18] A significant advantage of the described mechanism is the fact that when controlling the distribution function of a source, it achieves also to control all the source traffic parameters with only one mechanism. On the contrary, one Leaky Bucket mechanism can control only one source traffic parameter and thus, in order to control two traffic parameters (as for example SCR and mean burst size) a dual Leaky Bucket mechanism is required.[7, 9]

8.5.2 Performance Evaluation on an Internodal Node

In Reference 18 the effect of the proposed methodology on an internodal node was also examined because the protection of the conforming users from the misbehaving ones is in fact the most important issue for a policing mechanism. In Reference 25 it was shown that even when a misbehaving user is detected by the Leaky Bucket and excess cells are dropped, a considerable amount of traffic passes through the mechanism and deteriorates the QoS parameters of the conforming users which share the same internodal link queue. For this reason, the performance of the mechanism described in Section 8.4 was examined in such a queue. As it can be concluded from Tables 8.3 and 8.4, the tighter and faster control of a source that this mechanism achieves leads to more statistical gain and thus, better guarantee of the QoS requirements of the conforming sources. Not only does the proposed mechanism achieve smaller CLR than the Leaky Bucket, but, moreover, the network resources that it saves from the misbehaving sources result in smaller CLR even than in the case that all the sources comply with negotiated values of the source traffic parameters (Table 8.3). Similar conclu-

TABLE 8.3

Three Bursty Sources (with the Traffic Parameters of Table 8.1)

	Three sources, two non-conforming with deviation of SCR $d = 1.4$		Three conforming sources Leaky Bucket, proposed scheme
	Leaky Bucket	Proposed scheme	
Average total lost cells of nominal source in the internodal node	44.9 ± 9.03	4.26 ± 1.56	14.287 ± 4.198
Cell loss probability of the nominal source in the internodal node	1.88×10^{-4} ± 0.4×10^{-4}	1.79×10^{-5} ± 0.66×10^{-5}	6.035×10^{-5} ± 1.77×10^{-5}
Average dropped cells of the non-nominal sources by policing mechanisms	4903.2 ± 620.53	261184 ± 2449.2	2.584 ± 5.1
Average delay introduced by the queue in the buffer of the node (msec)	0.017 ± 0.027	0.003 ± 0.012	0.012 ± 0.022

Note: Link capacity C = 20 Mb/sec, internodal buffer size B = 150 cells, average session length = 100 sec

TABLE 8.4

Five Video Sources, Three Nonconforming Sources with Deviation of SCR $d = 1.5$

	Leaky Bucket	Proposed scheme
Average total lost cells of nominal source in the internodal node	3082.01 ± 262.03	620.93 ± 118.78
Cell loss probability of the nominal source in the internodal node	1.663×10^{-3} ± 1.414×10^{-4}	3.35×10^{-4} ± 6.409×10^{-5}
Average dropped cells of the non-nominal sources by policing mechanisms	62178.03 ± 4849.142	2167572 ± 11690.14
Average delay introduced by the queue in the buffer of the node (msec)	1.031 ± 0.3	0.205 ± 0.064

Note: link capacity C = Mb/sec, internodal buffer size B = 1500 cells, average session length = 200 sec

sions can be obtained as far as the average delay in an internodal queue is concerned.

In summary, it was shown through simulation that the UPC mechanism that uses a Reinforcement Learning Algorithm to enhance the Leaky Bucket, detects much smaller deviations from the negotiated values and reacts faster than the Leaky Bucket. This results in better guarantee of the QoS parameters of the conforming sources in internodal nodes.

8.6   Hardware Implementation and Conclusions

In this chapter, an efficient UPC mechanism was described, which enhances the Leaky Bucket using a Reinforcement Learning Algorithm, called SELA. This mechanism achieves tight and fast control for every type of VBR traffic source. Consequently, the network resources as well as the conforming users are protected much better from the misbehaving sources. Numerical results obtained from a comparative performance study with the Leaky Bucket confirmed the above statements. The necessity to define the values of the parameters of SELA does not affect the performance and the effectiveness of the proposed methodology, since it can be done off-line without affecting the real-time execution of the algorithm. The ability to control not only one traffic parameter at a time but, instead, the distribution of the source, counterbalances the increased computational effort which is needed in comparison with the Leaky Bucket. Nevertheless, the proposed scheme does not involve excessive information storage or computational overhead, while the required updating time lies within an acceptable range for real-time applications. As far as its hardware implementation is concerned, learning algorithms generally are amenable to simple hardware implementation using basic stochastic computing elements, which have long proved their success in synthesizing LAs.[26] A simple estimate of the hardware requirements that was made in Reference 18 showed that few Kbytes of ROM and RAM as well as a simple 8-bit microprocessor is adequate for its implementation. This implementation is simpler and less costly than the implementation of other Computational Intelligence techniques such as Artificial Neural Networks and Fuzzy Systems.

Exercises

1. Which other function could be used instead of the distribution function in equation (8.4)? In this case how should we modify the form of the feedback?

2. Making a simulation model try to examine the effect of the variable q_{thr} to the performance of the algorithm in a single node. What do you notice?

3. Based on the theoretical explanation of Section 8.4.4, try to estimate analytically the optimal values of the variable q_{thr} for the examined scenarios in the simulation study. Compare these values with the ones used in this study. What do you notice? What is the reason for this?

4. Choosing the appropriate models for different types of traffic sources, try to apply their distribution functions as feedback to the learning algorithm.

5. Try to improve the performance of the learning algorithm choosing a resolution parameter N that it is not constant, but, instead, at any time instant it varies proportionally according to the difference between the current feedback (Equation 8.4) and the true estimate (Equation 8.2).

References

1. ITU-T, Draft Recommendation I.371: Traffic Control and Congestion Control in B-ISDN, Geneva, 1996.
2. The ATM Forum Technical Committee, Traffic Management Specification, version 4.0, 1996.
3. Lague B., Rosenberg C., Guillemin F., A Generalization of some Policing Mechanisms, *in Proc. IEEE INFOCOM*, 767, 1992.
4. Butto M., Cavallero E., Tonnietti A., Effectiveness of the Leaky Bucket Policing Mechanisms in ATM networks, *IEEE J-SAC*, 335, April 1991.
5. Rathgeb E. P., Modeling and Performance Comparison of Policing Mechanisms for ATM Networks, *IEEE J-SAC*, 325, April 1991.
6. Tutufor K., On Admission Control and Policing in an ATM Based Network, *in Proc. 7th ITC Seminar*, Morristown, NJ, October 1990, paper 5.4.
7. Monteiro J. A. S., Gerla M., Fratta L., The Leaky Bucket Input Rate Control in ATM Networks, *in Proc. ICCC '90*, New Delhi, India, 370, 1990.
8. Elwalid A.I., Mitra D., Analysis and Design of Rate-Based Congestion Control of High Speed Networks, I: Stochastic Fluid Models, Access Regulation, *Queuing Systems*, 9, 29, 1991.
9. Pancha P., Zarki M., The Leaky Bucket Access Control for VBR MPEG Video, *in Proc. IEEE INFOCOM*, 796, 1995.
10. Cheung K. et al., Fuzzy Logic Based ATM Policing, *in Proc. ICCS*, 535, 1994.
11. Catania V., Ficilli G., Palazzo S., Panno D., Using Fuzzy Logic in ATM Traffic Control: Lessons and Perspectives, *IEEE Communication Magazine*, 34, 70, 1996.
12. Douligeris C., Develekos G., A Fuzzy Logic Approach to Congestion Control in ATM Networks, *in Proc. ICC*, 1969, 1995.
13. Ndousse T., Fuzzy Neural Control of Voice Cells in ATM Networks, *IEEE J-SAC*, 12, 1488, 1994.
14. Douligeris C., Develekos G., Neuro-Fuzzy Control in ATM Networks, *IEEE Communication Magazine*, 154, May 1997.
15. Catania V., Ficili G., Palazzo S., Panno D., A Comparative Analysis of Fuzzy Versus Conventional Policing Mechanisms for ATM Networks, *IEEE/ACM Transactions on Networking*, 449, June 1996.

16. Tarraf A., Habib I., Saadawi T., A Novel Neural Network Traffic Enforcement Mechanism for ATM Networks, *IEEE J-SAC*, 1088, August 1994.
17. Atlasis A. F., Vasilakos A. V., LB-SELA: Rated-Based Access Control for ATM Networks, *Computer Networks and ISDN Systems*, 30, 963, 1998.
18. Atlasis A. F., Stassinopoulos G. I., Vasilakos A. V., The Leaky Bucket Mechanism with Learning Algorithm for ATM Traffic Policing, *in Proc. ISCC'97*, Alexandria, Egypt, 68–72, July 1997. It has also been submitted for publication to *Computer Communications*.
19. Vasilakos A. V. Papadimitriou G., A New Approach to the Design of Reinforcement Scheme for Learning Automata: Stochastic Estimator Learning Algorithms, *Neurocomputing*, 7, 275, April 1995.
20. Vasilakos A. V., Loukas N. H., Atlasis A. F., The Use of Learning Algorithms in ATM Networks Call Admission Control Problem: A Methodology, *in Proc. ICCC*, Seoul, Korea, August 1995. It has also been submitted for publication to *CN-ISDN Systems*.
21. Narendra K., Thathachar M., Learning Automata: a Survey, *IEEE TRANSACTION SMC*, 4, 4, 323, 1974.
22. Mark B., Ramamurthy G., Real-time Estimate of UPC Parameters for Arbitrary Traffic Sources in ATM Networks, *in Proc. IEEE INFOCOM*, San Francisco, March 1996.
23. Maglaris B., Anastassiou D., Sen P., Karlsson G., Robbins J., Performance Models of Statistical Multiplexing in Packet Video Communications, *IEEE Transactions on Communications*, 36, 834, July 1998.
24. Anick D., Mitra D., Sondhi M., Stochastic-Theory of Data Handling System with Multiple Sources, *Bell System Technical Journal*, 61, 1871, October 1982.
25. Hluchyj M., Yin N., On the Queuing Behaviour of Multiplexed Leaky Bucket Regulated Sources, *in Proc. IEEE INFOCOM*, 672, 1993.
26. Mars P., Poppelbaum W. J., *Stochastic and Deterministing Averaging Processors*, Peter Peregrinus, Stevenage, U.K. 1981.

9

Networking Algorithms and Computational Intelligence

Sumit Ghosh, Qutaiba Razouqi, P. Seshasayi,
Tony S. Lee, and Seong-Soon Joo

CONTENTS

0-8493-1075-X/01/$0.00+$.50
© 2001 by CRC Press LLC

ABSTRACT   A network is a system that enables (i) communication and (ii) automation and control, between two or more network terminating points that may refer to either users or machines. A network consists of geographically dispersed components—links and nodes, of which links transport traffic while the nodes provide the distributed computational intelligence necessary in the network. The computational demand stems from the key networking functions of route determination, routing of traffic cells, switching the traffic cells in the switch fabric, ensuring the authenticity and security of each user traffic, and buffer management. In turn, the nodes of a network are composed of computationally intelligent modules corresponding to each of the networking functions. This chapter focuses on the issue of computational intelligence in buffer management, especially the use of algorithms, fuzzy sets, and fuzzy thresholds. Buffers are incorporated into a switch fabric to temporarily store the excess cells of a bursty traffic and achieve the dual functions of minimizing cell loss and smoothing traffic. For details on the nature of the computational demand for the remainder of the networking functions, the reader is referred to Reference 1. The use of sophisticated fuzzy set oriented computational techniques in networking is a recent phenomena and is growing rapidly. The remainder of this chapter is organized as follows. Section 9.1 presents a brief overview of the state of the current research in the use of fuzzy sets in networking. Section 9.2 focuses on an important networking function, buffer management, and presents the role of computationally intelligent algorithms and fuzzy thresholds in minimizing cell loss from buffer overflow, under bursty traffic. Finally, Section 9.3 develops a recommendation for further studies that are needed to maintain continued growth in this field.

KEY WORDS: *Computational intelligence, computation in networks, attributes of networks, buffer management, input and output buffers, cell loss, cell delay, fuzzy thresholds, network intelligence, routing, quality of service, call processor, security.*

9.1   Current Research in Computational Intelligence in Networking

Present day complex networks are dynamic, there is great uncertainty associated with the input traffic and other environmental parameters, they are subject to unexpected overloads, failures, and perturbations, and they defy accurate analytical modeling. Fuzzy logic appears to be a promising approach to address many important aspects of networks.

Fuzzy sets[2-4] provide a robust mathematical framework for dealing with "real-world" imprecision and nonstatistical uncertainty. Qualitative, "linguistic" variables allow one to represent a range of numerical values as a single, descriptive term that is described by a fuzzy set. Given that the present day complex networks are dynamic, that there is great uncertainty associated with the input traffic[5] and other environmental parameters, that they are subject to unexpected overloads, failures, and perturbations, and that they defy accurate analytical modeling, fuzzy logic appears to be a promising approach to address key aspects of networks. The ability to model networks in the continuum mathematics of fuzzy sets, rather than with traditional, discrete values coupled with extensive simulation, offers a reasonable compromise between rigorous analytical modeling and purely qualitative simulation.

Applications of fuzzy logic in telecommunications networks is recent and relatively less extensive than in automatic control. A detailed search coupled with a thorough review of the literature reveals that current research in fuzzy logic in telecommunications networks extends from queuing, buffer management, distributed access control, and load management to routing, call acceptance, policing, congestion mitigation, bandwidth allocation, channel assignment, network management, and quantitative performance evaluation of networks. To facilitate comprehension, the fuzzy literature may be organized into four efforts—modeling, control, management and forecasting, and performance estimation.

Fishwick[6] presents the use of fuzzy sets in qualitative and natural language simulation. The use of fuzzy techniques to model queuing systems is reported in Li and Lee[7] and Prade.[8] Both Prade[8] and Li and Lee[7] stress that in practical situations, the mean of the arrival rate (represented by λ) and the mean service rate (represented by v) are frequently fuzzy, i.e., they cannot be expressed in exact terms. Also, the service rules are often imprecise, given the presence of ill-stated exceptions and perturbations. However, neither Prade nor Li and Lee extend their analysis to the issues of queue management. Bonde and Ghosh[9] utilize "soft" linguistic system variables to model buffer queues in cell-switching networks. They introduce the concept of fuzzy thresholds toward robust, adaptive buffer management in sharp contrast to the traditional, abrupt, inflexible, binary thresholds. This view features continuous set membership and implies a soft, gradual transition between completely "FULL" and completely "not FULL" (empty). The membership values are defined through an asymmetric "sigmoid"-shaped function. Bonde and Ghosh investigate a fundamental issue: the underlying principles that either refuse or permit the entry of transit[10] cells, i.e., from other switches, to the switch buffer, from the perspective of fuzzy modeling and thresholds. They demonstrate that fuzzy thresholds cause the buffer queue to exhibit "soft" behavior, i.e., greater ability to adapt to dynamic conditions, and robustness, i.e., resilience to rapid dynamic changes in network traffic, and favorably impact cell discarding. Scheffer and Kunicki[11] present an analysis of the modeling techniques utilized for packet data networks. The scope of their analysis includes analytic, nonanalytic, nonparametric techniques, such as neural networks, fuzzy logic, and fractal schemes, and it evaluates with respect to accuracy, robustness, and ease of implementation. Motivated by the well-known success of fuzzy logic in nonlinear control,

Beneke and Kunicki[12] present a design of a fuzzy system that aims to recognize and respond, in real time, to telephone traffic load fluctuations, and thereby improve the network's grade of service.

Recently, Celmins[13] has proposed the efficient, distributed access control of battlefield communications network utilizing fuzzy logic control schemes. In this scheme, every node of the network gathers information on the approximate status of the network and makes intelligent decisions relative to accessing the network, independently and concurrently. The viability of this approach is enabled by the fact that today's network nodes are sophisticated computing engines. Pithani and Sethi[14] note that the information utilized by a network in determining efficient routes is subject to dynamic traffic and network topology changes and other uncertainties. They model the uncertainties in latency delays through fuzzy sets to improve the performance of one class of routing algorithms. Tanaka and Hosaka[15] review the difficulties of obtaining appropriate membership functions for efficient network control for call acceptance and routing. They present a method of obtaining good membership functions through tuning wherein the input values and revenues of the network are first measured and then optimal values derived from these data and the past tuning history. Ascia, Ficili, and Panno[16] recognize that the issue of policing is complex, stemming from the random nature of the traffic and contention for network resources. They propose a fuzzy logic-based policing mechanism that is "soft" and outperforms traditional mechanisms in selectivity and dynamic response. They also present a VLSI fuzzy processor, organized as a cascade of pipeline stages, that executes the fuzzy inferences of their approach, in parallel. Douligeris and Develekos[17] present a fuzzy rule-based system, utilizing the leaky bucket and moving window mechanisms, to enforce policing and mitigate congestion. The approach is transparent to the sources, it responds fast to peak rate violations as well as simultaneous mean and peak rate violations, and it is effective in shutting down noncompliant sources. Cheng and Chang[18] present a fuzzy traffic controller that manages both congestion and call admission control for ATM networks. The controller is a fuzzy implementation of the two-threshold congestion control and the equivalent capacity admission control methods where the parameters of the membership functions and the fuzzy control rules are extracted through clustering analysis of analytical data. They report a 11% improvement in system utilization while maintaining the QoS contract comparable with that of the conventional equivalent capacity method. Catania, Ficili, Palazzo, and Panno[19] propose a fuzzy policing mechanism for ATM networks to detect violations of the negotiated parameters while reducing the probability of false alarm. The mechanism monitors the number of cells transmitted by the user since connection and utilizes fuzzy rules to increase the threshold for conformance and lower it during periods of nonconformance. To relieve congestion and achieve high server utilization in ATM-based networks, Pitsillides and Sekercioglu[20] utilize fuzzy logic control techniques to develop a novel backward congestion notification scheme

(FBCN). They state that the complexity of the ATM networks and their dynamic parameters renders their analytic modeling very difficult, if not impossible. Sekercioglu and Pitsillides[21] also show that, in ATM LANs, the FBCN scheme minimizes cell losses at the switch queues and prevents time-consuming packet retransmissions that can dramatically degrade network throughput.

Holtzman[5] discusses the use of fuzzy approaches in the forecasting of future telecommunications services, and in modeling aspects of uncertainty in broadband traffic. Millstrom, Bonde, and Grimaldi[22] propose a unique hybrid architecture for automated VHF frequency management that applies fuzzy logic for enhanced signal detection and decision-making. Two applications of fuzzy modeling techniques to network troubleshooting systems are described by Lirov[23] and Lewis and Dreo.[24] Chakraborty, Mansfield, and Noguchi[25] describe the development of a fuzzy system for network management and observe that fuzzy logic serves as a better tool to facilitate knowledge representation and inferencing. The system is observed to outperform conventional crisp rule-based systems. Maravall[26] observes that in the context of adaptive control and routing of calls in a telephone network, the fuzzy adaptive approach outperforms the usual fixed techniques that suffer from a clear deterioration under unexpected overloads and failures.

Abdul-Haleem, Cheung, and Chuang[27] study a fuzzy distributed dynamic channel assignment scheme that attempts to uncover an open radio channel for assignment. Their scheme is in polite mode when a perfectly feasible channel is available. Otherwise, it engages progressively in an aggressive mode. Their results indicate that, under heavy traffic, the fuzzy scheme even outperforms the globally optimal dynamic channel assignment scheme in terms of spectral efficiency, while suffering mild voice quality degradation. Edwards and Sankar[28] report encouraging results in a fuzzy algorithm driven hand-off operation for a cellular system where the cell size is reduced to 10–100 meters. The algorithm combines the strength of the received signal with distance measurements to yield a hand-off factor which decreases monotonically as a mobile unit moves away from the base station, thus tracking the actual signal closely. Lau, Cheung, and Chuang[29] report success at reducing the number of hand-offs through a fuzzy algorithm that dynamically adjusts the signal averaging interval and the hysteresis threshold. Gavrilov, Puzikova, and Pyl'kin[30] develop a linguistic model of the communication channel and a fuzzy set theory-based methodology to estimate the current state of a channel. Simulation results indicate that the average time to make decisions about the operability of a channel undergoes significant reduction. Yan[31] recognizes the problem of quantitatively evaluating computer networks, arguing that many of the evaluation elements cannot be quantified and, as a result, evaluation is primarily qualitative. Next, Yan presents three mathematical models, based on the analytic hierarchy process, Grey system theory, and fuzzy mathematics, toward network evaluation. Levin[32] presents a computation and performance analysis of computer networks with nondeterminate, fuzzy time parameters.

9.2 Computationally Intelligent Algorithms for Buffer Management

9.2.1 Buffer Management and the Need for Computational Intelligence

In sophisticated communications networks including the (i) store and for-ward computer networks and (ii) ATM networks, an incoming cell must be first intercepted at a switching node and then processed and forwarded to the subsequent node towards its ultimate destination. The interception and processing invariably requires storage which gives rise to the notion of buffers. In cell switching networks including ATM networks, the burst-iness of input traffic adds complications. Economic reasons dictate net-work designs that cater not to the peak burst rate but some intermediate, probably average, rate of traffic. Clearly, when a burst of cells arrive at a switching node through an incoming link, the latter must temporarily store those cells that it cannot propagate through its outgoing links imme-diately. The reason could either be its inability to process the cells or the inability of the outgoing links to handle the high traffic volume. The stor-age devices are termed buffers and the computational algorithm underly-ing their processing is termed buffer management. While buffers are filled upon the arrival of a large burst of cells, they are cleared when the input rate slows down. If a large burst of cells should arrive when the buffer is already full, one or more cells will have to be dropped, leading to cell loss. Thus, buffer management will be effective in that cell loss may be mini-mized only when the average rates of incoming and outgoing traffic are similar, for a given time period over which the average is computed. Where the average incoming traffic rate of all traffic sources at a switching node is consistently higher than the outgoing rate, buffer management will fail regardless of the size of the buffer, given that networks operate continuously and infinite buffer size is unachievable. Even where the input traffic is uniform, i.e., nonbursty, but the average incoming traffic rate of all the incoming links exceeds the outgoing rate, buffer manage-ment will be rendered ineffective and irrelevant.

Thus, fundamentally, the issue of buffer management arises due to the burstiness of input traffic and the desire to design and operate cost-effective networks with minimal cell loss. The key components of buffer management include the buffer usage policy, the size of the buffer, and the computational algorithm. While increasing the buffer size will very likely imply reduced cell loss, architectural considerations indicate that the delay may be increased leading to performance degradation. Kroner, Hebuterne, Boyer, and Gravey[33] note that interactive data and video communications are examples of bursty traffic. Yegani, Krunz, and Hughes[34] further note that emerging user applica-tions and services including multimedia, LAN interconnections, imaging and graphics are becoming increasingly bursty.

The literature on buffer management is rich. Traditional buffer management approaches for high-speed cell-switching networks utilized statically fixed thresholds in deciding whether to admit or block incoming traffic cells into the buffer from other switches and users. The aim was to strike an efficient balance between the network throughput, i.e., the number of cells that are propagated through the network without being dropped, and the network performance which is measured by the delay incurred in propagating the cells to their destinations. In addition, the traditional approaches proposed the use of space priorities to achieve efficient utilization of the limited buffer space. Given that ATM is increasingly being required to support different traffic classes, the use of priority[35] is natural. Priority schemes involve assigning higher priority to real-time, delay sensitive cells while lower priority is allocated to nonreal-time, delay insensitive cells. As a result, under heavy load, delay sensitive cells are expected to suffer lower cell loss. Rothermel[35] observes that priorities may manifest through time priorities, i.e., to decrease queuing delay of high priority cells, and loss priorities, i.e., to decrease the loss probability of high priority cells. While more complex hardware is required to implement priorities, a 15–20% decrease in buffer space is noted corresponding to the use of space priorities. According to the literature, buffer access or space priority plays a key role in improving network performance and a number of schemes are reported wherein the buffer space is shared between several competing ATM cells. Czachorski, Fourneau, and Pekergin[36] introduce the push-out scheme that consists of pushing out the lower priority cells, when the buffer is full, to create space in the buffer for the higher priority cells. Lin and Silvester[37] present complete buffer sharing with push out and head-of-the-line, partial buffer sharing, complete buffer partitioning yet complete bandwidth sharing, and complete partitioning. Kroner, Hebuterne, Boyer, and Gravey[33] note that there is no performance difference between push-out mechanism and partial buffer sharing. Tassiulas, Hung, and Panwar[38] present a study to determine an optimal method to reduce cell loss rate in an ATM network, especially for loss sensitive cells. Causey and Kim[39] compare the performance of three space priority schemes and report that while complete buffer sharing delivers the highest throughput and lowest loss rate for nonbursty traffic, it suffers from congestion and unfairness corresponding to bursty input traffic. While complete partitioning attempts to be fair, it is inefficient in utilizing the available buffer space. In contrast, the partial buffer sharing scheme strikes a balance between the remaining two schemes. Bhagwat, Tipper, Balakrishram, and Mahapatra[40] show that the complete sharing scheme is highly susceptible under unbalanced loads, a fact that is corroborated by Endo, Ohuchi, Kozaki, Kuwahara, and Mori.[41] Bhagwat, Tipper, Balakrishram, and Mahapatra also show that complete partitioning requires the largest buffer space among all of the schemes. They conclude that restricted buffer sharing offers the best performance through balancing fairness and efficiency. Liew[42] observes that despite restricting the offered load, bursty traffic may degrade the performance of a node in an ATM network. It proposes the use of large buffers or buffer sharing to effectively improve throughput. Yegani, Krunz, and Hughes[34] propose a combination of priority schemes and shared buffers for effective buffer management and claim that the common nested threshold cell discarding with multiple buffers scheme may provide guaranteed service quality through adjusting the parameters of the switching node.

The literature also reports novel algorithms and control schemes for buffer management in ATM networks. Sumita[43] proposes an analytical technique to synthesize control schemes for output buffer management. Badran and Mouftah[44] present a switch design with buffers at both input and output links and claim superior performance over input buffering alone. They note that the burst length severely affects the buffer size required to ensure a given cell loss probability, even with the switch operated at fairly low utilization. Badran and Mouftah[44] do not provide representative values for burst length in the real world. Furthermore, the proper balance in the input and output buffer size appears to be more effective than increasing the total buffer size. Chen and Mark[45] propose sharing the buffer among multiple switch output ports towards effective buffer management. Li[46] proposes a new analytical model for overload control in finite message storage buffers. Eckberg, Luan and Lucantoni[47] propose a core network congestion control, termed "Band-Width Management," wherein cells that correspond to excessive traffic are first tagged selectively based on traffic agreements and real-time traffic monitoring and then discarded where congestion is encountered.

Trajkovic and Halfin[48] note that the buffer size allocation at a node is a function of the bandwidth ratio of the incoming and outgoing links. They observe that the required buffer sizes at different nodes in an ATM network are unexpectedly high for zero loss Quality of Service despite well regulated and conforming input traffic and "Peak Load Allocation." They conclude that leaky bucket policing without additional congestion control mechanisms will fail to provide guaranteed Quality of Service and recommend introducing scheduling algorithms.[49] Kroner, Hebuterne, Boyer, and Gravey[33] suggest increasing the buffer size for higher "load improvement" and claim that future VLSI technologies will offer increased buffer sizes without any increase in the end-to-end delay. From the computer architecture point of view, buffers may be viewed as memory or general purpose registers. They may be organized either as (A) simple queues with head and tail access points, (B) a general structure where the elements may be accessed randomly, and (C) sophisticated queues with random access to every slot in addition to head and tail access points. Figure 9.1 illustrates the three organizations. For large buffer sizes, clearly the delay through the switch is high under organization (A). When organized according to (B), the increased buffer size implies a larger address space which is accompanied by a corresponding increase in the access time. The cause underlying the access time is fundamental and lies in the address decoding time. With organization (C), the random access to any slot in the queue is achieved through pointers which are addresses that must be decoded and are therefore subject to significant access delays. Thus, larger buffers will be invariably associated with greater delay and lowered perfor-

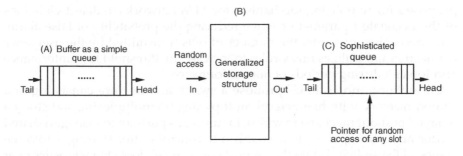

FIGURE 9.1
Buffer organization in switching nodes.

mance. Eckberg, Luan, and Lucantoni[47] corroborate by noting that large and expensive buffers add significantly to end-to-end delays. Zhang[50] stresses the importance of finite buffers in switching nodes and notes that the assumption of infinite buffers is unrealistic. Zhang states that the cell loss probability is only dependent on the ratio of the buffer capacity to the burst length. Bernabei et al.[51] states that accurate dimensioning of network resources, i.e., buffers, coupled with proper call admission and bandwidth assignment rules, constitute effective congestion control in ATM networks.

Buffer Management schemes without cell discard are able to remove short blocking periods, thus realizing low packet loss. However, they are ineffective for longer blocking periods and suffer from high packet loss rates.[46] Furthermore, schemes that utilize cell priorities and thresholds to determine cell discard or admission, also suffer from the limitations that they are generally unable to accurately and dynamically determine the effective threshold settings. While the aim is always to achieve a desired tradeoff between the number of cells carried through the network, propagation delays of cells, and the number of discarded cells, the fixed threshold approach has shown to be very restrictive and abrupt.[9] Bonde and Ghosh,[9] propose the notion of cell-blocking that utilizes fuzzy thresholding based on Zadeh's fuzzy set theory. Their comparative study reveals that fuzzy queue management adapts to sharp changes in cell arrival rates and maximum burstiness of bursty traffic sources, yielding lower cell discard rates. This fuzzy threshold approach[9] has introduced a buffer management scheme which exhibits a softer behavior, and has displayed greater ability to adapt to dynamic condition changes in network traffic.

Cheung and Chang[18] present a fuzzy traffic controller that manages both congestion and call admission control for ATM networks. The controller is a fuzzy implementation of the two-threshold congestion control and the equivalent capacity admission control methods where the parameters of the membership functions and the fuzzy control rules are extracted through clustering analysis of analytical data. They report a 11% improvement in system utilization while maintaining the QoS contract comparable with that of the conventional equivalent capacity method. Catania, Ficili, Palazzo, and Panno[19] propose a fuzzy policing mechanism for ATM networks to detect violations of the negotiated parameters while reducing the probability of false alarm. The mechanism monitors the number of cells transmitted by the user since connection and utilizes fuzzy rules to increase the threshold for conformance and lower it during periods of nonconformance.

Choudhury and Hahne[52] propose a switch architecture consisting of a shared memory with hierarchical multiplexing/demultiplexing and study a delayed pushout mechanism which, in turn, uses pushout to manage a shared buffer and a backpressure mechanism to control buffer sharing across the stages of the switch. While they report superior cell loss characteristics over the competitors, their analysis does not include the fuzzy logic-based buffer management techniques, published in the literature.

Both flow control mechanisms[53] and buffer management schemes share the same objective, namely to improve network performance. However, their differences, although subtle, are real. The credit based flow control approach,[54, 55] for instance, consists of a receiver generating "credits" based on the queue lengths and the sender responding by propagating an appropriate number of cells. Although credit generation is dependent on buffer occupancy, the primary aim[53] is to maximize the flow of cells from the sender to the receiver. In contrast, buffer management is concerned primarily with minimizing buffer overflow and cell loss.

The limitations with most of the traditional approaches, reviewed here, are twofold. First, they are susceptible to cell loss due to buffer overflow. Second, the results obtained are limited to a single switch and, thus, the techniques may not be generalized to a realistic switching network composed of multiple switches. While the results have focused on the cell loss, the impact of the buffer management strategies on the end-to-end cell delay characteristic—a key issue in performance, has never been reported for any of the traditional approaches. The remainder of this section is organized as follows. Section 9.2.2 presents a buffer management scheme in which fuzzy thresholding constitutes the computational intelligence while Section 9.2.3 details a competing approach where a computationally intelligent algorithm, GNCD, manages the buffers efficiently.

9.2.2   A Fuzzy Thresholding Scheme with Selective Blocking

This section introduces the notion of cell-blocking, wherein a fuzzy thresholding function, based on Zadeh's fuzzy set theory, is utilized to deliberately refuse entry to a fraction of incoming cells from other switches. The blocked cells must be rerouted by the sending switch to other switches and, in the process, they may incur delays. The fraction of blocked cells is a continuous function of the current buffer occupancy level unlike the abrupt, discrete thresholds in the traditional approaches. The thinking is that binary thresholds are excessively restrictive and that fuzzy thresholds will cause the buffer queue to exhibit "soft" behavior, i.e., greater ability to adapt to dynamic conditions, as opposed to the traditional, inflexible, binary thresholds. Also, fuzzy thresholds are expected to exhibit robustness, such as resilience to rapid dynamic changes in network traffic and favorably impact cell discarding. The fuzzy cell-blocking scheme is simulated on a computer and the simulation results for a given, realistic traffic stimulus, characterized by Poisson arrivals and exponentially distributed departures, are contrasted against those from the traditional, fixed thresholding, approach. A comparative analysis reveals that fuzzy queue management adapts superbly to sharp changes in cell arrival rates and maximum burstiness of bursty traffic sources, yielding lower cell discard rates, high throughput of cells through the network, and lower cell-blocking rates. The remainder of this subsection is organized as follows. Section 9.2.2.1 introduces the notion of cell-blocking and contrasts fuzzy threshold against fixed threshold schemes. Section 9.2.2.2 details the design of experiments to evaluate and contrast the two schemes. The experiments are simulated on a computer and the simulation results are analyzed to estimate the impact of the schemes on the performance of the cell-switching network.

9.2.2.1   Cell Blocking Strategies

The traditional notion of blocking,[56] defined in the context of circuit switched telephone networks, refers to the scenario when a telephone call cannot be completed due to resource limitations. Li[46] extends the notion of blocking to refer to the queue-full scenario when ATM voice packets are blocked. In contrast to this classical view of blocking as a negative artifact of queue management, this chapter introduces the notion of selective cell-blocking, where a switch, utilizing certain criteria and under specified conditions, deliberately refuses entry to one or more cells from a source switch, as opposed to all cells of a specific message, into the buffer queue. Here, a message consists of a number of individual cell-bursts where a cell-burst contains a variable number of cells. Where a cell-burst from a source switch, consisting of "N" cells, desires entry into a switch buffer, the switch may selectively block or admit this cell-burst or a fraction of the "N" cells. The remainder of the "N" cells must be routed by the source switch to another destination switch and, in the process, these cells may incur delays. In this spirit, the notion of cell-blocking combines potentially useful elements of both admission control and smoothing functions.

It may be observed that selective cell-blocking impacts the traditional view of virtual circuits. In particular, ATM networks, standardized for broadband-ISDN,[57] utilize a virtual-circuit connection-oriented switched architecture. This chapter proposes to expand the scope of the standard ATM architecture to include a selective cell-blocking option. Evidently, this feature implies some level of message fragmentation, may require the system to reorder cells at the destination, and raises the issue of algorithms to route blocked cells. However, given the focus on thresholding, these issues, as well as the extent and impact of fragmentation, are beyond the scope of this section and the reader is referred to Reference 1.

The threshold function determines, for each cell-burst, how many of the arriving cells to admit into the buffer. This function bears significant influence on the performance of the network including the fraction of cells lost due to dropping or excessive delays and the delay distribution of the cells. The traditional "fixed" scheme utilizes a binary threshold: admit or no-admit, depending on the occupancy of the buffer. In the fuzzy scheme, the admittance and no-admittance decision is based on a continuous threshold function, detailed subsequently. In both cases, admission or refusal to admit cells is expressed by a simple control rule:

> if (buffer is getting FULL)
>> refuse admittance to incoming cells
> else admit incoming cells

For both "fixed" and fuzzy schemes, only those cells that are already in the buffer are subject to being dropped when the buffer overflows. Conceivably, the choice of thresholds, relative to the maximum burst rate, in both schemes may be such that no cell is ever dropped. Such choices, however, are observed to constitute either poor buffer utilization or cause inefficient delays due to increased routing.

The dual issues of defining "getting FULL" and the manner in which it leads to a blocking decision are addressed subsequently.

9.2.2.1.1   Fixed Thresholds

The concept of fixed thresholds is not new[58, 59] and is described here only for the purpose of contrasting it with the use of fuzzy thresholding functions. In general, the choice of a threshold directly impacts the balance between the network throughput, i.e., the number of cells that are propagated through the network without being dropped, and the network performance, i.e., the delay incurred in propagating the cells to their destinations. For binary thresholds, this choice is especially critical since the choice is represented through a single value. For a very "conservative" choice, i.e., new cells are admitted only when the buffer occupancy is very low, say 25% of capacity, the utilization of the buffer may be very poor. That is, the buffer although largely unfilled, will mostly reject entrance to new cells. In contrast, a "liberal" threshold, say 90% of capacity, may cause severe problems under "bursty" traffic scenario, wherein the buffer may fail to "absorb" the sudden influx of cells.

The choice of binary thresholds is further aggravated by bursty traffic in two ways. By definition, a bursty traffic scenario implies that cells may be generated at near-peak rate for a while followed by zero cell generation. First, the value of the threshold must be determined *a priori*, i.e., before the occurrence of the events. Events are dynamic, they occur extremely fast, and manually determining and setting threshold values before every event is impractical. Second, the number of cells asserted at a switch is dynamic and

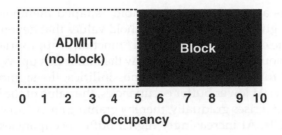

FIGURE 9.2
Conceptual partitioning of the buffer created by binary threshold.

adjusting the threshold value corresponding to every event is not pragmatic. An event corresponds to the generation and assertion of a cell-burst at a switch.

The second reason also implies that the switch must use the knowledge of the maximum "burstiness," i.e., the maximum number of cells possible within a cell-burst, to determine a threshold value that eliminates cell discard. For instance, consider a 10 cell-capacity buffer with a binary threshold value of 50%. Clearly, this scheme is "safe," i.e., it will absorb a cell-burst, where the burstiness ranges from 0–5 cells. For larger cell-bursts, the buffer may overflow and discard cells. The concern for not dropping cells has inevitably injected inefficiency into the approach.

Conceptually, a binary threshold partitions the buffer into two states: admit (no block) and block, as shown in Figure 9.2. Where the occupancy level ranges from 0 through 50% at a simulation time instant, all cells of a cell-burst are admitted. However, where the occupancy level exceeds 50%, all cell bursts are refused entry. The sharp change in the decision from admit to block for a change of occupancy level from 49% to 51%, in this approach, is very abrupt. It appears counter-intuitive that the abrupt swing in the decision making for a small change in the buffer occupancy level of 2%, i.e., from 49% to 51%, is realistic and logically sound.

9.2.2.1.2 Fuzzy Thresholds

Fuzzy sets[2] were introduced in the literature to explicitly deal with the "imprecisely defined classes" that one encounters in the real physical world, and to replace binary class assignments with continuous set membership. Here, to develop a more realistic view of buffer occupancy and to provide a robust decision making, the traditional notion of "FULLness" which leads to the binary decision classes, admit and block in Figure 9.2, are replaced respectively with two fuzzy states—"getting FULL" and "not getting FULL." This view features continuous set membership and implies a soft, gradual transition between completely "FULL" and completely "not FULL" (empty). The membership values are defined through a function and specific values are represented through the symbol, μ_{FULL}. In contrast to the use of triangular or trapezoidal symmetric membership functions in many fuzzy systems, this section utilizes an asymmetric "sigmoid"-shaped membership function which, in turn, gives rise to fuzzy threshold values that depend on the occupancy level. The choice of the "sigmoid" function is appropriate since it naturally captures our own intuition of how the buffer fills up. While triangular or trapezoidal functions are definite possibilities, the sigmoid function is more expressive. At lower buffer occupancies, the value of the blocking function is small and it rises gradually, thereby granting entry to a greater fraction of incoming cells. At increasingly higher buffer occupancies, the function rises more steeply through a "transition region." At high occupancy levels, the function again rises gradually but its value is large, thereby permitting entry to a smaller fraction of incoming cells.

Given that the cell-blocking strategy is expressed by a simple, single antecedent (if the buffer is getting FULL), single consequent (then block) control rule, the degree to which the buffer is full directly controls the level of cell-blocking. This simple rule results in a continuous-valued cell-blocking strategy, as shown in Figure 9.3. For instance, at 30% occupancy level, the value of μ_{FULL} is approximately 0.1, which results in a cell-blocking value of 0.1 or 10%. That is, 10% of the arriving cells would be refused entry into the buffer and they must be rerouted by the sending switch to other switches. Note that the degree of admittance given by (1 − blocking) is 90%. Thus, unlike the "fixed" cell-blocking scheme, represented by a single threshold value, X, and expressed through "fixed at value X," the fuzzy threshold function covers a specified range of buffer occupancy values. In the remainder of this section, while fixed cell-blocking is represented through "fixed at value X," fuzzy cell-blocking is represented by "fuzzy (A, B)," where A and B are respectively the "zero-valued" and "unit-valued" end points.

9.2.2.2 Performance of Fuzzy Thresholding

This section reports the design of simulation experiments to evaluate the "fixed" and fuzzy cell-blocking schemes under simulated congested traffic conditions. The measures, selected for the evaluation, include the (i) throughput, i.e., the number of cells serviced by the buffer, (ii) the number of cells refused entry into the buffer, and (iii) the number of cells dropped due to buffer overflow. The experiment assumes a finite-sized buffer of capacity 8 cells. Should the number of cells in the buffer equal or exceed 9, the excess cells are discarded. Otherwise, cells are not lost. The measures are obtained for (i) two sets of traffic volumes, 500 cells and 5000 cells, (ii) a number of fixed blocking schemes with different values of binary thresholds at 2, 3, and 4, respectively, (iii) six sets of traffic with maximum burstiness values of 5, 6, 7, 8, 9, and 10, respectively, and (iv) three sets of traffic with different mean arrival rates, $\lambda = 4$, $\lambda = 2$, and $\lambda = 1$, respectively. The mean service rate, ν is assumed equal to 1.

The simulations utilize a single-server queueing model[60] with the input traffic stimulus characterized by Poisson arrivals, mean arrival rate given by

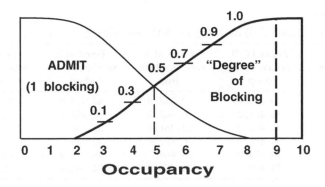

FIGURE 9.3
Fuzzy (2,8) cell-blocking scheme: Asymmetric, continuous set membership.

λ, and exponentially distributed departures, mean service rate given by v. Burstiness in the input traffic stimulus is modeled by generating a random number, N, between 0 and a user-specified maximum, and permitting N cells to be asserted at the buffer. The user-specified maximum is at least equal to or greater than 50% of the buffer capacity, i.e., 4. Corresponding to an event, the blocking scheme, in effect, will cause a fraction of the cells asserted to the buffer to be admitted into the queue with the remaining cells being blocked. The throughput is measured by incrementing the "service" counter each time a cell departs from the queue. The software is written in C and executed on a Macintosh IIci.

Tables 9.1 and 9.2 record the simulation results for 500 cells, $v = 1$, fuzzy threshold function with end points at occupancy levels 1 cell and 7 cells respectively, and maximum burstiness values ranging from 5 through 10. Tables 9.1 and 9.2 present three measures: number of cells serviced, number of cells blocked, and number of cells dropped, for three fixed blocking schemes, with binary thresholds at 4, 3, and 2, respectively and one fuzzy blocking scheme. Binary threshold choices of 5 and higher are observed to be too restrictive and are, therefore, not pursued. While Table 9.1 corresponds to faster arriving cells, given by a mean arrival rate of $\lambda = 4$, Table 9.2 relates to slower arriving cells, given by $\lambda = 2$. It may be noted that the total number of cells serviced, blocked, and dropped adds to slightly less than 500 in each case and this refers to a few cells that remain in the buffer at the completion of an experiment. The high percentages of blocked cells in Tables 9.1 and 9.2 do not imply inefficiency in queue management. It merely reflects a deliberate choice of parameters to create extreme congestion and contrast the performance of the two schemes.

As expected, with an increase in the maximum burstiness from 5 through 10, the three fixed blocking schemes drop more and more cells. The loss is reduced by successively selecting conservative thresholds, from fixed at 4 to fixed at 3, to fixed at 2. At the same time, the buffer is increasingly sparsely populated causing a decrease in the throughput, increase in blocking, and

TABLE 9.1

Fuzzy vs. "Fixed" Cell-Blocking for 500 Cells with $\lambda = 4$, $\nu = 1$

Maximum burst	Fuzzy (1, 7) service, block, drop	Fixed at 4 service, block, drop	Fixed at 3 service, block, drop	Fixed at 2 service, block, drop
5	40, 455, 0	40, 458, 0	40, 459, 0	40, 460, 0
6	36, 461, 0	36, 462, 2	36, 463, 0	34, 463, 0
7	32, 463, 0	32, 461, 5	32, 464, 0	32, 465, 0
8	30, 466, 0	30, 462, 6	30, 464, 2	30, 466, 0
9	26, 474, 0	26, 468, 7	26, 474, 2	26, 476, 0
10	24, 469, 0	24, 459, 9	24, 463, 6	24, 467, 2

TABLE 9.2

Fuzzy vs. "Fixed" Cell-Blocking for 500 Cells with $\lambda = 2$, $\nu = 1$

Maximum burst	Fuzzy (1, 7) service, block, drop	Fixed at 4 service, block, drop	Fixed at 3 service, block, drop	Fixed at 2 service, block, drop
5	80, 413, 0	80, 415, 0	79, 419, 0	73, 425, 0
6	68, 427, 0	68, 427, 3	67, 430, 0	63, 434, 0
7	63, 430, 0	63, 428, 5	63, 426, 6	62, 432, 0
8	58, 435, 0	58, 428, 10	58, 430, 6	56, 440, 0
9	56, 436, 2	56, 424, 15	55, 428, 13	55, 440, 4
10	54, 440, 5	54, 426, 19	54, 432, 13	54, 443, 6

consequent increase in the delay of cells. In contrast, the fuzzy scheme appears relatively unaffected even as the maximum burstiness is increased. The fuzzy approach drops fewer cells than any of the fixed thresholds, yet achieves a level of throughput equal to the most liberal fixed blocking scheme—fixed at 4. In addition, a decrease in the traffic arrival rate, from $\lambda = 4$ (Table 9.1) to $\lambda = 2$ (Table 9.2) is accompanied by an increase in the number of cells serviced and reduction in blocking.

To explore the trends exhibited in Tables 9.1 and 9.2 in greater detail, three experiments are simulated, each for a total of 5000 cells. Two maximum burstiness values are utilized—nominal, represented by value 5, and the extreme case, represented by value 8. Tables 9.3, 9.4, and 9.5 record the simulation results for three scenarios—fuzzy (1, 7), fixed at 3, and fixed at 2, and correspond respectively to arrival rates given by $\lambda = 4$, $\lambda = 2$, and $\lambda = 1$. It may be observed that the fuzzy (1, 7) scheme exhibits higher throughput, lower blocking, and reduced cell discard, relative to the fixed blocking schemes. In 17 of the 18 experiments, fuzzy(1, 7) scheme results in fewer cell discards. As expected, a decrease in the cell arrival rate, from $\lambda = 4$ to $\lambda = 2$ to $\lambda = 1$ is associated with increased number of cells serviced and reduced number of cells blocked.

Clearly, within the scope of the experiments, the fuzzy cell-blocking scheme appears to adapt better than traditional, fixed, cell-blocking schemes to changes in maximum burstiness, cell arrival rates, and traffic volume.

TABLE 9.3

Fuzzy vs. "Fixed" Cell-Blocking for 5000 Cells with $\lambda = 4$, $\nu = 1$

Maximum burst	Fuzzy (1, 7) service, block, drop	Fixed at 3 service, block, drop	Fixed at 2 service, block, drop
5	425,4568,0	425,4572,0	418,4580,0
8	269,4728,0	269,4719,13	264,4731,2

TABLE 9.4

Fuzzy vs. "Fixed" Cell-Blocking for 5000 Cells with $\lambda = 2$, $\nu = 1$

Maximum burst	Fuzzy (1, 7) service, block, drop	Fixed at 3 service, block, drop	Fixed at 2 service, block, drop
5	850, 4143, 0	837, 4158, 0	801, 4200, 0
8	548, 4449, 0	532, 4443, 28	513, 4478, 3

TABLE 9.5

Fuzzy vs. "Fixed" Cell-Blocking for 5000 Cells with $\lambda = \nu = 1$

Maximum burst	Fuzzy (1, 7) service, block, drop	Fixed at 3 service, block, drop	Fixed at 2 service, block, drop
5	1657, 3338, 0	1526, 3472, 0	1438, 3559, 0
8	1117, 3860, 20	1054, 3907, 37	998, 3981, 14

Figure 9.4 presents a graphical view of buffer occupancy, blocking function, and cell arrivals, as functions of simulation time, for a limited length of the entire simulation run. The schemes presented include fuzzy(1, 7) and fixed at 3. Figure 9.4(c) represents the stochastic distribution of the discrete number of cells asserted to the buffer. For this given distribution, unlike the graph corresponding to fixed blocking in Figure 9.4(a) which reveals sharp spikes, the graph corresponding to fuzzy blocking varies much less abruptly. In contrast, the fixed blocking scheme only admits cells if the buffer occupancy is less than three, and "oscillates" abruptly, causing both overflow and missed service opportunities. Figure 9.4(a) reveals instances of cell discard where the graph corresponding to fixed blocking equals 9. Also shown on Figure 9.4(a) is an instance of missed service when the occupancy, for the case of fixed blocking, falls to 0 just prior to a service time. Figure 9.4(b) presents the variation of the blocking function values, as a function of simulation time, for the fuzzy and fixed at 3 schemes. Once again, while the graph for fixed blocking swings abruptly between 0 and 1, that for fuzzy blocking varies relatively more smoothly. The observed superior performance of the fuzzy scheme agrees with common intuition. The fuzzy scheme more naturally captures the notion of fullness through the continuous-valued blocking function control.

9.2.3 A Guaranteed-No-Cells-Dropped Buffer Management Scheme

9.2.3.1 *A Guaranteed-No-Cells-Dropped (GNCD) Buffer Management Approach*

To quickly recapitulate, the principal weakness of the traditional fixed threshold scheme lies in the lack of a mechanism to determine the value of the threshold objectively and that the statically predetermined value fails to adapt to dynamic situations. The result is a combination of (1) cells discarded, i.e., refused admittance into the buffer, (2) cells dropped in the buffer due to buffer overflow, and (3) delays. In contrast, the fuzzy thresholding approach[9] constitutes, in essence, an adaptive approach wherein the threshold is adjusted continuously to reflect the dynamic state of the buffer. In Reference 9, the switching node first records the current occupancy level of the buffer and then computes a percentage value utilizing the fuzzy function, regardless of the size of the incoming cell-burst. Next, the percentage value is multiplied with the size of the cell-burst to determine the fraction of the cells admitted into the buffer, while the remainder of the cells are blocked selectively. Although the fuzzy thresholding approach adapts superbly to sharp changes in cell arrival rates and maximum burstiness of bursty traffic sources, yielding lower cell loss due to buffer overflow, it suffers from the weakness that, at times, one or more cells may be admitted into the buffer causing buffer overflow and consequent cell drop.

This section observes that at the time of computing the percentage value, while the current buffer occupancy is taken into consideration, the computation is unaware of the size of the input cell-burst. However, in the subsequent determination of the number of cells of the cell-burst that are blocked, the switch requires knowledge of the size of the cell-burst. This section argues that a combination of the knowledge of the current buffer occupancy and the size of the input cell-burst is likely to imply superior performance and proposes a new scheme, GNCD, wherein the buffer occupancy is first recorded and the number of empty buffer slots computed immediately. Then, an exact number of cells from the cell-burst are admitted that equals the number of empty slots. The remainder of the cells of the cell-burst are blocked at the sending switch or user. Thus, the total number of cells subject to blocking under GNCD correspond to the cells that are discarded plus the cells lost due to buffer overflow in the fixed thresholding scheme. The fraction of the blocked cells in GNCD that corresponds to that discarded in the fixed scheme, is caused by excessive traffic and its mitigation is beyond the scope of this chapter. While both the GNCD and fuzzy scheme incurs blocking, given that GNCD eliminates cell loss due to buffer overflow, it is logical to infer that in the worst case, the fraction of cells that are lost due to buffer overflow under fuzzy thresholding may be manifest as additional blocked cells in GNCD. These cells are rerouted by the sending switch to other switches and, in the process, they may cause the cells in the network to incur additional delays. Thus, the admission of cells of a cell-burst arriving at a switching node is defined by the absolute available space in the buffer. While it is computationally simple and fast, this scheme

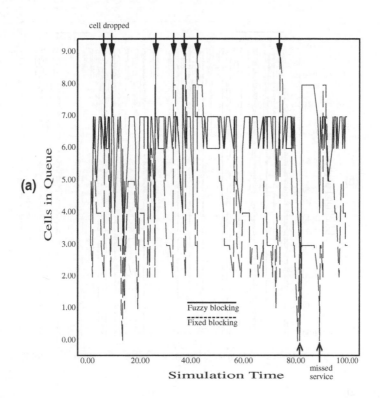

FIGURE 9.4a
Buffer occupancy (fuzzy vs. "fixed at 3" cell-blocking).

also guarantees the absence of cell loss due to buffer overflow which may be potentially invaluable to certain ATM traffic classes.

The key advantages of GNCD over fuzzy thresholding and other traditional approaches reported in the literature are two-fold. First, the computation involved in the decision to admit or refuse entry into the buffer is simple, fast, and relatively inexpensive which is critical in high-speed cell-switching networks. Second, GNCD eliminates cell loss due to buffer overflow, completely. Intuition, however, dictates that since the buffer utilization will be very high in GNCD, the cell delay in the network is likely to be significantly high. The design of experiments to uncover GNCD's true behavior is detailed in the subsequent sections.

9.2.3.2 Modeling GNCD for a Large-Scale Cell-Switching Network and Comparative Performance Analysis

9.2.3.2.1 Modeling GNCD for a Large-Scale Cell-Switching Network

In this research, the GNCD scheme is also modeled for a 50-switch, worldwide, representative cell-switching network,[61] shown in Figure 9.5. In the network, the node names correspond to major world cities and are expressed

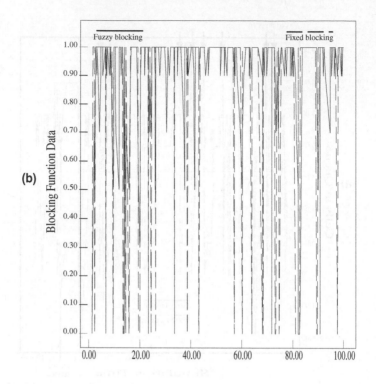

(b)

FIGURE 9.4b
Cell-blocking function (fuzzy vs. "fixed at 3" cell-blocking).

through acronyms. Associated with every link is the value of the propagation delay, computed from dividing the actual distance between the corresponding cities by the speed of electromagnetic transmission in optical fibers. The second quantity, associated with every link, refers to its capacity as a multiple of the basic 155.52 Mbits/s ATM link, and the capacity of a link between any two cities is derived based on the respective populations and subject to a normalizing factor. Every switch is characterized by a buffer which is governed by the GNCD buffer management scheme. The actual arrival of the cells of a cell burst at a switch is preceded by negotiation. The receiving switch measures the current buffer occupancy, determines the number of cells of the incoming burst that it may accept, and propagates it to the sending switch or user. In response, the sending switch or user transports the appropriate number of cells to the receiving switch and the remainder of the cells of the cell burst are considered blocked. Since the number of cells that is admitted into the buffer space is based on the exact available space in the buffer, no cell will be lost due to buffer overflow. Cell loss due to buffer overflow constitutes the most serious issue in buffer management since these cells may never be recovered. As with fuzzy thresholding, the GNCD approach incurs a two way exchange of information between the sending and receiving switches, which is likely to adversely impact on the throughput and delay characteristics.

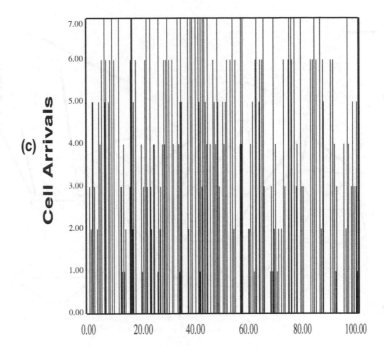

FIGURE 9.4c
Cell arrivals as a function of simulation time, for example traffic scenario with $\lambda = 2$, $\nu = 1$, maximum burstiness = 7.

At any time instant, the buffer may contain a number of cells destined for different outgoing links. The switch extracts each cell from the buffer and propagates it to its destination along the appropriate outgoing link when the latter is free. When an outgoing link is busy, the cells in the buffer, destined to travel along that link, may not be propagated immediately and will incur delays. Clearly, the utilization of a specific link is influenced by its capacity relative to the traffic scheduled for transport through it.

To achieve a systematic comparison of the GNCD and fuzzy thresholding schemes, this research also models the latter for the representative switching network. In the fuzzy threshold scheme, although the sending switch or user negotiates with the receiving switch, as in GNCD, cells may be lost due to buffer overflow. This section argues, as stated earlier, that, conceivably, the cost of eliminating cell loss due to buffer overflow in GNCD, unlike in the fuzzy scheme, may consist in increased selective blocking at the sending switch or user. Should GNCD experience higher throughput or lower blocking, relative to the fuzzy scheme, it will still be assumed that GNCD incurs additional blocking, equivalent to the fraction of the cells lost due to buffer overflow in the corresponding fuzzy scheme. This fraction is referred to as "special blocked" cells. Assuming that a cumulative total of D cells in the network are lost due to buffer overflow in the corresponding fuzzy thresholding scheme, it is hypothesized that the superiority of the GNCD scheme

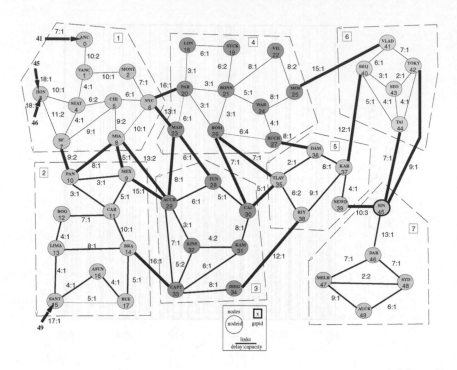

FIGURE 9.5
A 50-switch, representative cell-switching network.

is demonstrated only when D cells of the total number of blocked cells at the sending switches or users are successfully rerouted to their respective destinations. For every user traffic, asserted at a switch, the cell-switching network computes a virtual path from the origin to the destination switch, that constitutes the route for every cell burst of the traffic. The virtual path computation employs Dijkstra's shortest path algorithm which, in turn, utilizes the link propagation delays to constitute the cost function. In addition, the network computes a second, usually the next-best, virtual path for every possible node pair, to be used to reroute the "special blocked" cells.

9.2.3.2.2 Implementation Issues

The GNCD buffer management model of the representative cell-switching network is implemented in C and simulated on a network of 25+ Pentium workstations under linux, configured as a loosely-coupled parallel processor. The implementation constitutes an accurate, large-scale, asynchronous, distributed simulation that occupies 6000 lines of C code and executes for 12–14 hours of wall clock time. That is, the longest running of all of the processors requires 12–14 hours of wall clock time. A total of over 150 simulation runs are executed for different input parameters including traffic distributions. The cumulative size of the output data generated for each of the 50 switches, following a simulation run, approximates 6.6 Mbytes. A uniprocessor parser program, 200 lines of C code, processes the data to generate performance results that are presented in Section 9.4. The parser program requires approximately 1–3 hours of execution time.

The accuracy of the simulation is reflected by the fact that the value of the timestep, 2.53 μs, is determined by the fastest link in the network namely, 155.5 Mb/s, the switches process every individual cell, and that the underlying discrete event simulation algorithm ensures the correct order of execution of the events. In addition, the input traffic distributions utilized here correspond to the current traffic models in the literature, underscoring the significance of the results. In theory, users may assert traffic into the network at every switch. However, initial experiments reveal that when traffic, subject to the selected traffic model parameters, is asserted at every switch, the result is excessive congestion with extremely low, 15–20%, throughput. Since real-world, operational networks are likely to operate at throughput levels in the range 80–90% or higher, in this study, traffic is asserted into the network at arbitrarily selected switches. However, all of the switches in the network may participate in the routing of the traffic cells.

9.2.3.2.3 *Traffic Models and Input Traffic Distributions*

In this study, traffic distributions are generated utilizing three traffic models, one for each of video, audio, and data. The net traffic asserted at the switches reflect a stochastic combination of the three traffic models achieved through multiplexing. While the traffic models have been published in the literature, they are significant in that their parameters are derived from actual traffic. The video traffic generator utilizes the parameters provided in Reference 62 that, in turn, are derived from the movie, "Last Action Hero," compressed using MPEG-1 algorithm. A total of 238,626 frames of dimension 320 × 240 pixels, resolution 8 bits/pixel, and a frame rate of 30/s constitute the net duration of 2 h, 12 m, and 36 s of the movie. A total of 2000 frames are selected for this study and the choice of the frames correspond to continuous scene changes, implying frames of different sizes and, therefore, highly bursty traffic. The maximum burst length is set at 243 cells. The audio traffic generator utilizes an ON/OFF model with the key parameters derived from actual voice traffic.[63] The mean of the active period interval (ON) is 352 ms while that of the silent period interval (OFF) is 650 ms. The call holding time is 90 s, the intercell duration within the ON period is 2112 μs, and the maximum burst length is 1420 cells. The data traffic generator utilizes an ON/OFF Markov chain model, subject to traffic shaping.[64, 65] The maximum burst length is 64 cells.

9.2.3.2.4 *Simulation Results and Comparative Performance Analysis*

In this study, the simulation experiments are organized into two parts. Part 1 focuses on simulating GNCD in the absence of rerouting for the 50-switch network, measures the throughput, cell loss from buffer overflow, blocking, and end-to-end delay characteristics, and compares them with the corresponding

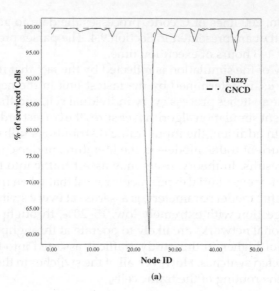

FIGURE 9.6a
Cumulative number of serviced cells.

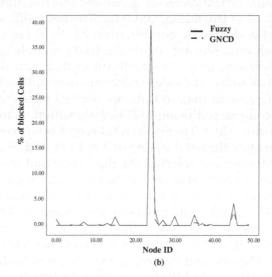

FIGURE 9.6b
Cumulative number of blocked cells.

measures from fuzzy thresholding, for different input traffic distributions. Part 2 of the study will be elaborated later in this section.

Given that the maximum burst length is approximately 1500 cells, the buffer size for both GNCD and fuzzy schemes is set at 1500. For the fuzzy scheme, a sigmoid function[9, 66] is utilized where the abruptness or slope parameter is set at 10.5 and the displacement parameter value at 5.4.

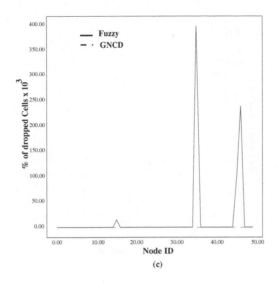

FIGURE 9.6c
Cumulative number of cells dropped, for GNCD and fuzzy schemes, in percentage, as a function of the switching nodes, for moderate traffic load.

Under Part 1, the first experiment utilizes a traffic distribution character-ized by moderate input cell arrival rate and a total of 1 million cells, which yields the net throughput for the network at 88.1% for GNCD and 87.09% for fuzzy. Figure 9.6(a) presents the cumulative number of serviced cells, in percentage, as a function of each of the 50 switching nodes. Clearly, the throughput is slightly higher for the GNCD scheme, even at the individual switches.

The total number of blocked cells in the entire network equals 129,078 for GNCD and 139,069 for the fuzzy scheme. Figure 9.6(b) presents the cumula-tive number of blocked cells, in percentage, as a function of the switching nodes. The reason underlying a lower value for GNCD is that the utilization of the buffers is very high, implying higher throughput and lower blocking. Nevertheless, as hypothesized earlier, the "special blocked" cells constitute a subset of the total number of blocked cells under GNCD. The total number of cells dropped in the entire network equals 339 cells for the fuzzy scheme. Fig-ure 9.6(c) presents the cumulative number of cells dropped, in percentage, as a function of the 50 switching nodes. Clearly, cell loss is nil in the GNCD scheme.

In each of Figures 9.6(a), 9.6(b), and 9.6(c), for node 24, while the percentage of cells serviced is relatively lower, the percentage of cells blocked and dropped are higher. The underlying reason is that node 24 encounters a sig-nificant number of cells, relative to other nodes in the network, as reflected in Figure 9.7(a). Figure 9.7(b) presents the buffer occupancy as a function of sim-ulation time, for a select set of nodes—24, 11, and 35, from different peer groups 4, 2, and 5, respectively, and corroborates the reasoning.

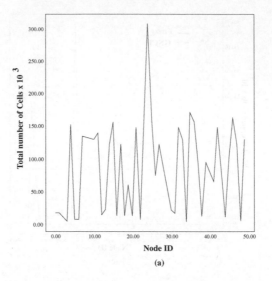

(a)

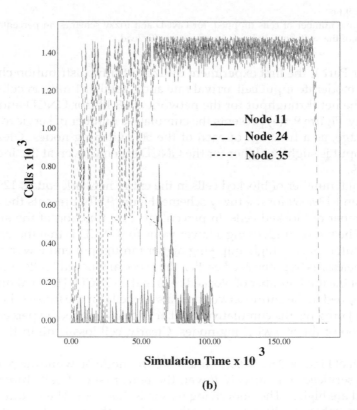

(b)

FIGURE 9.7
(a) Total number of cells intercepted by the switching nodes, (b) buffer occupancy, in cells, as a function of simulation time for switching nodes 11, 24, and 35.

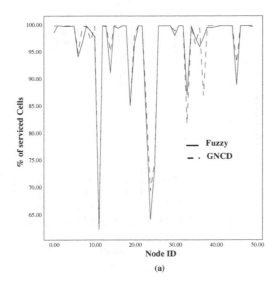

(a)

FIGURE 9.8a

Cumulative number of serviced cells.

The second and third experiments under Part 1 utilize two different traffic distributions. The aim is to examine the comparative performance of GNCD versus fuzzy scheme, for different traffic loads. Experiment 2 utilizes a high cell arrival rate that constitutes a very heavy traffic load with 1.5 million cells and yields a throughput of 73.99% and 73.76% for GNCD and fuzzy thresholding, respectively. Experiment 3 utilizes a low input cell arrival rate constituting a less dense traffic with less than 1 million cells.

Figures 9.8(a), 9.8(b), and 9.8(c), present the throughput, cell-blocking, and cell-dropping characteristics for GNCD and fuzzy schemes, under heavy traffic load. For the 50-switch network, the total number of cells lost due to buffer overflow equals 1457 for the fuzzy scheme and 0 for the GNCD. In addition, relative to the fuzzy approach, while the throughout is higher, blocking is lower for the GNCD. The significant increase in congestion is reflected in Figure 9.9 that plots the buffer occupancy distribution as a function of time for the same set of switches—11, 24, and 35.

The third experiment under Part 1 yields 91.9% and 91.35% throughputs for the GNCD and fuzzy scheme and Figures 9.10(a), 9.10(b), and 9.10(c) present the throughput, cell-blocking, and cell-dropping characteristics. The number of cells lost due to buffer overflow equals 267 for the fuzzy approach.

The results of experiments 1 through 3 under Part 1 clearly imply that for a representative, 50-switch network and under different traffic loads, ranging from light to moderate to heavy, despite eliminating cell loss due to buffer overflow, GNCD consistently yields higher throughput and lower cell blocking, relative to the fuzzy approach. The difference of throughput between GNCD and the fuzzy scheme, shown in Figure 9.11(a), is highest for moderate input cell arrival rate and decreases for both higher and lower input cell

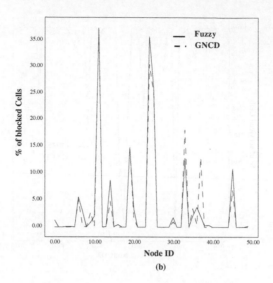

FIGURE 9.8b
Cumulative number of blocked cells.

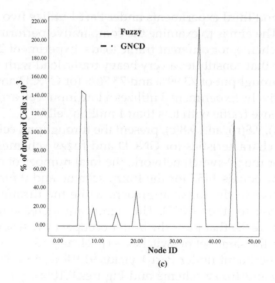

FIGURE 9.8c
Cumulative number of cells dropped, for GNCD and fuzzy schemes, in percentage, as a function of the switching nodes, for heavy traffic load.

arrival rates. Evidently, at low traffic density values, the buffer is not subject to strain and the behaviors of both approaches are similar. Under very high traffic densities, the buffer occupancy is consistently very high for both approaches and GNCD's insistence on eliminating cell loss due to buffer

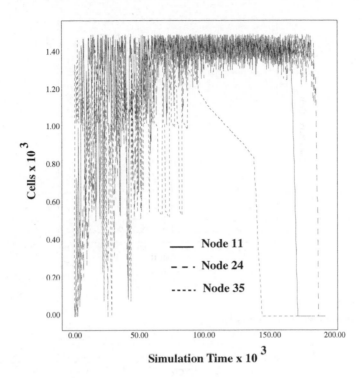

FIGURE 9.9
Buffer occupancy, in cells, as a function of simulation time for switching nodes 11, 24, and 35.

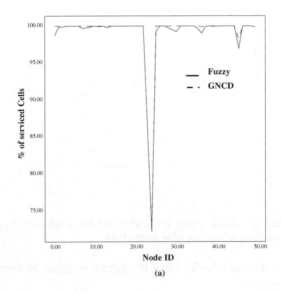

FIGURE 9.10a
Cumulative number of serviced cells.

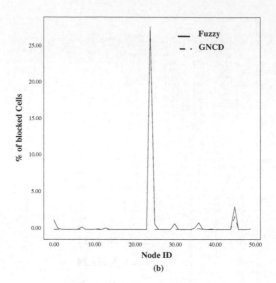

FIGURE 9.10b
Cumulative number of blocked cells.

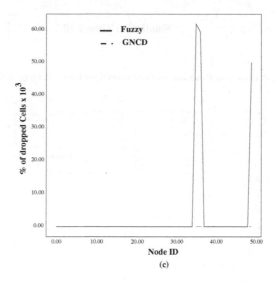

FIGURE 9.10c
Cumulative number of cells dropped, for GNCD and fuzzy schemes, in percentage, as a function of the switching nodes, for light traffic load.

overflow causes its superiority over the fuzzy scheme, in terms of throughput, to rapidly erode.

For each of the three experiments under Part 1, the cell bursts received at every destination node of the network, are marked. For each cell burst, the delay in its arrival at the destination switch is expressed as the following

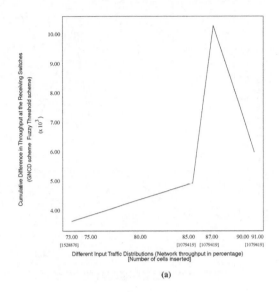

(a)

FIGURE 9.11a
Network performance as a function of different input traffic distributions, cumulative throughput difference, GNCD minus fuzzy scheme.

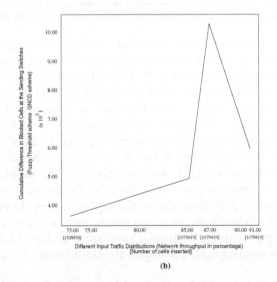

(b)

FIGURE 9.11b
Network performance as a function of different input traffic distributions, cumulative blocking difference, fuzzy scheme minus GNCD.

ratio: The transit time of the cell burst from source to this destination, divided by the propagation delay of the corresponding virtual path from source to destination. This ratio is referred to as the cell delay ratio and it may be noted that the reference, namely the cumulative propagation delay of the virtual

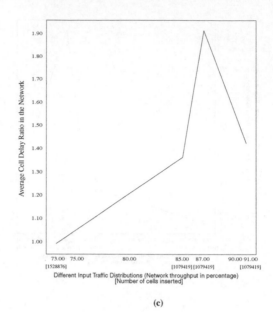

(c)

FIGURE 9.11c

Network performance as a function of different input traffic distributions, average cell delay ratios—average network cell delay under GNCD divided by that under the fuzzy scheme.

path, does not take into consideration the processing delays at the switches. The average of the ratios over all cell bursts for each of the input traffic distributions and for both GNCD and fuzzy schemes are computed, recorded, and termed average cell delay ratio for a given destination node. Figure 9.11(c) presents a plot of the ratio of the average cell delay ratio for GNCD to the average cell delay ratio for the fuzzy scheme, for the entire network, i.e., for all relevant nodes, corresponding to each of the three input traffic distributions. The ratio starts at 1.43 for low input cell arrival, increases to 1.92 for moderately dense traffic, and then drops to 0.93 for high density traffic. For light to moderate input traffic, the GNCD scheme performs superior buffer management, realizing higher throughput, despite eliminating cell loss through buffer overflow, and, as a result, incurs higher cell delays. For high density input traffic, the throughput superiority of GNCD over the fuzzy scheme is eroded and, consequently, both incur similar average cell delay values.

To examine the presence of a relationship between the buffer size and the cells lost due to buffer overflow condition in the fuzzy scheme, an experiment is designed. In this experiment, labeled 4, different buffer sizes, ranging from 400 to 3000 cells, are selected for every switch and the network is simulated with moderate density traffic as used in experiment 1. Figure 9.12 presents the cumulative cell dropping characteristic, throughout the entire network, as a function of the buffer size and reveals that the cell loss due to buffer overflow is significantly high for low buffer size in the range of 500 cells. The cell

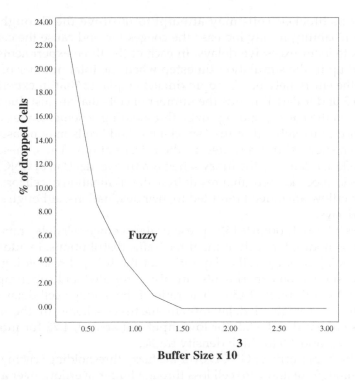

FIGURE 9.12
Cumulative number of cells dropped for fuzzy scheme, in percentage, as a function of the buffer size.

loss drops off sharply for a buffer size of 1500 cells which equals the maximum burst length of the incoming traffic. For buffer sizes beyond 1500 cells, the cell loss behavior is asymptotic, implying that significantly large buffer sizes, relative to the maximum burst length, may not yield an appreciable return.

Part 2 of the study is concerned with the rerouting of the fraction of blocked cells viewed as "special blocked" cells here. The throughput under GNCD is higher than in the fuzzy scheme and one may not necessarily conclude that GNCD blocks those cells that are otherwise lost through buffer overflow in the fuzzy scheme. As stated earlier, however, this section adopts a conservative stand and assumes that GNCD incurs blocking corresponding to the fraction of cells lost through buffer overflow in the fuzzy scheme. A total of three experiments are designed, one for each of the three traffic distributions used in experiments 1, 2, and 3 in Part 1. All of the experiments correspond to the rerouting approach integrated into GNCD. In each of the experiments, the switches attempt to reroute the "special blocked" cells to their destinations, subject to the constraint that for every cell burst, in the course of its transit through the network, its blocked cells may be subject to rerouting only once. The underlying reason is that while repeated rerouting of a cell burst or its blocked cells may attempt to improve the throughout and increase reliability, it may increase the congestion and cause the cells in the network to incur excessive delays. In each of the three experiments, data is obtained up to the simulation timestep when the total number of blocked cells in the entire network is approximately equal to that in experiments 1 through 3 under Part 1, minus the number of cells that are lost due to buffer overflow in the corresponding fuzzy thresholding scheme. Under these circumstances, virtually all of the "special blocked" cells may be assumed to have been successfully rerouted to their destinations. Thus, these experiments reflect, relative to the fuzzy scheme, a true realization of GNCD in that the "special blocked" cells that result from the elimination of cell loss through buffer overflow are indeed rerouted to their destinations, although subject to greater delays.

Figures 9.13(a) through 9.13(c) presents the average cell delay ratios for the destination nodes, for each of the three traffic distributions. In addition, the corresponding average cell delay ratios for the fuzzy thresholding without rerouting simulation experiments, are also computed and superimposed on Figures 9.13(a) through 9.13(c). The ratio of the average cell delay ratio for GNCD to the average cell delay ratio for the fuzzy scheme, for the entire network, is observed to be 2.5 for low input cell arrival, 4.72 for moderately dense traffic, and 1.6 for high density traffic.

Thus, the superiority of GNCD over the fuzzy thresholding scheme includes higher throughput and zero cell loss through buffer overflow, over a range of input cell arrival rates. However, cells incur higher delays in the network under GNCD even in the absence of rerouting. The increase in the number of blocked cells at the sending switches, under the GNCD scheme, may be successfully rerouted to their destinations at a cost of a further increase in the average cell delay in the network.

9.3   Proposed Research in Computational Intelligence in Networking

The field of networks is in critical need of intuitive and innovative approaches and novel algorithms to address the growing complexity in every one of its different aspects—performance, stability, security, connectivity, efficiency, routing, etc. There is the need to extend the research efforts outlined in Pithani and Sethi,[14] and Tanaka and Hosaka[15] to understand and improve routing in high-speed networks. The policing and congestion reduction efforts outlined in Ascia, Ficili, and Panno,[16] Douligeris and Develekos,[17] and Pitsillides and Sekercioglu[20] and need further research and elaboration. There is the need to extend the research effort in Bonde and Ghosh[9] to study the impact of fuzzy thresholding in buffer management in a representative cell-switching network consisting of 50–100 switches, the impact of utilizing

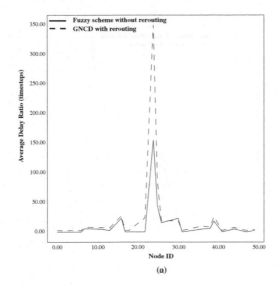

(a)

FIGURE 9.13a

Average cell delay ratios as a function of the destination nodes of the network for lightly loaded traffic.

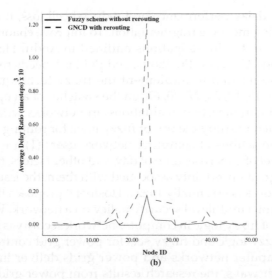

(b)

FIGURE 9.13b

Average cell delay ratios as a function of the destination nodes of the network for moderate traffic.

priorities on network performance, and the degree of message fragmentation, i.e., splitting of a message due to selective blocking, in the network. Although ATM networks, by definition, require all cells of a message to travel the same virtual path, it may be worthwhile to explore whether the gains

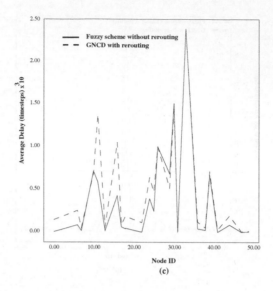

FIGURE 9.13c

Average cell delay ratios as a function of the destination nodes of the network for heavy traffic, for GNCD with rerouting and the fuzzy scheme without rerouting.

from improved delay performance and reduced cell loss, if any, may offset the accompanying message fragmentation. To improve channel assignments under heavy load, the investigations outlined in Abdul-Haleem, Cheung, and Chuang[27] and Gavrilov, Puzikova, and Pyl'kin[30] needs further attention. Research is also needed to implement the fuzzy techniques outlined in References 9, 14, 15, 17, 20, 27, 30, 67, on the switches of an operational ATM network and obtain objective evaluations. In networks with mobile hosts, research is needed to study the use of fuzzy logic for routing.

Given that the actions of computer network users (1) are defined by the type of user, level of expertise, time of day, and other factors, (2) are dynamic, and (3) have a great uncertainty associated with them, the issue of security in computer networks is essentially fuzzy. Hosmer[68] proposes the use of fuzzy sets to implement a multilevel security policy on a network. While the literature on the use of fuzzy logic in computer network security is sparse, it is rich in the use of fuzzy logic and fuzzy sets for power grid control and security. Given that computer networks and power grids deliver information and power in similar ways, the research results from power grids may provide valuable insights into the network security domain. Nijmura and Yokoyama[69] estimate security in power grid control through the use of fuzzy sets, thereby preventing catastrophic failures and degradation of service. Kurihara and Takahashi[70] propose an autonomous decentralized control system toward improved fault tolerance. Chang[71] presents a methodology for utilizing fuzzy models to provide static and dynamic security in power grid systems. Sinha[72] generates fuzzy estimations through pattern recognition for power grids. Further, field-related research needs to be directed in this area.

There is great uncertainty and imprecision relative to many of the real-world parameters connected with such issues in networks as (1) call setup in ATM and other high-speed networks, (2) policing, (3) bandwidth allocation, (4) virtual path and channel assignment, (5) route selection, and (6) security measures. Furthermore, while the complexity of today's networks defies analytic modeling, pure simulations with arbitrary choices of specific values for the different parameters convey very little insight into network behavior. Under these circumstances, it is theorized that a combination of (1) modeling networks in the continuum mathematics of fuzzy sets, and (2) extensive simulations, where each simulation is set up with a specific choice of values from the fuzzy sets corresponding to the fuzzy parameter variables, will result in a quasi-mathematical understanding of complex network behavior. The quantitative performance modeling efforts in Yan[31] and Levin[32] are noteworthy and the subject deserves greater attention.

References

1. S. Ghosh. *Fundamentals of High-Speed Networks: ATM and Future Networks.* Under Preparation, 1999.
2. L.A. Zadeh. Fuzzy Sets. *Information and Control,* Vol. 8:338–353, 1965.
3. L.A. Zadeh. Outline of a New Approach to the Analysis of Complex Systems. *IEEE Transactions on Systems, Man and Cybernetics,* Vol. 3:–, 1973.
4. D.E. Thomas and B. Armstrong-Helouvry. Fuzzy Logic Control—A Taxonomy of Demonstrated Benefits. *Proceedings of the IEEE,* Vol. 83(3):407–421, March 1995.
5. J.M. Holtzman. Coping with broadband traffic uncertainties: statistical uncertainty, fuzziness, neural networks. In *Proceedings of the IEEE Globecomm '90,* volume 1, pages 7–11, San Diego, CA, December 1990.
6. P.A. Fishwick. Fuzzy set methods for qualitative and natural language oriented simulation. In *Proceedings of the 1990 Winter Simulation Conference,* pages 513–519, 1990.
7. R.-J. Li and E.S. Lee. Analysis of Fuzzy Queues. *Computers and Mathematics Application,* Vol. 17(7):1143–147, 1989.
8. H.M. Prade. An outline of fuzzy or possibilistic models for queuing systems. In *Proceedings of the Symposium on Policy Analysis and Information Systems,* pages 147–153, Durham, NC, December 1980.
9. A. Bonde, S. Ghosh. A comparative study of fuzzy versus "fixed" thresholds for a robust queue management in cell-switching networks. *IEEE/ACM Transactions on Networking,* Vol. 2 (No. 4):337–344, August 1994.
10. M. Schwartz. *Telecommunication Networks: Protocols Modeling and Analysis.* Addison-Wesley Publishing Company, MA, 1988.
11. M.F. Scheffer and J.S. Kunicki. Comparative analysis of modeling techniques for packetized data. In *Proceedings of the IEE St. Petersburg International Teletraffic Seminar: New Telecommunication Services for Developing Networks,* pages 209–219, St. Petersburg, Russia, June–July 1995.

12. J.J.P. Beneke and J.S. Kunicki. Prediction of telephone traffic load using fuzzy systems. In *Proceedings of the IEE St. Petersburg International Teletraffic Seminar: New Telecommunication Services for Developing Networks*, pages 270–280, St. Petersburg, Russia, June–July 1995.

13. A. Celmins. Distributed fuzzy control of communications. In *Proceedings of the ISUMA-NAFIPS '95 The Third International Symposium on Uncertainty Modeling and Analysis and Annual Conference of the North American Fuzzy Information Processing Society*, pages 258–262, 1995.

14. S. Pithani and A.S. Sethi. A fuzzy set delay representation for computer network routing algorithms. In *Proceedings of the Second International Symposium on Uncertainty Modeling and Analysis*, pages 286–293, College Park, MD, April 1993.

15. Y. Tanaka and S. Hosaka. Fuzzy control of telecommunications networks using learning technique. *Electronics and Communications in Japan*, Vol. 76, Part 1 (No. 12): 41–51, Dec 1993.

16. G. Ascia, G. Ficili, and D. Panno. Design of a VLSI processor for ATM traffic sources management routing algorithms. In *Proceedings of the 20th Conference on Local Computer Networks*, pages 62–71, Minneapolis, MN, Oct 1995.

17. C. Douligeris and G. Develekos. A fuzzy logic approach to congestion control in ATM networks routing algorithms. In *Proceedings of the 1995 IEEE International Conference on Communications*, volume 3, pages 1969–1973, Seattle, WA, June 1995.

18. R-G. Cheng and C-J. Chang. Design of a fuzzy traffic controller for ATM Networks. *IEEE/ACM Transactions on Networking*, Vol. 4(No. 3):460–469, June 1996.

19. V. Catania, G. Ficili, S. Palazzo, and D. Panno. A Comparative Analysis of Fuzzy Versus Conventional Policing Mechanisms for ATM Networks. *IEEE/ACM Transactions on Networking*, Vol. 4(No. 3):449–459, June 1996.

20. A. Pitsillides and Y.A. Sekercioglu. Fuzzy logic control of cell flow in asynchronous transfer mode (ATM). In *Proceedings of the Australian Telecommunications Networks and Applications Conference*, pages 249–254, Clayton, Australia, Dec 1994.

21. Y.A. Sekercioglu and A. Pitsillides. Fuzzy logic control of ABR traffic flow in ATM LANs. In *Proceedings of the Australian Telecommunications Networks and Applications Conference*, pages 227–232, Clayton, Australia, Dec 1994.

22. N. Millstrom, A. Bonde, Jr., and M. Grimaldi. A hybrid neural-fuzzy approach to VHF frequency management. In *Proceedings of the 1993 IEEE Workshop— Neural Networks for Signal Processing 3*, pages 441–449. IEEE Press, 1993.

23. Y. Lirov. Fuzzy logic for distributed systems troubleshooting. In *Proceedings of the Second IEEE International Conference on Fuzzy Systems*, Volume 2, pages 986–991, San Francisco, CA, 1993.

24. L. Lewis and G. Dreo. Extending trouble ticket systems to fault diagnostics. *IEEE Network*, pages 44–51, Nov 1993.

25. B. Chakraborty, G. Mansfield, and S. Noguchi. Fuzzy technique in network management expert system. In *Proceedings of the Second International workshop on industrial fuzzy control and intelligent systems*, pages 40–48, College Station, TX, Dec 1992.

26. D. Maravall. Probabilistic and fuzzy learning automata for the optimal management of a communication system. In *Proceedings of the 12th European Meeting on Cybernetics and Systems Research*, pages 1385–1392, Vienna, Austria, April 1994.

27. M. Abdul-Haleem, K.F. Cheung, and J.C.I. Chuang. Aggressive fuzzy distributed dynamic channel assignment algorithm. In *Proceedings of the 1995 IEEE International Conference on Communications*, pages 423–427, Seattle, WA, June 1995.

28. G. Edwards and R. Sankar. Hand-off using fuzzy logic. In *Proceedings of the IEEE Globecom*, volume 1, pages 524–528, Singapore, Nov 1995.

29. S.S-F. Lau, K-F. Cheung, and J.C.I. Chuang. Fuzzy logic adaptive handoff algorithm. In *Proceedings of the IEEE Globecom*, volume 1, pages 509–513, Singapore, Nov 1995.

30. A.N. Gavrilov, L.A. Puzikova, and A.N. Pyl'kin. A sequential procedure for making decisions about the state of a communication channel by verifying fuzzy hypothesis. *Journal of Computer and Systems Sciences International*, Vol. 33(4):115–122, July–Aug 1995.

31. B-P Yan. An HGF integrated evaluation method for computer network. In *Proceedings of the IEEE Singapore ICCS '94 Conference*, volume 1, pages 86–89, Singapore, Nov 1994.

32. V.I. Levin. An Analysis of computer networks with non-determinate parameters using non-deterministic logic. *Automatic Control and Computer Sciences*, Vol. 25(5):17–24, 1991.

33. H. Kroner, G. Hebuterne, P. Boyer, and A. Gravey. Priority management in ATM switching nodes. *IEEE Journal on Selected Areas in Communications*, Vol. 9(No. 3):418–427, April 1991.

34. P. Yagani, M. Krunz, and H. Hughes. Congestion control schemes in prioritized ATM network. *IEEE International Conference on Communications*, Vol. 2:1169–1173, 1994.

35. Karl Rothermel. Priority mechanisms in ATM networks. *Proceedings of GLOBECOM '90*, Vol. 2:847–851, 1990.

36. T. Czachorski, Jean-Michael Fourneau, and F. Pekergin. Diffusion model of the push-out buffer management policy. *Proceedings of IEEE INFOCOM '92*, Vol. 1:252–261, 1992.

37. A. Lin and J. Silvester. Priority queuing strategies and buffer allocation protocols for traffic control at an ATM integrated broadBand switching system. *IEEE Journal on Selected Areas in Communications*, Vol. 9 (No. 9):1524–1536, Dec 1991.

38. L. Tassiulas, Y. Hung, and S. Panwar. Optimal buffer control during congestion in an ATM network node. *Proceedings of IEEE INFOCOM '93*, Vol. 3:1059–1066, 1993.

39. J. Causey and H. Kim. Comparison of buffer allocation schemes in ATM switches: complete sharing, partial sharing and dedicated allocation. *IEEE International Conference on Communications*, Vol. 2:1164–1168, 1994.

40. S. Bhagwat, D. Tipper, K. Balakrishram, and A. Mahapatra. Comparative evaluation of output buffer management schemes in ATM networks. *IEEE International Conference on Communications*, Vol. 2:1174–1178, 1994.

41. N. Endo, T. Ohuchi, T. Kozaki, H. Kuwahara, and M. Mori. Traffic characteristics evaluation of a shared buffer ATM switch. *Proceedings of GLOBECOM '90*, Vol. 3:1913–1918, 1990.

42. Soung Liew. Performance of various input-buffered and output-buffered ATM switch design principles under bursty traffic: simulation study. *IEEE Transactions on Communications*, Vol. 42(No. 2):1371–1379, Feb 1994.

43. S. Sumita. Synthesis of an output buffer management scheme in a switching system for multimedia communications. *Proceedings of IEEE INFOCOM '90*, Vol. 3:1226–1233, 1990.

44. H.F. Badran and H.T. Mouftah. ATM switch architecture with input-ouput-buffering: effect of input traffic correlation, contention resolution policies, buffer allocation strategies and delay in backpressure signal. *Computer Networks and ISDN Systems*, Vol. 26:1187–1213, 1994.

45. D.X. Chen and J.W. Mark. A buffer management scheme for the SCOQ switch under nonuniform traffic loading. *Proceedings of IEEE INFOCOM '92*, Vol. 1:132–140, 1992.

46. S-q. Li. Study of information loss in packet voice systems. *IEEE Transactions on Communications*, Vol. 37(11):1192–1202, Nov 1989.

47. A.E. Eckberg, D.T. Luan, and D.M. Lucantoni. Meeting the challenge: congestion and flow control strategies for broadband information transport. *Proceedings of GLOBECOM '89*. Vol. 3:1769–1773, Nov 1989.

48. L. Trajkovic and S. Halfin. Buffer requirements in ATM networks with leaky buckets. *IEEE International Conference on Communications*, Vol. 3:1616–1620, 1994.

49. R. Chipalkatti, J. Kurose, and D. Towsley. Scheduling policies for real-time and non-real-time traffic in a statistical multiplexer. *IEEE International Telecommunications Symposium*, pages 774–783, 1989.

50. Zhensheng Zhang. Finite buffer discrete- time queue with multiple Markovian arrivals and services in ATM network. *Proceedings of IEEE INFOCOM '92*, Vol. 3:2026–2035, 1992.

51. F. Bernabei, R. Ferretti, M. Listanti, and G. Zingrillo. ATM system buffer design under very low cell loss probability constraints. *Proceedings of IEEE INFOCOM '91*, Vol. 2:929–938, 1991.

52. A.K. Choudhury and E.L. Hahne. A new buffer management scheme for hierarchical shared memory switches. *IEEE/ACM Transactions on Networking*, Vol. 5 (No. 5):728–738, Oct 1997.

53. R. Jain. Congestion control and traffic management in ATM networks: recent advances and a survey. *Computer Networks and ISDN Systems*, Vol. 28:1723–1738, 1996.

54. H.T. Kung et al. Flow controlled virtual connections proposal for ATM traffic management. *AT-TM 94-0632R2*, Sept 1994.

55. J. Scott et al. Link by link, per VC credit based flow control. *AF-TM 94-0168*, March 1994.

56. Jean Walrand. *Communication Networks: A First Course*. Irwin Publishers, Boston, 1991.

57. D. Bertsekas and R. Gallagher. *Data Networks*. Prentice Hall, NJ, 1992.

58. S-q. Li. Overload control in a finite message storage buffer. *IEEE Transactions on Communications*, Vol. 37(12):1330–1338, Dec 1989.

59. D.W. Petr and V.S. Frost. Nested threshold cell discarding for ATM overload control: optimization under cell loss constraints. *IEEE INFOCOM 1991*, Vol. 3:1403–1412, April 1991.

60. F. Solomon. *Probability and Stochastic Processes*. Englewood Cliffs, NJ: Prentice-Hall, 1987.

61. Arthur Chai and Sumit Ghosh. Modeling and distributed simulation of broadband-ISDNetwork on a network of Sun workstations configured as a loosely-coupled parallel processor system. *IEEE Computer*, Vol. 26 (No. 9):37–51, Sept 1993.

62. C. Huang, M. Devetsikiotis, I. Lambadarism, and A. Kaye. Modeling and simulation of self-similar variable bit rate compressed video: a unified approach. *SIGComn*, pages 114–123, 1995.

63. C. Courcoubetis, G. Fouskas, and R. Weber. On the performance of an effective band-width formula. *Proceedings of ITC14*, pages 201–212, 1994.
64. C.A. Cooper and T. Eliazov. A study of the statistical multiplexing efficiencies achievable with variable bit rate traffic on a BISDN. Technical report, Bell Communications Research, 1991.
65. C.A. Cooper. A reasonable solution to the B-ISDN congestion control problem. Technical report, Bell Communications Research, 1990.
66. L.H. Tsoukalas and R.E. Uhrig. *Fuzzy and Neural Approaches in Engineering.* John Wiley & Sons, New York, 1997.
67. G. Ascia and V. Catania. A VLSI parallel architecture for fuzzy expert systems. *International Journal of Pattern Recognition and Artificial Intelligence*, Vol. 8(2): 1995.
68. H. Hosmer. Using fuzzy logic to represent security policies in the multipolicy paradigm. *SIG Security Audit and Control Review*, Vol. 10(4):12–21, Fall 1992.
69. T. Nijmura and R. Yokoyama. Security enhancement of electric power systems by approximate reasoning. In *Proceedings of the IEEE International Conference on Fuzzy Systems*, volume 1, pages 205–210, 1995.
70. I. Kurihara and K. Takahashi. Preventive control in an autonomous decentralized system. *Electrical Engineering in Japan*, Vol. 112(7):63–76, 1992.
71. C.S. Chang. Model-based fuzzy control of power system static/dynamic security using the pattern recogition approach. *IEE Proceedings Generation, Transmission, Distribution*, Vol. 141(4):270–278, July 1994.
72. A.K. Sinha. Power system security assessment using pattern recognition and fuzzy estimation. *Electrical Power and Energy Systems*, Vol. 17(1):11–19, 1995.

10

QoS-Based Hierarchical Routing in ATM Networks Using Reinforcement Learning Algorithms: A Methodology

Marios P. Saltouros, Antonios F. Atlasis, Athanasios V. Vasilakos, and Witold Pedrycz

CONTENTS

ABSTRACT In this chapter the use of a Reinforcement Learning Algorithm for optimizing the routing in *Asynchronous Transfer Mode* (ATM) networks based on the *Private Network-to-Network Interface* (PNNI) standard is proposed. This algorithm aims at maximizing the network throughput (allocating efficiently the network resources) while guaranteeing the Quality of

0-8493-1075-X/01/$0.00+$.50
© 2001 by CRC Press LLC

Service (QoS) requirements for each connection. In this study, large-scale networks are considered where it becomes necessary to be organized hierarchically so that a scale in terms of computation, communication, and storage requirements will be achieved. A comparative performance study of the proposed and other well-known routing schemes is demonstrated by means of simulation on an existing commercial network. Simulation results over a wide range of uniform, time-varying, and skewed loading conditions show the effectiveness of the proposed routing algorithm, and disclose the strength and weakness of the various schemes.

KEY WORDS: *ATM Networks, QoS Routing, PNNI Protocol, Reinforcement Learning Algorithms.*

10.1 Introduction

The implementation of the emerging high-speed communication networks, such as Broadband ISDNs, has been proposed by the ITU-T[1] to be based on the ATM technology. In ATM networks, various kinds of sources (such as voice, video, or data) with different traffic features and *Quality of Service* (QoS) requirements are statistically multiplexed at very high rates. However, the integrated character of ATM networks causes a number of serious congestion problems characterized by cell losses and excessive delays making the statistical multiplexing ineffective without control.

The QoS parameters of a connection are the *Cell Loss Ratio* (CLR), the *maximum Cell Transfer Delay* (maxCTD), and the *Cell Delay Variation* (CDV) incurred by the cells belonging to that connection. In ATM networks, the QoS of a connection is closely linked to the allocated network resources. A connection request for a given call is accepted only when sufficient resources are available to carry the new connection through the entire network at its requested QoS while maintaining the agreed QoS of the already established connections. It is the *Connection Admission Control* (CAC) which is responsible for deciding whether a new connection request should be accepted or not. To operate efficiently, CAC takes into account the connection traffic descriptor and the requested QoS for both the incoming call and the already established connections, as well as the available network resources. Once the connection has been established, the requested QoS should be provided as long as the connection is compliant with the negotiated traffic contract.

Part of the CAC actions is the routing algorithm,[1] whose objective is the identification and selection of a suitable route between the source-destination pair of an incoming call. A routing algorithm aims at maximizing the network throughput, which implies minimization of the call blocking rate, while guaranteeing the QoS constraints. Thus, for a QoS-based routing algorithm, not only is it important to select a route that satisfies all the QoS requirements, but also it is significant to manage efficiently the network resources, e.g., buffer space and bandwidth. So, it is clear that an efficient scheme should aim at reducing the consumption of resources and at balancing the traffic load across the network when necessary. The later property is particularly helpful for maximizing the ability to accommodate future calls. Furthermore, a high-performance ATM routing protocol should be dynamic, being capable of selecting the routes in response to variation of the traffic load and network failure, contributing in this way to the better exploitation of the network resources.

In this chapter, a computational intelligence approach to the ATM hierarchical (PNNI) routing issue is presented based on the *Stochastic Estimator Learning Algorithm* (SELA), an efficient Reinforcement Learning Algorithm. The learning approach presented in this study treats routing as a stochastic allocation problem amenable to a dynamic decentralized approach. Based on real-time feedback of network status, learning algorithms operating at various network switches determine how the traffic is to be routed. It is evident that the power of the learning algorithm approach is best realized in complex systems when several learning algorithms operate in a distributed fashion with each having a small number of actions. Such situations arise frequently in large-scale systems, where practical constraints make distributed decentralized control necessary. Traffic routing in telecommunication networks is an example of such systems.

Moreover, the need for learning in such systems depends on the information that is available regarding the system, the characteristics of the present noise, and the constraints that exist on the inputs of the learning algorithm. For low levels of uncertainty, learning may not be the most effective approach. On the contrary, for highly uncertain systems it may be essential for adequate system performance. Consequently, the effectiveness of learning routing techniques on telecommunication networks is justified by the fact that the states of these networks, which vary with time, are regarded as nonstationary environments and the information concerning these states at a particular instant may no longer be valid at the subsequent instant. Moreover, in large-scale networks it is impractical for any single entity (i.e., nodes and links) to have access to detailed state information about all the other network entities. This leads to the need of topology aggregation for reasons of scaling in terms of computation, communication, and storage requirements, but also for security reasons since the internal topology of a network may have to be hidden. As a result, the selection of a route across such large networks will typically be performed based on partial or inaccurate information derived from the process of topology aggregation performed. In general, the accuracy of the provided state information decreases as the amount of the aggregation that is carried out increases. This inaccuracy can lead to wrong routing decisions.[2] Learning algorithms techniques, choosing a path probabilistically, attempt to minimize the effects of inaccuracies on decision making. The greatest potential of a *Reinforcement Learning Algorithm* methodology is that it permits the analysis of very complex dynamic systems and enables an optimum global operation. Even when little information is available, it tends to stabilize a nonstationary system to predict its behavior.

Simulation results on an existing commercial topology, configured as a two-level hierarchical network, demonstrate the effectiveness of the proposed methodology for hierarchical QoS routing. The cases of connections with specific CLR and maxCTD requirements are considered. This effectiveness is highlighted when compared in terms of call blocking ratios and network throughput with other well-known routing algorithms over a wide range of uniform, time-varying, and skewed loading conditions. Moreover, the proposed algorithm is simple with no computational overhead and it can be implemented in hardware easily according to the procedures described in Reference 3.

The rest of the chapter is structured as follows. Since the route optimization methodology is proposed for ATM networks based on the PNNI specification, at first background information on the PNNI standards is provided, and prior work on hierarchical routing is summarized in Section 10.2. In Section 10.3, a concise description of the SELA algorithm is given. The proposed routing methodology based on SELA is developed in Section 10.4, while in Section 10.5 the simulation model is described and the extracted numerical results from the comparative performance study between the various routing schemes are exhibited. Finally, Section 10.6 concludes the chapter and intimates some interesting aspects of future work.

10.2 Background

In this section, at first a brief overview of the PNNI standards for ATM networks is presented, and then prior work on hierarchical routing is summarized.

10.2.1 Overview of PNNI Standards

Private Network-to-Network Interface (PNNI) is the interface between two switching systems which can be either a simple switch or a set of switches that are managed by the same manager.[4] One of the categories of protocols that PNNI includes is used for routing. These protocols determine the distribution of the information between switches and a cluster of them, as well as a way of organizing large-scale networks hierarchically making their management easier.

At the lowest level of hierarchy, each node corresponds to a switch, while each link corresponds to a physical link. At the other levels, each node (which is called *logical node*) corresponds to a cluster of one or more nodes of the lower level, while each link (which is called *logical link*) corresponds to a logical connection that consists of one or more links of the lower level in tandem or/and in parallel. A *peer group* is the cluster of the nodes in a specific level of hierarchy in which each node exchange topology information (consisting of *PNNI Topology State Elements* called PTSEs) with the other nodes of the same cluster. Of course, a peer-group can consist of a cluster of peer-groups of the lower level. Finally, a *border node* in a peer group is a node that is attached to at least one link that connects it with a corresponding node in another peer group at the same level of hierarchy (Figure 10.1).

In PNNI, the first (originating) switching system that receives the connection setup request, routes the incoming call to the destination (*source routing*) according to the *topology information* that each one of the candidate switching systems advertises. This information can be the state of the links attached to these switching systems (either physical or logical), the QoS parameters that can be guaranteed, as well as the ability to transmit various types of traffic. In other words, the originating switching system determines the path to the destination comprising all the detail of the hierarchy known to it. This is called a *hierarchically complete source route*. Such a path is not a fully-detailed source route because some portions of it are abstracted as a sequence of logical group nodes to be traversed. Thus, this source route does not hold the details of the path outside the originator's peer group and is considered as a *logical path*. The actual selection of a physical path outside the source switching system peer group has to be made by each border node of peer groups along the logical path. When the call setup arrives at the border node of a peer group, this node is responsible for selecting a lower level source route describing the crossing of that peer group.

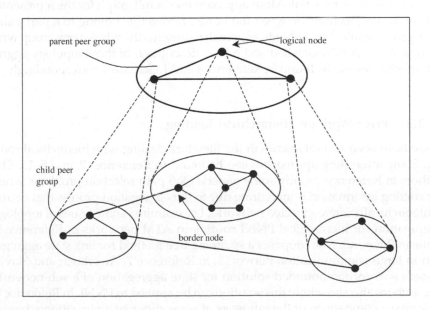

FIGURE 10.1
Hierarchical structure of ATM networks according to PNNI.

There are significant reasons that induce the compression and the aggregation of the information that is exchanged in a peer-group and is advertised to the higher level (*parent peer-group*). First of all, a decrease in the amount of the exchanged information decreases the resources that are consumed for their distribution, which is very important for WAN. Besides, it is not practical for every entity of the network (e.g., every node) to maintain detailed information about the state of all the other entities. Moreover, there are cases that an internal topology of a network should be concealed from nodes outside it for security reasons. However, the compression and the aggregation of the information (which is called *topology aggregation*) should be implemented in such a manner that the real network topology be represented satisfactorily so that efficient routing and resource allocation will be achieved.

The problem regards what information each switching system should advertise and also how often this information should be distributed to the other switching systems. The problem of having fast updating so that each switching system will have up-to-date information about all the other switching systems is particularly difficult due to the large propagation delay that large-scale ATM networks introduce. An increase in the frequency of updating and in the amount of the exchanged information increases the probability of choosing a more suitable route, but it also increases both the processing complexity and the resources that are consumed, and vice versa.

Several schemes for topology aggregation have been proposed; the best known are the *Star*, the *Full-Mesh*, and the *Simple-Node* approaches. The Star approach, which has been adopted by PNNI,[4] is an efficient compromise between the other two. Full-Mesh approach uses a full graph for the representation of the connections between the border nodes that belong to a particular peer-group, while Simple-Node approach represents the whole peer-group with a single node. A fully-detailed and a brief description of these topology aggregation schemes can be found in Reference 4 and Reference 5 correspondingly.

10.2.2 Prior Work on Hierarchical Routing

There have been several studies in the literature dealing with hierarchical routing. Some interesting approaches can be found in References 2, and 5–13. The authors in Reference 2 study some general QoS path selection problems when the routing information is inaccurate due to the aggregation process that occurs in hierarchically interconnected networks. Lee examines some issues of topology aggregation for hierarchical PNNI routing in ATM networks in Reference 5, while in Reference 6 he proposes a spanning tree method for link state aggregation in large communication networks. In Reference 7, Awerbuch and Slavitt present a distortion-bounded solution for state aggregation of a sub-network. The authors also show how this solution can be applied to PNNI. In Reference 8, they give a comparison of the influence of some different aggregation schemes on network performance, by means of simulation. However, in the model used only one source-destination pair is considered. In References 9 and 10 Baras et al.

present a method based on a hierarchical reduced load approximation to evaluate PNNI routing. Yet, no simulation results are presented. Orda suggests approximation methods for the fundamental problem of constrained path optimization (typical of QoS routing) in Reference 11, which exploit the structure of hierarchical aggregated topologies to give ε-optimal solutions. In Reference 12, the same author proposes routing algorithms that base their decision on a probability vector (probabilistic routing) for networks characterized by uncertain parameters. Finally, Kim et al. suggest a scalable inter-domain routing scheme which addresses the issue of selecting a path with multiple metrics for building QoS in a homogeneous high speed WAN.[13]

10.3   Stochastic Estimator Learning Algorithm

Generally, a Learning Automaton (LA) is a finite-state machine that interacts with a stochastic environment, trying to learn the optimal action the environment offers through a learning process (Figure 10.2). At any iteration the automaton chooses an action, according to a probability vector, using an output function. This function stimulates the environment which responds with an answer (reward or penalty). The automaton takes into account this answer and jumps, if necessary, to a new state using a transition function.

The Stochastic Estimator Learning Algorithm (SELA) used in the proposed methodology is a powerful and flexible ergodic Learning Automaton, especially when operating in a nonstationary stochastic environment.[14] Ergodic means that it converges at the optimal action with a distribution independent of the initial state. SELA has already been successfully used in other problems, e.g., Routing,[15] Connection Admission Control (CAC)[16] and UPC.[17] For reasons of completeness of our presentation, after some necessary definitions

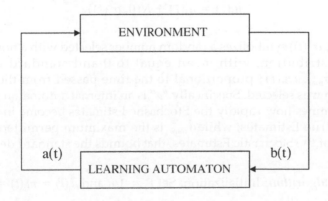

FIGURE 10.2
A learning automaton that interacts with a stochastic environment.

a concise description of SELA algorithm is given as described in Reference 14: $A = \{a_1, a_2, \ldots, a_n\}$ is the set of the n actions $(2 \le n < \infty)$ offered by the environment. The action selected at the time t is symbolized as: $a(t) = a_k \in A$. B is the set of the possible environment responses (feedback). The feedback of action a_i $(1 \le i \le n)$ at time instant t is symbolized as $b_i(t) \in B$. We denote by $P = \{P_1, P_2, \ldots, P_n\}$ the probability vector of choosing each action, i.e., P_i is the probability of choosing action a_i. The Estimator $E(t)$ contains at any time instant t the estimated environment characteristics. We define: $E(t) = \{D(t), M(t), U(t)\}$ where:

$D(t) = \{d_1(t), d_2(t), \ldots, d_n(t)\}$ is the True Estimate Vector. The True Estimate d_k of the selected action a_k is the mean reward which this action received the last W times that it was selected. It is computed as:

$$d_k(t) = \frac{\sum_{i=1}^{w} b_k(i)}{W} \tag{10.1}$$

where $\sum_{i=1}^{w} b_k(i)$ is the total reward received by the automaton during the last W times that action a_k was selected. The parameter $W \in N^*$ (where $N^* = \{1, 2, 3, \ldots\}$) is an integer internal automaton parameter called "learning window" and is used for ignoring old—and probably invalid—environmental responses.

$M(t) = \{m_1(t), m_2(t), \ldots, m_n(t)\}$ is the Oldness Vector; $m_i(t)$ of action a_i at any time instant t is a nonnegative integer number which expresses the time passed (counted in number of iterations) from the last time that action a_i was selected.

$U(t) = \{u_1(t), u_2(t), \ldots, u_n(t)\}$ is the Stochastic Estimator Vector. The Stochastic Estimate $u_i(t)$ of action a_i is defined as:

$$u_i(t) = d_i(t) + N(0, \sigma_i^2(t)) \tag{10.2}$$

where $N(0, \sigma_i^2(t))$ symbolizes a random number selected with a normal probability distribution, with mean equal to 0 and standard deviation $\sigma_i = \min\{\sigma_{max}, am_i(t)\}$ proportional to the time passed from the last time each action was selected. Specifically "a" is an internal automaton parameter that determines how rapidly the Stochastic Estimates become independent from the True Estimates, while σ_{max} is the maximum permitted standard deviation of the Stochastic Estimates, that bounds the standard deviation.

The SELA algorithm: Initialization: Set $P_i = 1/n$ and $d_i(t) = m_i(t) = u_i(t) = 0$, $\forall i \in \{1, 2, \ldots, n\}$.

Step 1: Select an action $a(t) = a_k$ according to the probability vector.

Step 2: Receive the feedback $b_k(t)$ of action a_k from the environment.

Step 3: Compute the new True Estimate $d_k(t)$ of the selected action a_k according to (1).

Step 4: Update the Oldness Vector by setting $m_k(t) = 0$, and $m_i(t) = m_i(t-1) +1$ $\forall i \neq k$.

Step 5: Compute the new Stochastic Estimate $u_i(t)$ $\forall i$ according to (2).

Step 6: Sort the set of the n actions in increasing order of their stochastic estimate of mean reward so that the first element of that classification ("1st-optimal" action a_m) will be the one with the highest value and the last one ("nth-optimal" action a_r) that with the lowest value. Thus, $u_m(t) = \max\{u_i(t)\}$ and $u_r(t) = \min\{u_i(t)\}$.

Step 7: Update the probability vector in the following way:

For every action a_i ($i = 1, 2, \ldots m-1, m+1, \ldots, n$) with $P_i(t) > 0$ set: $P_i(t+1) = P_i(t) - 1/N$ where N is a parameter called "resolution parameter" and determines the step size $\Delta(\Delta = 1/N)$ of the probability updating.
For the "optimal" action a_m set:

$$ P_m(t+1) = 1 - \sum_{i \neq m} P_i(t+1), \quad \text{with} \quad 0 \leq P_i \leq 1 \quad \text{and} \quad \sum_{i=1}^{n} P_i = 1 $$

Step 8: Go to step 1

A complete formal description of SELA automaton can be found in Reference 14. It should be noted that the choice of automaton's parameters (a, $\sigma_{\max}$, N) is a critical issue, relative to the automaton's performance under various switching environments. A learning automaton is called ϵ-*optimal* (see Reference 18), if there is an internal parameter N such that:

$$ \lim_{N \to \infty} (\lim_{t \to \infty} E\{P_m(t)\}) = 1 $$

(The symbolism $E\{\ldots\}$ stands for the expected value). We have:

THEOREM: The SELA learning automaton is ϵ-*optimal* in every stochastic environment that offers symmetrically distributed noise. Let $d_1, d_2, \ldots, d_n$ be the mean rewards offered by the environment to the actions $a_1, a_2, \ldots, a_n$ respectively. If action a_m is the optimal one $d_m = \max_i \{d_i\}$ for $i = 1, \ldots, n$) and $P_m(t) = [Pa(t) = a_m]$, then for every value $N \geq N_0$ ($N_0 > 0$) of the resolution parameter there is instant $t_0 < \infty$ such that for every $t \geq t_0$ it holds that $E\{P_m(t)\} = 1$.

The proof of the theorem is given in Reference 14. The assumption of symmetrically distributed noise is not unreasonable. The noise of all known stochastic environments is symmetrically distributed about the mean rewards of the actions.

10.4 Routing Scheme Based on SELA

In this section a QoS-based, dynamic source routing algorithm is presented, in compliance with the PNNI routing protocol, aiming at an efficient resource allocation and at a maximization of the network throughput. In particular, each source node maintains a topology database consisting of a collection of all PTSEs received from its neighbors that represents that node's present view of the PNNI routing domain. Moreover, as it has been shown in several studies dealing with nonhierarchical (flat) QoS routing (e.g., Reference 19), efficient resource exploitation can be realized when routing is restricted to short paths, even though there are longer candidate paths that may satisfy the QoS requirements. Hence, in this study such a restriction on candidate paths is applied deliberately. Specifically, at each source node a routing table (database) of precomputed hierarchically complete k-shortest path routes connecting this source node to every possible destination node is constructed and maintained. This routing table needs to be updated whenever the network topology is modified. One of the algorithms described in Reference 20 can be used off-line to compute the k-shortest path (minimum-hop) routes for each source-destination pair. Upon receiving a request to establish a new connection with specific traffic parameters and QoS requirements, the source node selects one of the precomputed k-shortest path routes that can be accepted by the CAC using its path selection routing algorithm. If no such a route exists, then the incoming call is rejected.

Below the definition of the SELA actions, the aggregated state parameters of the links, the computation of the environmental feedback, and finally the description of the SELA routing algorithm are given.

10.4.1 The Definition of the Actions of the SELA

In this study a dynamic decentralized learning approach is used where one SELA learning algorithm operates at each source node, determining how each incoming call is to be routed from this node to every possible destination node. The SELA algorithm uses k actions that represent the k-shortest path (minimum-hop) routes for each source-destination pair. k depends on the network topology size and the relative traffic flows.

10.4.2 Environmental Feedback Received by SELA

In the proposed methodology, the utilization of the links (either logical or physical) are used in order to compute the feedback of SELA. Therefore, the algorithm requires that the nodes retain a current link utilization table (database) for the entire network in their level of hierarchy. It is also reminded that path selection is performed at the source node as well as at each entry border node. Hence, the feedback of the algorithm at the source node is computed using the actual metrics inside the source peer group and the aggregated metrics outside the source peer group. On the other hand, the feedback of the algorithm at each entry border node is derived from the actual metrics inside its peer group.

Furthermore, in order to enhance the performance of the SELA routing algorithm in heavy loading conditions—where uncontrolled alternate routing can lead to increased call blocking rate[15]—we developed it to include the concept of *trunk reservation* which is a technique suitable for achieving this objective. Its effectiveness is based on the fact that at heavy traffic load, where the saving of network resources is very critical, the highly utilized links are used only for minimum-hop routes resulting in even better usage of the network resources and so, in increased throughput. When the expected route utilization $\rho_{\exp}^{\text{route}}$ (which is defined as the maximum expected utilization of the links that comprise this route) is greater than a predefined utilization threshold ρ_{TRT}, the call is established at this route only if it is a minimum-hop one. Otherwise, another route is tried. It is noted that the expected utilization of the links that comprise a route is defined as the utilization the links will have assuming that the route will be accepted by the CAC and accommodate the connection request.

Thus, if n links comprise the route that corresponds to the action a_k chosen by SELA, and $\rho_{\exp}^{i}$ is the expected utilization of the ith link, the feedback of the automaton $b_k(t)$ for this action is computed as:

$$b(t) = 1.0 \times \text{MAX\_NO\_OF\_HOPS} - \sum_{i=0}^{n} \rho_i \qquad (10.3)$$

where,

$$\rho_i = \begin{cases} 1, & (\text{if } \rho_{\exp}^{i} > \rho_{\text{TRT}}) \\ \rho_{\exp}^{i}, & \text{otherwise} \end{cases}$$

and MAX_NO_OF_HOPS is the maximum number of hops that a route between any source-destination pair can have.

It should be noted that the first term on the right hand side of the Eq. (10.3) represents a constant, predefined, maximum theoretical bound for the sum of the utilization of the links that a route between any source-destination pair can have. This term indicates the maximum feedback that a route can take, which occurs when all the links that comprise the specific route are not utilized at all ($\rho_i = 0$).

Equation (10.3) indicates that as the sum increases, which means that the previously selected route becomes congested, the feedback of SELA deteriorates; such operation seems to be quite natural. Moreover, this sum also increases implicitly when the number of the hops of the route increases, which implies an alternate (not a minimum-hop) route. Thus, the algorithm generally favors the shortest path routes. It favors alternate routes only when the congestion level of the shortest routes is much greater than the one of the alternate routes. In addition to this, when an alternate route is selected, its corresponding environmental feedback strongly deteriorates if that route consists of links which have expected utilization greater than the trunk reservation threshold ρ_{TRT}. In that way the over-utilization of the alternate routes is avoided when the network becomes congested (inherent characteristic of the trunk reservation mechanism).

10.4.3   The SELA Routing Algorithm

The SELA routing algorithm is summarized as follows:

Off-Line operation: At each source node:

1. Find the k-shortest (minimum-hop) routes for each source-destination pair using one of the algorithms described in Reference 20 and inform the routing tables (databases) at these nodes.

2. Correspond each candidate route to one of the k actions of SELA.

On-Line operation:

1. Whenever there is a new call-request at the source node, estimate the expected utilization of the links of the candidate routes.

2. Run the SELA algorithm computing the environmental feedback as described in Section 10.4.2.

3. Select the 1st-optimal action.

4. IF that action corresponds to a shortest (minimum-hop) route AND this route is accepted by the CAC, THEN choose that action as the selected one and establish the call over the corresponding route.

   ELSE IF that action corresponds to an alternate route which is accepted by the CAC AND whose utilization $\rho_{exp}^{route} \leq \rho_{TRT}$, THEN choose that action as the selected one and establish the call over the corresponding route.

   ELSE IF nothing of the above stands, THEN repeat Step 4 concerning the next optimal action.

5. IF all actions have been tried without any of them being chosen as the selected one, THEN the call request is rejected.

10.5 Simulation Performance Study

The simulator was written in C++ programming language and executed on an Ultra Spark Sun workstation. The simulation runs were carried out at call level. We avoided performing the simulation runs at cell level because of the extremely long computation time that was required. Even for the call level execution of the simulator, more than 1000 minutes of CPU time on average were needed for four minutes simulation run. The confidence intervals of our measurements are 95% constructed with the method of independent replications. The Welch's procedure was used to determine the warm-up (initial transient) period so as to discard possible misleading data. The length of the runs was estimated using the Replication/Deletion approach[21] in order to gather reliable statistics. All the routing algorithms were performed using the same stream of random numbers (used for the generation of call requests), so that a fair comparison among the algorithms can be achieved, because, in this way any differences regarding the calls arrival pattern are eliminated.

The candidate paths that a source node or an entry border node can choose belong to the minimum-hop+2 restricted candidate paths set. It is clarified that the minimum-hop+2 set contains the minimum-hop paths, and additionally these paths which are only one and two hops longer. This is desirable because using a longer path to route a connection ties up resources at more intermediate nodes, and subsequently decreases the network throughput, especially at heavy traffic load.[22]

10.5.1 Network Model

Simulation runs were performed on the NSFNET-backbone topology (existing commercial network) shown in Figure 10.3, which consists of 14 switches with average degree 3, and 22 bidirectional links. For hierarchical routing the network was arbitrarily divided into four peer-groups, each with three or four switches, corresponding in this way to a two level hierarchical network. The Full-Mesh topology aggregation scheme was adopted and performed. All the links have identical bandwidth of 45 Mb/s. Distances between two neighboring switches correspond to 1500 Km. Assuming high-speed networks, where the propagation delay depends on the distance between both endpoints, we choose the propagation speed to be two thirds the speed of light. Thus, the link propagation delay is 7.5 ms. The network parameters are summarized in Table 10.1.

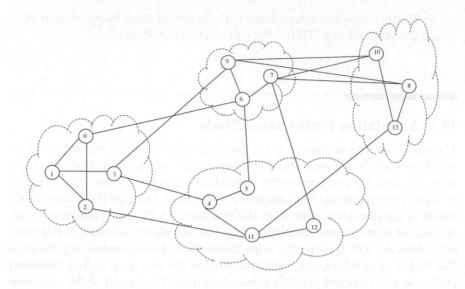

FIGURE 10.3
The simulated network which consists of 14 nodes, 22 bi-directional links, average degree 3, and is divided into four PGs for hierarchical routing.

TABLE 10.1

Network Parameters

Link rate	45 Mbps
Link propagation delay	7.5 ms
Link transmission delay	9.422 μs
Local link delay	negligible
Buffer size at source end system	infinite
Buffer size at switch	500 cells
Processing time at switch	negligible

10.5.2 Topology Aggregation

At first the aggregated (logical) links between two adjacent peer groups are defined as *inter-group links*, while the logical links between every two border nodes in the same peer group are defined as *intra-group links*. In general, the topology aggregation consists of the aggregation of nodal and link information.

In this simulation study, the aggregation of links hop-count, maxCTD (topology metrics), and utilization (topology attribute) are considered. In particular, for the *Full-Mesh* topology aggregation scheme that is adopted and applied in this study, all inter-group and intra-group links should advertise their metrics and attributes. It is noted that a logical intra-group link may correspond to a number of low level (physical) paths connecting the same source and destination border nodes. So, the hop-count, and the maxCTD metrics of an intra-group link are defined as the maximum number of hops

TABLE 10.2

Traffic Characteristics

Peak cell rate	10.6 Mbps
Sustainable cell rate	106 Kbps
Average bursty period	400 μs
Average call length	0.25 s

and the maximum end-to-end delay of its corresponding candidate low level paths, respectively. As far as the utilization topology attribute of an intra-group link is concerned, we regarded it as the maximum utilization of the links that constitute the corresponding candidate low level paths.

Similarly, an inter-group link may correspond to several physical links between two adjoining peer groups and consequently its aggregated utilization and maxCTD can be defined in the same manner. No nodal information is advertised.

10.5.3 Traffic Model

As far as the traffic model is concerned, real-time Variable Bit Rate (rt-VBR) traffic was used. Specifically, bursty (ON-OFF) sources were assumed with exponentially distributed burst and silent periods. The duration of the established calls and the inter-arrival time of new call requests were also assumed to be exponentially distributed (Poisson arrival process). The traffic parameters of the offered calls are summarized in Table 10.2. As far as their QoS requirements are concerned, the CLR was set to 10^{-6}, while the maxCTD constraint values ranged from 15 ms to 65 ms.

In our study, we consider uniform, time-varying, and skewed workloads. In *uniform workload*, the same call arrival rate is used for each source-destination pair. In this case, for light loading conditions, the offered call arrival rates were chosen to be $\lambda_I = 120$ calls/s and $\lambda_{II} = 140$ calls/s. Under moderate loading scenario the offered loads were fixed at $\lambda_I = 180$ calls/s and $\lambda_{II} = 240$ calls/s, while for heavy loading scenario they were set to $\lambda_I = 420$ calls/s and $\lambda_{II} = 540$ calls/s, respectively. Moreover, the performance of the routing algorithms was examined under changing traffic conditions. In this scenario, during the simulation the network load varied changing the call arrival rates for each source-destination pair every ten seconds of the simulated time. In this case, the call arrival rate is uniformly distributed in the range 0 to 540 calls/sec. Finally, in *skewed workload*, some switches are selected as the destination of the majority of the calls and consequently are regarded as 'hot-spots.' Specifically, for each source-destination pair the $\lambda = 120$ calls/s call arrival rate was used, except for the pairs where the 'hot-spots' switches were the destination nodes. In this later case, the $\lambda = 540$ calls/s call arrival rate was used. In this simulation study, the switches 1 and 13 were selected as the 'hot-spots' switches.

10.5.4   Connection Admission Control Scheme

The CAC scheme we adopted is the one proposed by Guerin et al. in Reference 23, which is based on the fluid-flow approximation model of Reference 24. In summary, the equivalent capacity of N multiplexed sources is given by

$$\bar{c}(N) = \frac{\sum_{i=1}^{N} [\lambda_i + \mu_i + r_{\text{peak}}^{(i)} z_0 - \sqrt{(\lambda_i + \mu_i + R_{\text{peak}}^{(i)} z_0)^2 - 4\lambda_i R_{\text{peak}}^{(i)} z_0}]}{2z_0}$$

In this equation, $1/\lambda_i$ and $1/\mu_i$ are the mean idle and burst periods of source i, respectively, and $R_{\text{peak}}^{(i)}$ is its peak cell rate (PCR). Furthermore, for a buffer size x and a buffer overflow probability requirement (CLR) smaller than ϵ, the factor z_0 is given by

$$z_0 = -\frac{\alpha}{x}, \quad \text{where } \alpha = \ln 1 / \epsilon$$

Using the above equivalent bandwidth approximation, the expected utilization of the links that constitute the candidate route can be easily estimated by the relation

$$\rho_{\text{exp}}^i = \frac{\bar{c}(N+1)}{C}$$

where C is the link capacity.

In order to satisfy the CLR requirements in a single node in case of a new call set-up request, the CAC accepts the new candidate connection only if $\bar{c}(N+1) \leq C$. Otherwise, the CAC rejects the candidate connection.

As far as the maxCTD requirements are concerned, when a connection i is routed over a path p which consists of n links with link delay d_l, the end-to-end delay $D^i(p)$ is computed by the equation:

$$D^i(p) = \sum_{l \in p} d_l, \quad \text{where}$$

$$d_l = d_{\text{propagation}} + d_{\text{transmission}} + d_{\text{queuing}} + d_{\text{processing}} \tag{10.4}$$

where $d_{\text{propagation}}$ and $d_{\text{transmission}}$ are the *propagation* and *transmission* delay, respectively, at link l, while $d_{\text{processing}}$ and d_{queuing} are the *processing* and the *queuing* delay, respectively, that a cell experiences at the previous switch where link l is attached. It is noted that, while the first three of them are constant, the queuing delay depends on the queue at the buffer of the switch. In this study, for the estimation of the queuing delay, we consider the worst-case where a cell enters a buffer with size B as the Bth cell and thus, it experiences the maximum possible queuing delay. In this case, $D^i(p)$ of Eq. (10.4) is an upper bound of the maximum end-to-end delay. Consequently, a candidate new connection i with end-to-end delay constraint D^i will be accepted by the CAC if $D^i(p) \leq D^i$.

10.5.5   The Compared Routing Algorithms

In this section, the six routing algorithms that were compared in the simulation study are briefly described.

- **SELA-routing** algorithm described in Section 10.4.
- **Least loaded-shortest path routing (LSR):** a feasible path (i.e., a path that satisfies the QoS requirements of the specific call request) with the minimum hop count (*minimum-hop path*) is tried. If there are several such paths, the one with the maximum available bandwidth (*least loaded path*) is chosen. If several such paths exist, one of them is randomly chosen. The call request is rejected if there is not any minimum-hop path that satisfies the QoS requirements.
- **Shortest-least loaded path routing (SLR):** a feasible path with the maximum available bandwidth is tried. If there are several such paths, the one with the minimum hop count is chosen. If several such paths exist, one of them is randomly chosen. A call request is rejected only if there is not any feasible path.
- **Shortest-least loaded path routing with trunk reservation (TSLR):** a shortest-least loaded path routing algorithm that uses the trunk reservation concept.
- **Least loaded-alternate path routing (LALT):** a feasible least loaded minimum-hop path is tried. If no feasible minimum-hop path exists, try the least loaded alternate (other than the minimum-hop) paths in order of preference (i.e., first the paths which are one hop longer, then the two hop longer and so on). If several such paths exist, one is randomly chosen. A call request is rejected only if there is not any feasible path.
- **Least loaded-alternate path routing with trunk reservation (TLALT):** a least loaded alternate path routing that uses the trunk reservation concept.

It should be mentioned that the values of the SELA parameters that were used during the simulation are tabulated in Table 10.3. These parameters were set after pilot simulation runs in order to achieve the best possible performance. Finally, the trunk reservation threshold was set to 0.81 for all the routing which make use of the trunk reservation concept.

TABLE 10.3

SELA's Parameter Values

Learning window W	14
Maximum variance σ_{max}	0.35
Internal parameter a	0.35
Resolution parameter N	30
MAX_NO_OF_HOPS	5

TABLE 10.4

Performance of the Routing Algorithms in Terms of the Call Blocking Ratios Under Various Uniform Loading Conditions

	Uniform light load		Uniform moderate load		Uniform heavy load	
Arrival rate λ calls/s	120	140	180	240	420	540
SELA	0.0695148	0.144309	0.274498	0.409072	0.619823	0.690002
LSR	0.071441	0.144383	0.276819	0.410835	0.620377	0.690372
SLR	0.172396	0.27767	0.410458	0.530282	0.696933	0.750723
TSLR	0.132162	0.250849	0.395951	0.522085	0.694297	0.749092
LALT	0.0684507	0.17145	0.322406	0.471278	0.664976	0.726298
TLALT	0.0698267	0.14437	0.276746	0.410839	0.619888	0.690061

TABLE 10.5

Performance of the Routing Algorithms in Terms of the Call Blocking Rates Under Time-Varying and Skewed Loading Conditions

	Time-varying load	Skewed load
SELA	0.470423	0.328653
LSR	0.471754	0.329544
SLR	0.577272	0.439432
TSLR	0.570539	0.427765
LALT	0.528636	0.368378
TLALT	0.471731	0.328674

10.5.6 Simulation Results

In this section the numerical results obtained from the comparative simulation performance study among the proposed methodology and the routing algorithms mentioned above on the NSFNET-backbone topology are presented. The measures of comparison include the call blocking ratios and the achieved throughput (expressed as the number of the successfully established calls) that the algorithms exhibit over a wide range of loading scenarios. The performance results are tabulated in Tables 10.4–10.7. Next, some general observations and also the details of our evaluation according to the simulation results are given:

TABLE 10.6

Performance of the Routing Algorithms in Terms of the Achieved Throughput (Expressed as the Number of the Successfully Established Calls) Under Various Uniform Loading Conditions

	Uniform light load		Uniform moderate load		Uniform heavy load	
Arrival rate λ calls/s	120	140	180	240	420	540
SELA	2,439,453	2,616,625	2,852,210	3,098,150	3,487,880	3,656,430
	±479.958	±1417.23	±1727.64	±2314.36	±1711.88	±2811.86
LSR	2,434,403	2,616,399	2,843,080	3,088,910	3,482,800	3,652,060
	±517.102	±1460.68	±1573.65	±1860.52	±2212.33	±3449.82
SLR	2,169,728	2,208,821	2,317,700	2,462,670	2,780,450	2,940,220
	±1419.77	±1726.21	±2106.6	±2432.66	±1952.24	±2677.28
TSLR	2,275,210	2,290,837	2,374,730	2,505,640	2,804,630	2,959,450
	±1253.95	±2198.27	±2055.84	±2461.72	±1410.43	±2632.66
LALT	2,442,243	2,533,630	2,663,870	2,772,020	3,073,630	3,228,310
	±722.975	±1859.85	±2108.24	±2164.81	±2058.05	±3070.15
TLALT	2,438,635	2,616,438	2,843,370	3,088,890	3,487,290	3,655,730
	±450.463	±1407.32	±1582.59	±2088.6	±2363.42	±3074.91

TABLE 10.7

Performance of the Routing Algorithms in Terms of the Achieved Throughput (Expressed as the Number of the Successfully Established Calls) Under Time-Varying and Skewed Loading Conditions

	Time-varying load	Skewed load
SELA	3,115,680	2,639,310
	±58705	±1520
LSR	3,109,180	2,635,810
	±61331.2	±1335.89
SLR	2,488,000	2,203,800
	±58423	±1612.51
TSLR	2,527,290	2,249,670
	±59543.1	±1912.8
LALT	2,773,540	2,483,140
	±59946.2	±1656.95
TLALT	3,108,660	2,639,230
	±59848.4	±1414.67

1. Giving priority to shortest paths, a routing scheme always performs better. This is due to the fact that the use of long paths is undesirable since they tie up network resources at more intermediate nodes than minimum hop paths. Thus, since the amount of the consumed resources increases, the number of the connections that can be accepted decreases.

2. In light loading conditions, even though a routing scheme should favorably select shortest paths for use, it should also balance the load of the network since there are plenty of resources. As the network load increases, it should use alternate routes more selectively decreasing the load balancing, so that waste of network resources and consequently, premature network congestion will be avoided.

3. Generally, the algorithms that use trunk reservation (SELA, TSLR, TLALT) exhibit better performance than the others (LSR, SLR, LALT) in every loading scenario except for the case where the network is loaded very lightly. In this latter case, LALT shows the best performance establishing more connections than the other algorithms.

4. The SELA routing algorithm achieves the best performance over all the other routing algorithms, in almost all the loading conditions. Specifically, it shows the lower blocking ratio in seven out of the eight examined loading scenarios, allowing more connections to be established. The only case that one of the other algorithms—specifically the LALT—blocks fewer calls than SELA is under very light loading conditions (when the arrival rate is λ_I). In this case SELA exhibits the second lower blocking ratio.

Generally, it could be concluded that among the compared routing schemes, the SELA routing algorithm always has the best (or the second best) performance irrespective of the loading conditions. Thus, it manages to admit more calls into the network than the other algorithms, while guaranteeing the QoS requirements (since the CAC algorithm rejects the calls when the QoS constraints cannot be satisfied). The SELA routing algorithm achieves to route the offered traffic effectively because it avoids unnecessary use of alternate routes while balancing the load when required. By predicting an oncoming congestion, it routes the offered loads through the less loaded paths. The simulation results demonstrate the effective compromise that the SELA routing algorithm attains between the use of minimum-hop routes and the load balancing, which is the key of its efficiency.

10.6 Conclusions and Future Work

In this chapter, a dynamic call routing methodology was proposed that uses an efficient reinforcement learning algorithm (SELA) suitable for real-time application in ATM networks. This algorithm uses hierarchical route information of the connection, aggregated topology, and loading information of the network to select feasible paths that satisfy the QoS constraints while achieving high resource efficiency.

In particular, it is demonstrated that for the implementation of the algorithm two sets of information are required. The first one regards the topology information that is necessary for the computation of the k-shortest path hierarchically complete routes, while the other includes the attribute information of the utilization of the links that comprise these routes. This latter information is used by the proposed algorithm for the selection of the most suitable among these k-shortest path routes. Since the PNNI protocol adopted by the ATM Forum[4] includes mechanisms for distributing the information as well as hierarchy mechanisms which provide an aggregated view of clusters of switching elements, it is clear that the proposed routing methodology can be implemented in the framework of the PNNI.

A comparative performance study among the proposed and five other well-known path selection routing algorithms was demonstrated by means of simulation on an existing commercial network. The used performance metrics are the call blocking ratios and the achieved network throughput that the various algorithms exhibited under uniform, time-varying, and skewed traffic workloads. Simulation results showed the effectiveness of the proposed routing algorithm and disclosed the strength and weakness of the various schemes.

Our future research includes studying other aggregation schemes and route selection policies. We are now engaged in developing the *Star* and *Simple Node* topology aggregation approaches. Our objective is to compare the performance of the various aggregation schemes and study the influence that the various workload distributions exercise on that performance. Besides, we also plan to extend the scheme to handle point-to-multipoint and multipoint-to-multipoint routing. Moreover, as far as the SELA algorithm development is concerned, a much promising approach seems to be the use of fuzzy sets for the determination of its parameters based on the traffic characteristics and network parameters.

10.7 Problems

1. Try to prove that the SELA learning automaton is ϵ-optimal in every stochastic environment that offers symmetrically distributed noise (Theorem in Section 10.3).

2. Define the following parameters for a communication network:

   - ρ_{exp}^i : the expected utilization of the ith link.
   - W^i : the equivalent bandwidth already allocated at the ith link.
   - W_{exp}^i : the expected equivalent bandwidth of the ith link in case that the new call request be accepted.

- $W_{exp}^i - W^i$: the additional equivalent bandwidth that is required to establish the new call at the specific ith link.

Thus, if n links comprise the route that corresponds to the action a_k chosen by SELA the feedback of the automaton $b_k(t)$ for this action can be defined similar to Equation (10.3) as:

$$b(t) = 1.0 \times \text{MAX\_NO\_OF\_HOPS} - \sum_{i=0}^{n} [\rho_{exp}^i (W_{exp}^i - W^i)]$$

    a. Considering the above environmental feedback function for the automaton, run the SELA routing algorithm and observe its performance.

    b. One can try many other expressions for the calculation of the automaton environmental feedback $b_k(t)$ that corresponds to the action a_k chosen by SELA. Give examples of alternative feedback expressions that can be used and provide reasoning for each choice.

    3. We can modify the Step 7 of the SELA algorithm (shown in Section 10.3), and describe a new method to update the probability vector of choosing each action of the automaton. Specifically:

- Assume that the probabilities P_i of choosing actions i always total 100.
- Let x be the amount of the probability of an action that we wish to adapt.
- Multiply the vector of probabilities P_i a constant factor so they total $100 - x$, that is

$$P'_i = \frac{P_i(100 - x)}{100} \text{ for } x \in [0, 100] \text{ and } 1 \leq i \leq n$$

where n is the number of actions offered by the environment. Call this new vector of probabilities P'_i the *base probabilities*.

- Without loss of generality, assume that the set of the n actions in increasing order of their probabilities' values is: $P_1, P_2, \ldots, P_n$.
- For every action a_i $(1 \leq i \leq n)$ with $P_i(t) > 0$ set:

$$P_i(t + 1) = P'_i(t) + \frac{[P_i(t) - P_n(t)]x}{\sum_{i=1}^{n-1} P_i(t) - (n - 1)P_n(t)}$$

a. Considering the above modification of the SELA algorithm, run the SELA routing algorithm and observe its performance for various values of parameter x (e.g., $x = 40$, $x = 100$, etc.).

b. Suggest alternative solutions to determine the step size Δ (see Step 7 of the SELA algorithm) of the probability updating.

4. Learning automata routing can operate by replacing the alternate routing schemes at each switching center by automata. In other words the SELA routing algorithm (Section 10.4.3) can be modified to refer to a routing practice supported by many researchers where direct routing (minimum-hop routing) should be always used before applying the learning automata routing scheme which is only responsible for alternate routing. Consequently, we can correspond the k shortest alternate routes to the k actions of the SELA learning algorithm. In case that the minimum hop routes are blocked due to congestion, SELA chooses one of its actions, i.e., one of the k shortest alternate routes.

a. Based on the aforementioned routing practice, modify the pseudo-code of the SELA routing algorithm shown in Section 10.4.3. For the convenience of the reader, one can refer to Reference 25 for more details.

b. Considering this modification, realize and run the SELA routing algorithm observing its performance.

c. Each action of the SELA automaton can correspond to either the selection of a specific alternate route or to a particular sequence of selections of alternate routes. Obviously, for r allowable candidate alternate routes, $r!$ sequences are possible, and this can represent a fundamental restriction on the viable size of the automaton to be used. However, we urge the reader to correspond the sequences of selections of candidate alternate routes to the actions of the SELA automaton, and run the SELA routing algorithm observing its performance.

References

1. ITU-T, Recommendation I.371, Traffic Control and Congestion Control in B-ISDN, Geneva, 1996.
2. Guerin, R., and Orda, A., QoS-based routing in networks with inaccurate information: theory and algorithms, *IBM Research Report*, RC 20515, 1996.
3. Mars, P., and Poppelbaum, W. J., *Stochastic and Deterministic Averaging Processors*, Peter Peregrinus Ltd., Wheaton, A., London, 1981.

4. The ATM Forum Technical Committee, Private Network-Network Interface Specification Version 1.0 (PNNI 1.0), ATM Forum af-pnni-0055.000, 1996.
5. Lee, W., Topology aggregation for hierarchical routing in ATM networks, *Computer Communication Review: Special Issue on ATM*, pp. 82–92, 1995.
6. Lee, W., Spanning tree method for link state aggregation in large communication networks, *in Proc. IEEE INFOCOM'95*, 297, 1995.
7. Awerbuch, B., and Shavitt, Y., Topology aggregation for directed graph, http://www.cnds.jhu.edu/publications.
8. Awerbuch, B., Du, Y., Khan, B., and Shavitt, Y., Routing through teranode networks with topology aggregation, http://www.cnds.jhu.edu/publications.
9. Xie, H., and Baras, J., Performance analysis of PNNI routing in ATM networks: hierarchical reduced load approximation, *Institute of System Research Technical Report*, TR 97–65, University of Maryland, 1997.
10. Baras, J., and Corson, M., Tactical and strategic communication network simulation and performance analysis, in *Proc. 1st Annual Conf. on Advanced Telecommunications and Information Distribution Research Program (ATIRP)*, 1997.
11. Orda, A., Routing with end-to-end QoS guarantees in broadband networks, in *Proc. IEEE INFOCOM'98*, 1998.
12. Lorenz, D., and Orda, A., QoS routing in networks with uncertain parameters, in *Proc. IEEE INFOCOM'98*, 1998.
13. Kim, S. H., Lim, K., and Kim, Ch., A scalable QoS-based inter-domain routing scheme in a high speed wide area network, *Computer Communications*, 21, 390, 1998.
14. Vasilakos, A. V., and Papadimitriou, G., A new approach to the design of reinforcement scheme for learning automata: stochastic estimator learning algorithms, *Neurocomputing*, 7, 275, 1995.
15. Atlasis, A. F., Saltouros, M. P., and Vasilakos, A. V., On the use of a stochastic estimator learning algorithm to the ATM routing problem: a methodology, *Computer Communications*, 21, 538, 1998.
16. Vasilakos, A. V., Loukas, N. H., and Atlasis, A. F., The use of learning algorithms in ATM networks call admission control problem: a methodology, in *Proc. ICCC'95*, Seoul, Korea, 1995.
17. Atlasis A. F., and Vasilakos A. V., LB-SELA: rated-based access control for ATM networks, *Computer Networks and ISDN Systems*, 30, 963, 1998.
18. Narendra, K., and Thathachar, M., Learning automata: a survey, *IEEE Transactions on Systems Man and Cybernetics*, SMC-4, 323, 1974.
19. Ma, Q., and Steenkiste, P., On path selection for traffic with bandwidth guarantees, in *Proc. IEEE Int. Conf. on Network Protocols*, Atlanta, Georgia, 1997.
20. Shier, D. R., On algorithms for finding the k shortest paths in a network, *Networks*, 9, 1979.
21. Law, A. M., and Kelton, W. D., *Simulation Modeling and Analysis*, 2nd edition, Munson, E. M., and Margery, L., ed., McGraw-Hill International Editions, Singapore, 1991.
22. Matta, I., and Shankar, A. U., Dynamic routing of real-time circuits, in *Proc. IEEE Int. Conf. on Network Protocols*, 1996.
23. Guerin, R., Ahmadi, H., and Naghshineh, M., Equivalent capacity and its application to bandwidth allocation in high-speed networks, *IEEE Journal on Selected Areas in Communications*, 9, 968, 1991.
24. Anick, D., Mitra, D., and Sondhi, M. M., Stochastic theory of a data-handling system with multiple sources, *Bell Systems Technical Journal*, 61, 1871, 1982.

11

Network Routing with the Use of Evolutionary Methods

Masaharu Munetomo

CONTENTS

KEY WORDS: *Routing algorithms, genetic algorithms, classifier systems, path genetic operators, path crossover, path mutation, Internet, routing tables, circuit-switching networks, packet-switching networks, genetic-based routing algorithm.*

11.1 Introduction

This chapter gives an introduction to current status of researches in applications of evolutionary methods to network routing algorithms. There are not

0-8493-1075-X/01/$0.00+$.50
© 2001 by CRC Press LLC

many papers that have been published in this area; however, it is expected to be an active research area in the near future because of the following reasons:

1. Increasing importance of adaptive routing algorithms in large computer networks such as the Internet.

2. Difficulty of routing algorithms to be adaptive and at the same time efficient, caused by inherent uncertainty of the observation in a network that has communication latency. This problem becomes more serious in large networks.

3. Effectiveness of evolutionary methods in adaptation to environment with uncertainty, and their inherent parallelism.

Rapid expansion of the Internet increases demands for scalable routing algorithms that perform effectively in large networks. It is, however, essentially difficult to realize an adaptive algorithm which is also scalable. For a routing algorithm to be adaptive, each node needs to send packets frequently to observe network load status. Frequent observations inevitably produce excessive overheads especially in large networks. In addition, there exists uncertainty in the observations because they cannot avoid communication delay through the network and an observation itself changes network load status.

Evolutionary methods are expected to be a promising answer of this difficult adaptation problem. In natural evolution, information from the environment is inherently uncertain, noisy, unstable,etc. Evolutionary methods such as genetic algorithms (GAs) are considered robust to such environments.[10] Another possible advantage of evolutionary methods are their inherent parallel mechanism. In network routing in large networks, any centralized algorithm is considered unrealistic because the control node must be extremely heavily loaded. That is the reason why conventional routing protocols in the Internet employ simple distributed algorithms instead of using sophisticated centralized ones.

This chapter continues as follows: after giving brief introductions to network routing algorithms and genetic algorithms, the following three case studies concerning applications of GA to routing algorithms are given: first, an application of order-based genetic operators to the routing in circuit switching networks, second, an adaptive routing for packet switching networks using a fuzzy classifier system, and finally, an adaptive routing algorithm in the Internet which employs path genetic operators to generate alternative routes.

11.2 Network Routing

Network routing algorithms determine routes from source node to destination for communication. In *circuit-switching networks* such as telephone networks, a *circuit* is allocated between source and destination nodes before starting communication and then all the data are transmitted through the circuit. Routing algorithms for such networks generate a route for the circuit based on the network status such as topological information, load status of links, and so on. On the other hand, in *packet-switching networks* such as the Internet, data to be sent are decomposed into communication *packets*, each of which is independently transferred from its source node to its destination. In this approach, instead of allocating a route for all the data, routing algorithms determine a route for each packet according to the current status of the network.

A routing algorithm can be *static* or *dynamic*. A static algorithm determines routes in advance by using statistical information on user traffic. A dynamic algorithm allocates routes based on current observation of network status. To observe network status, a *centralized* algorithm collects information and controls whole routing decisions in a single node. On the other hand, a *distributed* algorithm performs observation in each node in parallel. In centralized algorithms, all the communication for the observations concentrate on the control node, which leads to extremely heavy load in that node. This makes centralized algorithms practically impossible to work effectively in large networks. Therefore, to design a routing algorithm for large networks, we need to take a distributed approach. The following three routing algorithms, by using GAs, are all based on distributed control in which each node performs the same routing algorithm independently.

11.3 Genetic Algorithms

A simple GA consists of three genetic operators: *crossover, mutation,* and *selection* applied repeatedly to a population of strings in order to find optimal or near optimal solutions. Crossover exchanges substrings between a pair of strings. Mutation makes a small perturbation to a string. By *selection*, well-performed strings are selected to survive. The search power of GAs lies in their schema processing by crossover. Mutations are considered to play a minor role that supplies schemata which are not existent in an initial population. By combining *building blocks* (BBs) of a problem, GAs can find an optimal solution efficiently. Therefore, the design of crossover operator is one of the most important factors in designing a GA. A classical one-point crossover works well when short, well-performed schemata lead to optimal solutions. When we have long BBs, they are easily disrupted by the crossover. For crossover to work effectively, it is necessary to encode strings that make BBs short, or necessary to design a tailored crossover operator that preserves BBs of the problem. In the following applications of GAs to network routing algorithms, both approaches are taken: problem-specific encoding and tailored version of genetic operators for the routing algorithms. By taking such approaches, we can employ both the power of genetic search and the problem specific knowledge to find solutions. In general, employing genetic algorithms does not prevent conventional problem-specific techniques from working effectively.

11.4   An Anticipatory Routing Algorithm in Circuit-Switching Networks

In *Handbook of Genetic Algorithms*,[11] Cox et al.[12] proposed a dynamic anticipatory routing algorithm in circuit-switching networks by using an order-based genetic algorithm in its optimization of allocating call requests to a network. This algorithm allocates bandwidth of the network to call requests $x(t)$ at time t. The purpose of the optimization is to find a feasible schedule u that minimizes its cost $C[x(t), u]$.

In real time call scheduling, a call request consists of at least the following five attributes: a source-destination pair, a bandwidth required, a desired start time, a duration, and a priority class.[12] Under the framework, their routing algorithm is divided into a master and a slave strategy. A master strategy calculates a set of alternative routes which are the few shortest paths between each pair of nodes by using Dijkstra's shortest path algorithm. A slave strategy allocates call requests based on the alternative routes as follows: first, it tests whether overall traffic pattern is significantly changed or not. If it has changed, a master process is invoked to renew alternative routes. Second, when a new call request is received, one of the alternative routes is assigned to it and the call is stored in a pending list until it finishes. If it fails to assign, the algorithm seeks for a new assignment that allows the call to be accepted by changing the allocation in the pending list. The search for a new set of assignment is accomplished by an order-based genetic algorithm.[12]

The order-based genetic algorithm is based on a permutation representation of strings, each of which represents the order of call request assignment. Call requests are allocated as follows: take first call of the permutation and assign it to the shortest feasible path among the calculated shortest paths by master strategy. Then take the second call and assign it to the shortest feasible paths. Repeat this assignment procedure until no feasible route exists or until all the requests have been assigned. To obtain an optimal permutation, the genetic algorithm employs a uniform order-based crossover and a scramble sublist mutation.[12]

The uniform order-based crossover reorders a random subset of elements in a permutation according to the order in the other permutation. This crossover operator is inspired by the uniform crossover. First, a binary string is randomly generated as a template (like uniform crossover). Based on the template, this crossover operator applied to a pair of strings (p1, p2) performs the following to generate an offspring c1:

1. On all bitpositions where the template contains a "1", fill in elements of c1 by copying them from p1.
2. Make a list of the elements from p1 associated with a "0" in the template.
3. Permute the elements as they appear in the same order in p2.
4. Fill the permuted elements in the gaps on c1.

To generate another child (c2), carry out a similar process.

The scramble sublist mutation selects a sublist of a parent string and permutes the selected sublist to be an offspring while the rest of the string is directly inherited from the parent. Fitness value of a string is evaluated by the sum of loss from all call requests not assigned under the permutation, which is to be minimized.

It was reported that the assignment algorithm based on the GA outperforms those with random search and 2-opt algorithm except in the case that there is not enough time for the optimization, and also reported that a hybrid algorithm consisting of the GA and 2-opt achieved the best results.

11.5 An Adaptive Routing Algorithm for Packet-Switching Networks by Fuzzy Classifier Systems

In a packet-switching network which transmits data with communication packets, a routing algorithm determines where to forward packets to deliver them to their destination nodes. Conventional routing protocols such as the RIP and the OSPF forward packets to the next hop of the shortest path based on hop-count metric. To minimize communication latency by considering load status of the network, it is sometimes necessary to forward packets along other than the shortest paths. It is simply because the shortest path based on hop-count metric is not necessarily an optimal route that minimizes communication latency which is dynamically changing.

Carse et al. applied a Pittsburgh-style fuzzy classifier system[13] to determine whether to forward packets along the shortest path or not. In this classifier system, fuzzy rules and membership functions are optimized simultaneously by using a GA with a new n-dimensional crossover operator and a fuzzy cover operator. Each rule R_k is represented by the following n-input and m-output system:

$$(x_1^c(k), x_1^w(k)); \ldots (x_n^c(k), x_n^w(k)) \Rightarrow (y_1^c(k), y_1^w(k)); \ldots (y_m^c(k), y_m^w(k)) \quad (11.1)$$

where x^c and x^w are the centers and the widths of the fuzzy set membership functions over the range of input variables, and y^c and y^w are those of the output. The n-dimensional crossover operator exchanges membership functions $(x_i^c(k), x_i^w(k))$ that satisfies $x_i^c(k) > C_i$ between a pair of rules, where $C_i (i = 1 \ldots n)$ is a randomly selected crosspoint vector. The crosspoint vector C_i is calculated by $C_i = MIN_i + (MAX_i - MIN_i)(r_c)^{1/n}$, where $[MIN_i, MAX_i]$ is the range of the input variable and r_c is a random number uniformly distributed in $[0, 1]$. A fuzzy cover operator was implemented as follows: if a set of inputs do not match any rules in the rule-base, a new rule is created with fuzzy set membership function of inputs whose centers are equal to the unmatched input vector; those for output membership functions are set randomly in the allowed range.[13] Fitness is obtained by calculating inverse value of the average of measured packet delays.[14]

In their experiments,[13] a 3-node network is employed and the routing algorithm makes decisions on whether it selects a link directly connected to the destination or not in sending a packet. In this simple example, inputs to the routing algorithm in each node are the following four variables:

DelayLeftDirect   The observed packet delay routed directly from the source node to the destination node which is located left of the source.

DelayLeftIndirect   The observed packet delay routed indirectly (passing through another node located right) to the destination node located left of the source node.

DelayRightDirect   The observed delay routed directly to the destination located right of the source node.

DelayRightIndirect   The observed delay routed indirectly to the destination located right of the source node.

This is because the network has only 3 nodes each of which has two links (named Left and Right) to be connected to the other 2 nodes. The packet delay is measured at the destination node and averaged over the last N packets. The actual input to the classifier system is the logarithm of the measured delay, which enables the algorithms to be sensitive to smaller delay.

The outputs of the routing algorithm are the following:

PLeftDirect   The probability for a packet destined to the left node to be routed directly to the node. Therefore, $(1 - PLeftDirect)$ is that to be routed indirectly.

PrightDirect   The probability for a packet destined to the right node to be routed directly to the node. Therefore, $(1 - PRightDirect)$ is that to be routed indirectly.

By distributing packets probabilistically between direct and indirect routes, this algorithm not only minimizes communication latencies of packets but also balances loads of links.

They also applied this algorithm to a larger network consisting of 19 nodes.[14] To utilize the same framework as employed in a 3-node network, the algorithm considers the shortest path and the second shortest path as candidates for decision. Instead of choosing direct or indirect routes, one of the shortest or the second shortest paths is selected in each node by using the fuzzy classifier system.

11.6 The Genetic-Based Routing Algorithm in the Internet

The Genetic-Based Routing (GBR) algorithm[15] is an adaptive routing algorithm in the Internet that employs a genetic algorithm with *path genetic operators*. The algorithm maintains a limited number of alternative routes to each destination and observes their communication latency adaptively. The path genetic operators which consist of *path crossover* and *path mutation* are designed to generate alternative routes according to topological information of the network. In the following, we will show an overview of the algorithm and its genetic operators. We also present results of simulation experiments to show the effectiveness of the algorithm.

11.6.1 Overview

Figure 11.1 illustrates the GBR algorithm. Each node has its routing table consisting of lists of alternative routes assigned to the destinations. A routing table does not necessarily contain routes to all destination nodes in a network; that is, it only contains alternative routes to destinations to which packets are sent frequently. For each route, delay along the route and its weight

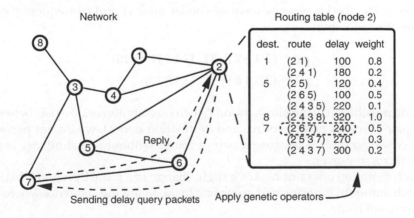

FIGURE 11.1
Overview of the GBR.

value are stored. Delay of a route is obtained by sending observation packets periodically along it. A weight of a route is calculated from its delay. Each route is encoded into a list of node IDs along it, for example, (2 4 8 5 9 11). Note that the route constrained to the network topology; that is, each step of a route must pass through a physical link of the network. To generate alternative routes, genetic operators called *path genetic operators*[15] are applied to routes in the routing table.

11.6.2   Path Genetic Operators

Path genetic operators consist of *path crossover* and *path mutation*. Path crossover operator exchanges sub-routes between a pair of routes. When we apply the crossover operator to a pair of routes r_1 and r_2, the operation proceeds according to the following sequence.[15]

1. List up a set of nodes N_c included in both r_1 and r_2 (excluding source and destination nodes) as potential crossing sites.
2. Select a node n_c as a crossing site from the N_c.
3. Crossing-over the routes by exchanging all the nodes after the crossing site n_c.

For example, when a path crossover is applied to following a pair of routes,

$$p1 = (2\ 4\ 5\ 7\ 8\ 11\ 15\ 17\ 20)$$
$$p2 = (2\ 3\ 4\ 6\ 9\ 11\ 12\ 14\ 17\ 19\ 20)$$

a crossing site is selected randomly from nodes that exist in both routes (in the above example, nodes 4, 11, and 17). When we select, say, node 11 as a crossing site, the crossover exchanges nodes after 11 and consequently generates the following:

$$p1 = (2\ 4\ 5\ 7\ 8\ \boxed{11}\ \mathbf{12\ 14\ 17\ 19\ 20})$$
$$p2 = (2\ 3\ 4\ 6\ 9\ \boxed{11}\ \mathbf{15\ 17\ 20})$$

If no possible crossing site is found (when no common node exists between the pair of routes except source and destination nodes), we do not perform this crossover. This is because a pair of routes without any similarity is not worth crossing-over.

Path mutation operator makes a slight change to a route by a perturbation. A path mutation is performed according to the following sequence where r is the original route.[15]

1. Select a mutation node n_m randomly from all nodes along the route r excluding its source and destination nodes.

2. Select a node n'_m from the neighbors of the mutation node, $n'_m \in \epsilon(n_m)$.

3. Generate a shortest path $r1$ from the source node to n'_m, and another shortest path $r2$ from n'_m to the destination.

4. If any duplication exists between $r1$ and $r2$, we discard the routes and do not perform a mutation. Otherwise, we connect the routes to have a mutated route $r' = r1 + r2$.

For example, when this mutation operator is applied to a route,

$$p1 = (2\ 4\ 5\ 7\ 8\ 11\ 15\ 17\ 20)$$

and node 8 is selected as a mutation node, we select another node randomly from neighbor of the mutation node. Assume that node 9 is a neighbor of node 8 and is selected, we connect source node 2 to node 9 and node 9 to destination (node 20) by using Dijkstra's algorithm and will generate the following offspring.

$$p1' = (2\ 4\ 6\ \boxed{9}\ 10\ 19\ 20)$$

In some cases, this mutation may create a route with duplication of nodes that passes through the mutation node twice. In such cases, we discard the offspring and do not continue the mutation any more because a duplication indicates that we cannot create meaningful alternative routes from the original route.

11.6.3 Fitness Evaluation

A fitness value of a route is calculated from communication latency along it. To observe delay, a delay query packet is sent to the destination and then sent back. This observation is performed at a specified interval of sending data packets. The delay of a route is calculated by averaging a delay from source to the destination and that back from the destination to the source. Using the delay obtained, we calculate weight w_i of route i by the following equations:

$$w_i = \frac{1/\eta_i}{\sum_{j \in S} 1/\eta_j}, \quad \text{where} \quad \eta_i = \frac{d_i}{\sum_{j \in S} d_j} \tag{11.2}$$

where d_i is the delay along route i and S is a set of routes to the same destination. By the above equations, a route with smaller delay is assigned a larger weight to be selected among the alternative routes. The sum of weights for routes to a destination is kept to one by the above equations.

11.6.4   Execution Flow

In the GBR, each node performs the same algorithm independently according to the following sequence:

1. Even when a packet is created at a node, the node determines a route of the packet based on its routing table. Concerning packets arrived from other nodes, the node forwards it according to its route, or receives it when the node is the destination of the packet. Initially, a routing table is empty. If there is no route for the destination of a packet, a default route is generated by employing Dijkstra's shortest path algorithm and is inserted to the table.

2. After a specified number of packets are sent along a route, a delay query packet is sent to observe communication latency of the route. If the packet arrived at its destination, another packet is sent back to notify its answer. After receiving the answer, communication latency of the route is obtained by calculating the average of the time to send the query packet and that receive its answer.

3. Once obtaining delay of a route, weights of routes to the same destination are calculated according to the equations (11.2). After every evaluation of weights, genetic operators are invoked at a specified probability to create alternative routes in the routing table.

4. After performing genetic operators, if the size of a routing table exceeds its limit, selection operators are performed to reduce its size. There are two selection operators: (I) *Local selection* which deletes a route with the smallest weight among routes of a same destination, and (II) *Global selection* which deletes all the routes of a destination which the sum of frequencies is the smallest among all the destinations in a routing table.

11.6.5   Comparison with Conventional Algorithms

Here, a theoretical and some empirical comparisons of the GBR algorithm with conventional algorithms such as the RIP[5] and the SPF[6] widely employed in the Internet. In Table 11.1, orders of the number of messages, the message size, and the overall communication overheads for the observations are shown, where n is the number of nodes in a network. The RIP needs $O(n^3)$ communication overheads because they employ broadcasts which needs to send $O(n^2)$ messages, each of which contains a whole routing table whose size is $O(n)$. The SPF also employs broadcasts that need $O(n^2)$ messages. Since it broadcasts only link status of the node, its message size is dependent upon the degree of a node, which ranges from $O(1)$ (in mesh, etc.) to less than $O(n)$ ($O(n)$ is for perfectly connected networks; however, we do not need to use a routing algorithm in such networks).

TABLE 11.1

Communication Overheads for Routing Algorithm

	# of messages	Message size	Total overheads
RIP	$O(n^2)$	$O(n)$	$O(n^3)$
SPF	$O(n^2)$	$O(1) \sim O(n)$	$O(n^2) \sim O(n^3)$
GBR	$O(n)$	$O(1)$	$O(n)$

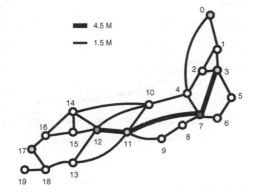

FIGURE 11.2
A 19-node network for simulation experiments.

On the other hand, the GBR needs to send only $O(n)$ messages to observe load status of routes, because each node sends delay request messages for a limited number of routes frequently used, which do not increase when n increases. Since overheads for genetic operators are much smaller than communication overheads, the GBR will apparently outperform conventional algorithms in large networks.

Note that the GBR needs to employ SPF-like algorithm to initialize topological database of the network in each node. Only to collect topological information which is not frequently changed, the conventional algorithms work effectively. The above results indicate that we cannot use algorithms based on the RIP and the SPF to collect communication latency adaptively in large networks because such algorithms need to send broadcast messages frequently to observe accurate delay of routes, which causes $O(n^2) \sim O(n^3)$ overheads. The GBR realizes observation with only $O(n)$ communication overheads.

Empirically, some simulation experiments are performed in Reference 15 to show the effectiveness of the GBR. To estimate overall performance of the routing algorithm, mean response time of data packets to arrive were observed for the RIP, the SPF, and the GBR. In the following experiments, a 20-node network in Figure 11.2 is employed, which is generated from geographic location of Japanese major cities. Data packets are generated in each node and their destinations are limited to nodes 0, 3, 7, 11, 12, and 17 (circles with gray). Bandwidth

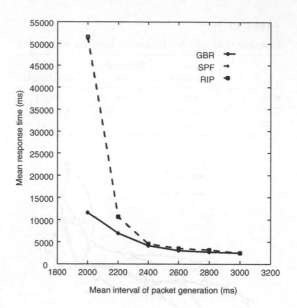

FIGURE 11.3
Mean response time of data packets.

of the thinner link is 1.5 Mbps and that of the thicker one is 4.5 Mbps. We perform simulation experiments for 3000 seconds. Conditions for the GBR are the following: fitness of a route is evaluated after every 10 packets sent along the route. The probability to apply a path mutation after the evaluation is 0.1 and that to apply a path crossover is 0.05. The limit of the population size is 100.

The result is shown in Figure 11.3 The *x*-axis of the figure is the mean interval of packet generation which is exponentially distributed (smaller interval leads to heavier load of the network). We plot the mean response time of data packets by changing the mean generation interval. From this result, we can easily see that we have much smaller response time of sending data packets by the GBR compared with the other two algorithms. This is not only because the GBR can observe communication latency of routes with less communication overheads but also because the algorithm realizes a load balancing mechanism among alternative routes by distributing packets probabilistically.

To see the effects of the load balancing mechanism, we show the result of load status of links in Figure 11.4. The thickness of each link stands for the log-scaled number of packets in packet queue of the link. Since it is log-scaled, a thick line (for example, from node 11 to node 13 in RIP) shows a link extremely heavily-loaded.

From the results, the RIP causes some extremely heavily loaded links (such as a link from 11 to 13) because of its inefficiency and overheads caused by broadcasting whole routing tables. The SPF can reduce load of links slightly,

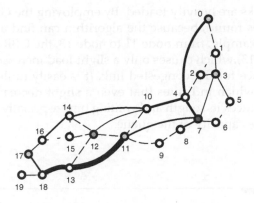

RIP

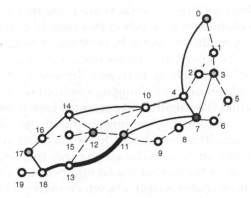

SPF

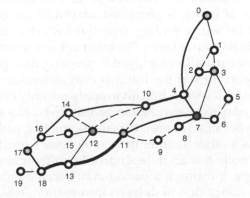

GBR

FIGURE 11.4
Load status of links during simulation.

but still some links are heavily loaded. By employing the GBR, no such congestion of links is found, because the algorithm can find alternative routes effectively. For example, from node 11 to node 13, the GBR finds an alternative route (11 12 13),which causes only a slight load increase along the route and a great reduce for the congested link. It is easily understood from the queuing theory which indicates that even a slight decrease of arrival to a queue greatly reduces its length and waiting time especially when the arrival rate is close to the capacity of the queue.

11.7　Conclusion

In this chapter, three routing algorithms are presented as case studies for applications of evolutionary methods in this area. In circuit-switching networks, evolutionary methods seem to be not difficult to apply because many of the problems can be solved in a framework of combinatorial optimization problems. For packet-switching networks, however, routing algorithms need to perform adaptations to changing environments without enough supply of information. The algorithms presented here tried to manage their routing decisions by introducing learning procedures based on evolutionary technique. In the algorithm based on a fuzzy classifier system, one of the direct or indirect routs are selected based on fuzzy rule sets. This algorithm only determines one of the best or the second best and cannot generate a number of alternative routes adaptively, which makes this algorithm difficult to employ in actual, large networks. On the other hand, the GBR algorithm does not have such limitation for the number of alternative routes. The algorithm tries to maintain a set of well-performed alternative routes in a routing table each of which is generated adaptively by genetic operators. The GBR is based on source routing algorithms which determine the entire route of a packet in its source node. The Internet has a source routing option; however, this is not usually employed in sending data packets. Conventional routing algorithms in the Internet only determine the next hop of packets in each node. It seems difficult to apply not only evolutionary methods but also any adaptive learning mechanism to realize a next-hop based routing algorithm because we need to maintain routing tables in nodes to be consistent with each other in order to send a packet correctly to its destination. When each node has an independent learning mechanism that only specifies next hops, a route of a packet may become inconsistent among routing tables in nodes due to delayed information, which causes loops, packet loss, etc. Therefore, designing an adaptive routing algorithm based on a nonsource routing approach is considered a challenging goal for the evolutionary routing algorithms, which is necessary for them to be employed in routing protocols in the Internet.

Exercises

1. Perform a scramble sublist mutation to a permutation (1 2 4 5 3 6 7 8). (You can select any sublist to be scrambled.)

2. Perform a uniform order-based crossover to a pair of permutation (1 2 4 5 3 6 7 8) and (8 6 4 2 7 5 3 1) under a template (0 1 1 0 1 1 0 0).

3. Design a genetic algorithm that solves static (without considering starting time in call requests) bandwidth allocation of links for call requests $R = \{(s_i, d_i, b_i)\}$ to a network (N, E), $N = \{n_i\}$, $E = \{(e_i, c_i)\}$ where R is a set of requests, N is a set of nodes, E is a set of links, n_i is a node, e_i is an edge, c_i is the capacity of e_i, $s_i \in N$ and $d_i \in N$ are source and destination of a call that needs bandwidth b_i.

4. Perform a path mutation to route (1 3 7 10 11 15) based on the network in Figure 11.2. (You can select any node as a mutation node.)

5. Perform a path crossover to a pair of routes (3 5 6 7 8 11 10) and (3 1 6 9 8 12 10). (You can select any crossing site from nodes that both routes contain.)

6. Design a string migration mechanism for the GBR that exchanges strings among nodes. (Hint: You need to modify migrated strings after their arrival.)

References

1. U. Black. *TCP/IP & Related Protocols*—Second Edition. McGraw-Hill, 1995.
2. D. E. Comer. *Internetworking with TCP/IP, Vol. I: Principles, Protocols, and Architecture*, Third Edition. Prentice-Hall, 1995.
3. L. L. Peterson and B. S. Davie. *Computer Networks—A Systems Approach*. Morgan Kaufmann, 1996.
4. A. S. Tanenbaum. *Computer Networks*, Second Edition. Prentice-Hall, 1988.
5. C. Hedrick. *RFC-1058: Routing Information Protocol*. Network Working Group, 1988.
6. J. Moy. *RFC-1131: The OSPF Specification*. Network Working Group, 1989.
7. E. W. Dijkstra. A note on two problems in connexion with graphs. *Numerische Mathematik*, Vol. 1, 269–271, 1959.
8. R. E. Bellman. *Dynamic Programming*. Princeton University Press, 1957.
9. L. R. Ford, J. and D. R. Fulkerson. *Flows in Networks*. Princeton University Press, 1962.
10. D. E. Goldberg, *Genetic Algorithms in Search, Optimization, and Machine Learning*. Addison-Wesley, 1989.
11. L. Davis, editor, *Handbook of Genetic Algorithms*. Van Nostrand Reinhold, 1991.
12. L. A. Cox, Jr., L. Davis, and Y. Qiu. Dynamic anticipatory routing in circuit-switched telecommunications networks. In Lawrence Davis, editor, *Handbook of Genetic Algorithms*, pages 124–143. Van Nostrand Reinhold, 1991.

13. B. Carse, T. C. Fogarty, and A. Munro. Adaptive distributed routing using evolutionary fuzzy control. In L. J. Eshelman, editor, *Proceedings of the Sixth International Conference on Genetic Algorithms*, pages 389–396. Morgan Kaufmann Publishers, 1995.

14. B. Carse, T. C. Fogarty, and A. Munro. Evolving temporal fuzzy rule-bases for distributed routing control in telecommunication networks. *Genetic Algorithms and Soft Computing*, pages 467–488. Physica-Verlag Heidelberg, 1996.

15. M. Munetomo, Y. Takai, and Y. Sato. An adaptive network routing algorithm employing path genetic operators. *Proceedings of the Seventh International Conference on Genetic Algorithms*, pages 643–649. Morgan Kaufmann Publishers, 1997.

16. M. Munetomo, Y. Takai, and Y. Sato. An adaptive routing algorithm with load blancing by a genetic algorithm. *Transactions of the Information Processing Society of Japan*, Vol. 38, No. 2 (in Japanese), pages 219–227. 1998.

12

Design and Use of Neural Network Applications in Telecommunications

R. J. Frank, N. Davey, and S. P. Hunt

CONTENTS

0-8493-1075-X/01/$0.00+$.50
© 2001 by CRC Press LLC

12.1   Introduction

This chapter describes the use of neural networks in the analysis of software systems. The development of large software systems over long periods of time, and the software crisis that this has produced,[2, 3] provides a rich source of problems. Many of these problems may be addressed, in part at least, with the aid of pattern matching and pattern classification tools such as artificial neural networks. We concentrate on two problems: analyzing the complexity of large software systems following the procedure set out in Reference 1 and detecting copies (clones) of source code procedures in large software systems.

12.2   Using Neural Networks to Analyze Software Complexity

This section describes how neural networks, in combination with software complexity measures, can be used to investigate the structure of the component code in a large software system. For any given unit of software, it is possible to measure its complexity using one of a variety of well-established software metrics. In such a way one block of code can be compared with another and, moreover, blocks of code of similar complexity can be identified.

Telecommunications software systems tend to be very large, with code produced in varying circumstances and over a considerable period of time. There are two problems with the use of software complexity measures in the analysis of such systems. The first problem is the number of code segments to be considered; the second is finding the appropriate measure of complexity for the varying styles of code that a large system may contain. Neural networks are well suited to analyzing large collections of data, and thereby suggest themselves for dealing with problems of this size. To deal with the second problem we follow the method of Sheppard and Simpson,[1] in which a variety of software complexity measures are amalgamated into a single feature vector. This facilitates the comparison of software units across a spectrum of different measures of similarity.

12.2.1   Software Metrics Employed in the Analysis

Twelve metrics were employed in Reference 1 and the same twelve were used for this work:

1. Lines of code (LOC) is the simplest of the measures, and provides information on the size of a program/procedure. It includes both the executable and the comment lines. This measure is not particularly useful on its own, as it will obviously vary according to the programmer's style, but it can give an indication of complexity.

2. McCabe's Cyclomatic Complexity (V_g) is based on the control-flow graph of a program or procedure. A simpler version of V_g, which is calculated by counting the decision points in a function (e.g., IF, FOR, WHILE, CASE) and adding 1, was used in Reference 1 and is, therefore, used for this work. V_g is not a sufficient measure of software complexity on its own as it ignores the effects of nesting; however, a set of possible standards have been supplied[8] to use when analyzing code with V_g.

3. Control Density (D) is a measure that combines both LOC and V_g. Its interpretation in Reference 1 proved to be somewhat uncertain. A high density should indicate a high complexity; however, a high value of D was found to be the significant feature of all the procedures that were classified as "Standard" (see below), implying that the opposite was true.

4. n1 number of unique operators

5. n2 number of unique operands

6. N1 total number of operators

7. N2 total number of operands

8. n = n1 + n2 the *vocabulary* of the procedure

9. N = N1 + N2 the *length* of the procedure

10. $V = N (\log_2 n)$, a measure of the *volume* (size) of the function—an estimate of the number of bits required to store the function in memory

11. $L = 2/n1 \times n2/N2$ the program *level*—intended to give an indication of how difficult it is to understand the function

12. E = V/L gives an indication of the *effort* required to write the function.

These measures include McCabe's Cyclomatic Complexity (V_g),[5] and Halstead's Software Science measures.[6, 7] Both McCabe's and Halstead's measures were developed with the intention of analyzing software independently of the programming language used, and thus the results should not differ between the "C" code used in Reference 1 and the PROTEL code used for this study. When taken together these twelve measures do not form an independent set; the last five measures are all calculated from the number of operators/operands in the code and D is the quotient of the two preceding measures in the table.

We include all twelve because, as observed by Moore,[9] all cluster tools that rely on one of the normal distance metrics to measure similarity of inputs can only ever find first order correlations between vectors. In order to find second or higher order correlations it is necessary to explicitly represent these in the training set. An implication of this is that blocks of code that had a similar D measure, for example, would not be clustered unless D was represented as a field in the input vector.

12.2.2 Recap of Sheppard and Simpson's Work

Sheppard and Simpson[1] used a simple competitive network comprising twelve input nodes and twelve output units. Over 4000 functions written in C were analyzed. These functions were taken from approximately 20 different programs, written by different programming teams. Nine of the output nodes were found not to be involved in the classification process. This may have been indicative of three natural clusters in the input space, but could have been caused by the limitations of the particular algorithm used. Simple competitive networks are prone to producing dead units. A second network was then constructed with twelve input nodes and three output nodes, and the results were the same.

Sheppard and Simpson's networks identified three clusters of functions. The first of these comprised 95.1% of the functions analyzed, the second cluster contained 4.75% of the functions, and the third cluster contained the remaining 0.15%. The functions in these clusters were termed "Standard", "Marginal", and "Non-Standard", respectively. Contrary to what might have been expected, Standard functions were characterized by a high value for D. Marginal functions were characterized by high n_1 and n values (large vocabulary), and Nonstandard functions had a high value for LOC and V_g.

12.2.3 The Data Used for the Current Study

The first data set contained 2236 procedures drawn from a single telecommunications software product. The second set contained 4456 procedures drawn from a variety of software products. All procedures were written in the proprietary language PROTEL, a block structured language designed for the control of telecommunication systems and used by Nortel. As in Reference 1, the method and metrics used were language independent.

12.2.4 Preprocessing the Data

This task accounted for over 50% of the effort expended on this work. Because PROTEL procedures can be of a user-defined type, it was necessary to develop a software tool to parse the code, identify all the procedures and apply the metrics. The tool applied six of the twelve metrics to each procedure: the remaining six values were calculated from the results. It should be noted that once the tool had been developed, the additional effort required to derive these remaining six measurements from the six recorded by the tool was trivial. See below for details of how the final vectors were produced.

12.2.5 The Data Vectors Produced

Each procedure was represented by a 12-ary vector, one field for each complexity measure. Within a data set vectors of raw measures are not used since they differ widely in magnitude; rather, each feature is represented as a proportion of the largest value for that feature. Specifically, for the set of raw vectors: $V_{raw} = \{\mathbf{v}_i$ the actual vectors used are:

$$V = \left\{ \frac{\mathbf{v}_i}{\max_j(\mathbf{v}_{ij})} \right\}$$

12.2.6 The Network Models Used

Like Sheppard and Simpson,[1] we chose to use self-organizing neural networks for this task. In these nets the output units, or more precisely the weight vectors of these units, move in the input space so as to minimize a cost function. The input vectors are then classified into clusters according to which output unit is closest. Problems arise when using such networks to find clusters in two respects:

- the number of output units in the network may influence the number of clusters that will be found in the input;
- the network may converge on solution to the problem for which some of the output units do not participate in the final classification: such units are said to be dead.

A variety of methods have been proposed to overcome these problems, including: feature mapping,[10] a conscience mechanism,[11] and Adaptive Resonance Theorem (ART).[12]

We used four different types of self-organizing networks with a variety of parameter settings. The networks chosen, with the motivation for their choice, were:

- *a simple competitive net* (SCN): to attempt to replicate the original results of Reference 1;

- a competitive net with a conscience mechanism, known as a *frequency sensitive competitive net* (FSCN): to examine whether the small number of nodes used in the SCN classification and concomitant dead units could be overcome;

- *a self-organizing map* (SOM): to investigate whether a global order was present in the representation space;

- a *fuzzy ART* network: to avoid prescribing the number of categorizing units used.

12.2.6.1 Simple Competitive Net

Initially the net consists of a set of nodes with small random initial weights. When an input is presented the winning node is selected as that with greatest similarity to the input. The winner's weight vector is then modified to become more similar to the input. We use Euclidean distance as the measure of similarity and, therefore, simply move the weight vector of the winner towards the input in Euclidean space. The size of the update is determined by a learning rate.

For input x and weight vectors $\{w_i\}_j$ the winner, $w_i{}^*$ is defined by:

$$\forall i \bullet |w_i{}^* - x| \leq |w_i - x|$$

The change to the weight vectors is then given by:

$$\text{if } i = i^* \quad \text{then} \quad \partial w_i = \rho(x - w_i) \quad \text{where } \rho \text{ is the learning rate}$$

else $\partial w_i = 0$

This type of network is prone to poor classifications, due partly to the problem of dead units—units that never win in the competition and thus never move towards the inputs.

12.2.6.2 Frequency Sensitive Competitive Net

In order to overcome the above problem one mechanism, originally suggested in Reference 14, is to make it harder for frequently winning units to win again.

This is done by subtracting a bias term from each unit, that is proportional to the number of times the unit has won. Stabilization can be a problem with this method and we follow the scheme suggested in Reference 14. Here the bias is defined by:

$$B_i = \gamma\left(F_i - \frac{1}{N}\right)$$

Where γ is a problem specific parameter, N is the total number of units, and F_i is the win count of unit i.

A potential problem with this type of network is that the biasing mechanism, while producing reasonable code vectors for the data set, may obscure natural clustering in the data.

12.2.6.3 Self-Organizing Map

In a SOM[10] the output units are arranged in a fixed topology, usually a two-dimensional grid. We use a grid wrapped around at the boundaries. Winning units pull their neighbors in the grid with them toward the input. The neighborhood of a unit should initially be set to a large fraction of the output space and decrease over time. A useful heuristic for setting the neighborhood is that it should start at roughly half the output space and decrease to unity, to ensure global order.[15] We use a ten by ten SOM and therefore initialize the neighborhood to five. We also include a conscience mechanism as described in the FSCN, with γ set to one initially, and reduced in line with the neighborhood reduction. A successfully trained SOM provides more information than the other networks used here. The SOM not only captures clustering in the input space but also the relative position of these clusters, which can be located from their position in the output space.

12.2.6.4 Fuzzy ART

Fuzzy ART[13] is an extension of the basic ART[11] algorithm to allow for continuous, rather than binary, data. In the original ART1 network, the proximity of two vectors is measured by taking their dot product or, equivalently, by performing a bit-wise Boolean AND on the two vectors and taking Norm 1 of the result. In Fuzzy ART the fuzzy logic AND connective, min, is used to extend the method to real values. This formulation of similarity is unusual and a little inconsistent, since the resulting measure is not the canonical Norm 1 metric. It has arisen here in order to maintain the biological inspiration of the original ART model.

The network begins with a set of units with their weights randomly allocated but all disabled. For each epoch and each input vector the enabled unit nearest to the input is selected; if the unit satisfies the vigilance criterion its weight vector is moved towards the input; if it does not, one of the disabled units has its weight vector set equal to the input.

Three details must be specified:

- Proximity: the closeness, $d(x, w)$ of the input x to the unit w is defined by:

$$d(x, w) = \frac{\|x \wedge w\|_1}{\epsilon + \|w\|_1}$$

where

$$\|x\|_1 = \sum x_j, \quad x \wedge w = (\min(x_j, w_j)) \quad \text{and } \epsilon \text{ is a small value to break ties.}$$

- Vigilance: the unit w, satisfies the vigilance test for input x, if:

$$\frac{\|x \wedge w\|_1}{\|x\|_1} \leq \delta$$

where δ is the vigilance parameter
- Movement: the winner, w, moves towards the input x, according to:

$$w' = x \wedge w$$

The behavior of the net is determined by the value of the vigilance parameter and the number of nodes in the initial disabled set. In our experiments we used a pool of nodes greater than required, and a variety of vigilance parameters. Fuzzy ART is appealing as the clusters are formed dynamically but the algorithm has some specific problems. As noted in Reference 9 the weight vectors drift toward the origin. This is due to the update rule: components of weight vectors are monotonically decreasing. As described in Reference 13 some of the difficulties of this can be overcome by complementary coding; but this has the unfortunate side effect of doubling the length of the input.

12.2.7   Results and Discussion

All results reported here for those nets with random initializations (all those except Fuzzy ART) are typical runs. No significant divergences from the reported behavior were observed.

12.2.7.1   Simple Competitive Net

This experiment was an attempt to replicate the results obtained in Reference 1, so nets with 12 units were used to analyze both data sets. In each case only four of the twelve units took part in the final classification. The clusters were identified as follows:

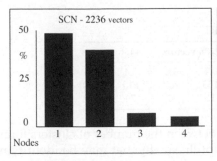 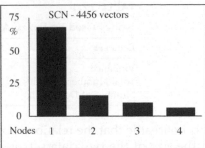

FIGURE 12.1
The proportion of vectors identified by each of the four active nodes of a simple competitive net.

Node 1 clustered "average" procedures (those with no particularly special features).

Node 2 clustered procedures with a large language content compared to their size.

Node 3 clustered very small procedures (those with fewer than 10 lines).

Node 4 clustered very large procedures (those with large values for all measures).

12.2.7.2 Frequency Sensitive Competitive Net

For both data sets the net was initialized with 100 units, all of which were used after training. For the smaller data set the resulting clusters ranged in size from 43 vectors to 8, with a uniform spread.

12.2.7.3 Self-Organizing Map

Again, 100 units were used, this time arranged in a ten by ten grid. The initial neighborhood was set to five and decreased over the training time to one. The system was trained with each of the two data sets and, in each case, two units did not participate in the classification. The classification produced was quite similar to the FSCN. However, with the SOM, additional proximity information could be inferred from the position of the classifying nodes in the SOM grid. For example, two vectors grouped by Fuzzy ART were found to be classified by adjacent nodes in the SOM.

12.2.7.4 Fuzzy ART

The net was trained over both data sets with two settings of the vigilance parameter: 0.4 and 0.6, and a pool of 2000 uncommitted nodes. As expected the net produced more clusters with the higher vigilance than lower. Table 12.1 details the results from the two data sets.

TABLE 12.1

Nodes Used by Fuzzy ART

Data set	2236 vectors		4456 vectors	
Vigilance	0.4	0.6	0.4	0.6
Total Number of nodes	155	227	619	930
Number of Dead Units	17	17	70	252

It is noticeable that the relationship between the number of clusters found and the size of the two data sets is not linear: the net may be identifying increased variation in the larger data set due to its multi-product source. A problem with the Fuzzy ART network becomes apparent when the weight vectors of the classifying units are examined—they are generally very small. For example, mean weight vector for the smaller data set and vigilance 0.4 is:

(0.026, 0.003, 0.120, 0.062, 0.031, 0.036, 0.079, 0.040, 0.024, 0.031, 0.031, 0.001) with length 0.178.

Another major difference between the earlier nets and Fuzzy ART is that the others are sensitive to frequency densities in the input space: if an area of the input space is heavily populated then many units will move there. This is not the case with Fuzzy ART, since a single unit will classify an area if all the vectors within it are sufficiently close and satisfy the vigilance test.

12.2.7.5 Comparison of the Four Network Types

The results obtained from the simple competitive net were radically different from those obtained from the other three nets. The SCN found 4 clusters in the input data (see Figure 12.1) while a more complex pattern emerged with the other networks: all found numerous clusters in the data. As demonstrated by the more sophisticated clusterers, this was the result of an inherent tendency of the SCN to not utilize some of its units; all the other networks have mechanisms for dealing with dead units. The FSCN and the SOM produced very similar clusters, while Fuzzy ART produced many more clusters than any of the other nets, and clustered the data in a different way from the FSCN/SOM. Comparative results are shown in Figure 12.2.

The results of these experiments illustrate the care with which unsupervised clustering must be undertaken. It is clear that the number and size of the clusters which are found is highly dependent on the architecture and the parameter settings used. In particular, SCNs may give unrepresentative results due to the problems of dead units.

Some of the limitations of using the ART type architecture are described above, and it is difficult to make a direct comparison with the other types of network due to the different distance metric used. The FSCN and the SOM gave similar results to one another, with the SOM providing additional information: the topological map.

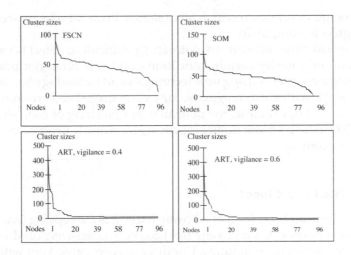

FIGURE 12.2
4456 vectors, the second data set, classified by FSCN, SOM, and Fuzzy ART.

The methods described here do not provide a simple way of partitioning software units on their complexity; however, the pattern of similarity could be potentially useful in locating procedures of *related* complexity. A similar application of neural nets to the location of software clones is described in the next section. Finally, it should be noted that the major activity of work such as this is the pre-processing of data.

12.3 Detecting Software Clones in Large Systems

In this section we discuss the use of neural networks as an analysis tool to find copies of procedures in a large software system. Our first attempt at producing a clone analysis tool, using a self-organizing map, is detailed in Reference 16.

The amount of code in most consumer products is doubling every two years: televisions may contain up to 500 kilobytes of software, and even an electric shaver may have 2 kilobytes.[17] A typical digital telephone exchange in the 1970s would have possessed a software system of a few hundred thousand lines of source code. Today a digital telephone exchange may incorporate tens of millions of lines of code. Such systems have evolved over the years, with current systems incorporating code from their very earliest forebears.

The development of these systems began in a time when software development methods were in their infancy. For example, structured analysis and design were not the norm, and the sophisticated reusability facilities provided by object-oriented implementation techniques were not available outside computer science research laboratories. Error rates increased as these systems grew in complexity.

Errors in real-time software are notoriously difficult to detect because they often occur only under certain conditions, and it is often not practical to rebuild existing systems using new techniques and technology because of the costs involved, with the average cost for well-managed code running at over $100 per line.[17] For example, the detection and patching of code sites where the year 2000 date problem occurs is proving immensely expensive—and this is a known problem.

12.3.1   What is a Clone?

Cloning is the most basic form of software reuse: it is the copying (and modifying) of blocks of code. Code cloning has been very extensively used within the software development community. Unofficial surveys carried out within large, long-term software development projects suggest that 25–30% of the modules in this kind of system may have been cloned. In some cases entire modules are cloned, though more usually part of a module is copied and used in another module.

The main advantage of cloning, as a means of software reuse, is that it is very simple to do. It can provide a development team with a quick start, and often a rapid solution to the problem. Copying procedures that are functionally similar to those which are required, and then modifying that code, allows customization without any software ownership negotiations. The designer copying the procedure does not have to enter into a discussion with the original designer about the procedure interface or functional modifications required, or wait for the original designer to update the procedure to incorporate the new functionality. It is obviously particularly tempting if contractors are paid for the number of lines of code they produce: a common employment contract strategy used in the past in an attempt to speed up production of code.

Clones can appear in unexpected places. The system used here has found a small number of relatively small clone procedures in the O/S of a large embedded system where much effort had been spent keeping the code small, clean, and fast.

12.3.2   The Problem of Finding and Documenting Clones

Large systems are organized hierarchically with code blocks defined within procedures, procedures gathered together into modules, modules into sections, sections into areas, etc. The names of the "units" may vary from language to language and system to system. The simplest unit to consider as a clone is the procedure.

As these large software systems have grown there is an increasing awareness that systems are becoming impossible to know. Quick bug fixes and new features written to tight deadlines all contribute to a general trend towards more disordered designs. Commercial considerations drive the system on as more investment is made. Even if clones are found they may not be removed "just in case": a "don't fix it if it ain't broke" philosophy. Complete rewrites become impossible, but periodically a need to reduce the number of system failures provides an opportunity to force a major investment simply to clean up the design. In such a redesign program a clone database tool that has already documented clones can be used to remove some or all of them. It is this approach we are using, and to do so we have built a tool to detect clones.

12.3.3   Design Objectives of a Clone Tool

We require a tool that can be used at the lowest possible level. Software engineers that have found a bug in a procedure should be able to find out whether that procedure is called by any other procedure in the system. Armed with this information they can check whether the proposed bug fix introduces other bugs into the system. A procedure database should provide this information, but may not provide information on whether the (incorrect) procedure has been copied. The clone database may then help in showing the software engineer other possible bug fix sites or lead him to opt for clone removal. Secondly, the tool should be applicable to very large systems (100,000 procedures) over a relatively long period of time, in order to catalog the whole system. Thirdly, the tool should be able to display the original procedure and a possible clone side by side for a visual check by the programmer.

12.3.4   Methods for Finding Clones

The simplest method of finding clones is a direct search. Obtain a pattern, such as a variable name, from the fixed procedure and search for that pattern throughout all the other procedures in the system. Sieves can be used: for example, ignore all procedures with an LOC value less than 90% or more than 140% of the original. The names of candidate clones may then be ranked in some way and programs such as UNIX diff and vdiff may be used to check the procedures side by side.

12.3.5   Representation of a Procedure

The direct search method is not very effective if we wish to document the clones of a system, since the whole text of every procedure will have to be searched each time we pick a new original procedure. Another method is to convert each procedure into a fixed length "feature" record using a fixed set of features. For example, many procedure clones that share the same functionality will also share similar names, so one feature that could be used is a procedure name string. Other possible "fingerprinting" techniques include

    i) Number of lines of code/comments in procedure

    ii) Keyword frequency counts

   iii) Number of parameters

   iv) Number of local variables

    v) Number of nested IFs, DOs, etc

   vi) Keyword pairs

  vii) Bigrams or trigrams of letters

 viii) Indentation

   ix) Some, or all, of the complexity measures described in Section 12.2 or combinations of the above.

12.3.6   Producing a Clone Database

This involves creating an associative memory. Given a test procedure, create a feature record and search for a procedure with a "similar" record in the database. A number of methods can be used to do this, e.g., Hash tables, k-means clustering, Trees, etc. we attempted to use a hashing algorithm on the procedure name but this was not very effective. Neural networks using records that could be compared with a Euclidean measure (in neural network terms vectors) were, however, found to be effective.

12.3.7   Representation of Software Units as Fixed Length Vectors

We chose to represent each source code module as a fixed length feature vector. The choice of representation is central to the success of an application such as this, since it is imperative that similar blocks of code have similar feature vectors.

More specifically we require that, if B is a clone of A then $|v_A - v_B|_2$ should be relatively small, where v_A and v_B are the vectors representing software units A and B, respectively.

In fact, the set of source code units can be thought of as a set of indexed taxonomic trees: one for each "original" code unit. Each node in such a tree corresponds to a set of identical copies of the same unit, with the root of the tree being the set of copies of the original unit. Branching points further down the tree correspond to instances of cloning *with modification*: the whole structure is analogous to a phylogenetic tree. The index of each node in the tree denotes the degree of divergence of the clone from its original parent, see Figure 12.3.

Due to the equivalence of indexed hierarchical trees and ultrametric spaces[18] it can be seen that the task of detecting clones becomes one of induc-

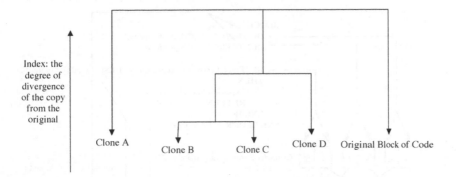

FIGURE 12.3
An indexed taxonomic tree for the clones of a single block of code. In this instance A and D were cloned and, subsequently, B and C were cloned from D.

ing an appropriate ultrametric in the space of representational vectors. Clustering the source code vectors with a SOM provides a similarity metric, whereas our newer model produces the desired ultrametric—the vectors are organized into a tree structure.

To represent all the information in a source code unit would not be feasible due to the resulting size of the vector and the need to represent a variable length structure in a fixed length format. Some degree of abstraction is therefore required. Our first simplification was that the user-chosen tokens (e.g., identifiers and operators), should be largely ignored; this gives a big reduction in the size of the problem. We were therefore left with the problem of capturing the information in the parse tree of the software unit. We do this in three ways: first, the frequencies of keywords in the unit are calculated; second, the indentation pattern is represented; and third, the length of each line is recorded. The method is applicable to any source code. The PROTEL language provides programmers with 96 keywords, including standard operators. The frequency of occurrence of each of the 96 keywords was calculated for every procedure, giving a 96-dimensional vector for each of the procedures under investigation.

As all the code is printed in a standard format (or can be easily filtered to a standard format), generated by the programming environment, the indentation pattern is isomorphic to the structure of the parse tree. The problem with representing it is that the number of lines in a unit of software, and therefore the number of indentations, is variable. What we were interested in was the *pattern* of indentation, so we drew a graph of indentation depth against line number for each procedure. To obtain fixed length vectors for the procedures we then sampled 100 points from each graph, using linear extrapolation where necessary. This gives a 100-dimensional vector for each procedure, and is relatively stable against minor modifications to the source code, such as the addition or removal of a line of code.

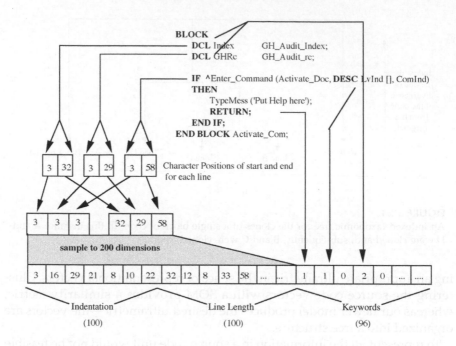

FIGURE 12.4
Illustrates how a feature vector is produced to represent a procedure.

Line length gives some indication of the usage of user-defined tokens. As with indentation, we were interested in the line length *pattern* for the procedures under investigation. This was represented on a graph in a similar way to the indentation pattern, which was again sampled at 100 points across the software unit to give a 100-dimensional vector.

Finally, each of the three vectors calculated for a procedure was normalized, so that each of them had roughly the same importance. Each keyword frequency was divided by the maximum frequency for that keyword averaged over a large set of source code units. Each indentation and line-length value was divided by a mean value calculated as before. That is, the j'th component of the i'th vector is calculated as:

$$x_{ij} = \frac{x_{ij}}{\underset{k}{mean}(x_{ik})}$$

The three vectors were then combined to form a single input vector for the clone detection tool. The overall structure of an input vector is seen in Figure 12.4. Once the procedures had been represented as a set of fixed length feature vectors it was then possible to use this set to train a neural net to identify clusters in the data. We used two different types of network: a SOM and, subsequently, a dynamic tree based network.

12.3.8 Potential Network Models

Self-organizing neural networks are powerful tools for clustering unlabelled data. They are particularly suited to the kind of problem described here as they scale up well to very large data sets and are a robust and relatively straightforward technology to use. After the network is trained the data set will be partitioned into a set of clusters of similar vectors. With most self-organizing neural networks the weight vector of the classifying output unit will be the centroid of its class, so that the network is performing a form of k-means clustering.[19]

Moreover, any novel vector can be presented to the trained net, which will identify the class in the training set to which the novel vector belongs, and therefore will identify the likely clones of this software module.

More formally, the set of N software modules: $\{S_i\}_{i \leq N}$ is first converted to a set of vectors: $\{v_i\}_{i \leq N}$.

An unsupervised neural network consisting of M output units (with $M << N$), $\{X_j\}_{j \leq M}$, is then trained on these vectors and will produce a classification of the training set. That is, a discrete partition of the training set is produced:

$$c : \{v_i\}_{i \leq N} \rightarrow \{X_j\}_{j \leq M}$$

so that the set of clusters: $\{c^{-1}(x_j)\}_{j \leq M}$ identifies the potential clones.

Having determined what the network was required to do, we were faced with a choice of neural net architectures. The available models were: the simple competitive net and its refinements, ART[11] type networks and self-organizing maps (SOMs).[10] Initially we implemented the tool using a SOM: a more complete description of this work is given in Reference 16.

We also examined the possibility of using dynamic competitive learning (DCL) to construct a "neural tree": the advantages of using such a system in place of a SOM being the speed with which the tree may be constructed and accessed.[19] A second prototype was developed using this alternative technology.

12.3.9 The Self Organizing Map (SOM)

The data set used was taken from a skim file of 1775 PROTEL procedures occupying approximately 5 Mbytes of source code. A clone detection rate was arrived at for each network as follows:
Set Clone Detection Rate to zero.
Select 100 test procedures from the data set at random.
For each of the 100 test procedures in turn:

> Find the procedure most similar to it in the clone database and make this the candidate clone.

Determine the similarity between the test procedure and the candidate clone by calculating the cosine of the angle between their representation vectors.

If the similarity of the two procedures is above a "cloniness threshold"

Have the two procedures examined side by side by a software engineer familiar with the PROTEL language and the telecommunication software, to determine whether or not the candidate clone is an actual clone.

Otherwise

The candidate clone is not an actual clone.

If the candidate clone is an actual clone

Add the similarity measure for the two procedures to the Clone Detection Rate.

Otherwise

Subtract the similarity measure from the Clone Detection Rate.

The final results were then scaled to be in the range 0 . . . 100. Table 12.2 shows the results of varying the input vector scheme with the size of the SOM.

12.3.10　The Dynamic Competitive Learning (DCL) Model

The limitations of the SOM-based tool are the long training times and the fixed number of clone classes that are created. Using the same input vector representation we have implemented a prototype tree-based dynamic neural network. As already stated the ideal classification structure for the type of data used here is a tree, rather than the flat structure induced by a SOM. As a simple example consider the tree in Figure 12.5.

TABLE 12.2

Clone Detection Rates for SOMs of Various Sizes

Input vectors used	SOM Size	Clone detection rate
Keywords	10 × 10	35.2
Keywords	15 × 15	36.0
Keywords	20 × 20	37.0
Indentation	10 × 10	41.3
Indentation	15 × 15	51.4
Indentation	20 × 20	55.0
Keywords + indentation	20 × 20	68.7

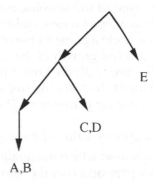

FIGURE 12.5
A simple tree structure representing the similarity of five vectors A . . . E.

Here the classification space has been divided into two clusters at the top level: one containing E only and the other, which is further subdivided, the other four vectors. The advantage of this structure is twofold:

- The depth in the tree of the nearest common ancestor of two vectors is a measure of their similarity. This is useful in the test phase—it is easy to specify the type of clone required, for example: look as far as second cousins.

- A search through such a data structure is significantly faster than a search through a flat data structure. Specifically, searching for an item in the tree is a task of complexity $O(\log bN)$, where N is the depth of item in the tree and b the average branching factor, whereas a search in a linear structure is of complexity $O(N)$. This means that both training and test phases can be completed more quickly.

12.3.10.1 The DCL Algorithm

A DCL network may be thought of as being organized like a tree, with a processing element (PE) at each node except the root. When an input vector is presented to the network it is passed to the PEs at each node in the first layer of the tree; standard competition takes place between the PEs at this level. If the weight vector of the winning PE is not sufficiently close to the input vector (the distance between them is said to be greater than a *quality* value), a new node is created at this level to classify the input.

The use of the quality value is analogous to the use of a vigilance parameter in ART. The quality value determines the radius of the hypersphere that defines the classification volume of a node; this value must therefore decrease, as the tree grows deeper, so that lower levels provide a finer classification than higher levels. If a new node is not required then the winner's weight vector is moved closer to the input vector, mediated by a learning rate, as in standard competitive learning. The learning process is then recursively applied to the children of the winner, until a leaf node of the tree is reached.

At any stage of the learning process a winning node which is currently a leaf node may spawn new children. This occurs if the relative frequency of wins of the node against its parent exceeds a *threshold* value. In order to prevent unbounded growth in the tree the threshold value is increased at each level in the tree, with the lowest thresholds closest to the root.

The shape of the resulting tree is determined by the input vectors and the choice of learning parameters. The following factors must be specified:

- the quality value of the top level nodes,
- the rate at which the quality factor decreases for successive layers, as a proportion of the previous value,
- the threshold value of the top level nodes,
- the rate at which the threshold value increases for successive layers, as a proportion of the previous value,
- the schedule for the learning rate—which should decrease after each epoch to promote stability as a proportion of the previous value.

The Quality Factor was arrived at with the aim of producing a reasonably large initial cluster. With the chosen setting the initial cluster contains 20 nodes with 4 developing a subclassification. So the initial space is divided up into 16 outlier clusters and 4 larger clusters which are subdivided. The Quality Reduction rate was chosen so that the network developed the appropriate amount of decomposition and clustering.

In the test phase the node closest to the input vector is found by a straightforward search of the tree:

- Set x to the input vector, and *current-node* to the root of the classification tree.
- Find the child of *current-node* that has a weight vector closest to x. If this node is a leaf then return it, else repeat with this child as *current-node*.

12.3.10.2   Training

With these parameters the network converges rapidly: with 10,257 of the 296-ary vectors described above only two or three epochs were required for reasonable convergence. The training time was less than forty minutes on a HP 712/80.

12.3.10.3   Results

A typical classification tree produced by the DCL is shown in Table 12.3. The leaf nodes of the tree classified, on average, about 20 input vectors. A fraction of the tree is shown in Figure 12.6.

TABLE 12.3

A Typical Classification Tree

No. of nodes	1500
Maximum ply	19
Average branching factor	2.84
Widest point	298 nodes, at depth 5

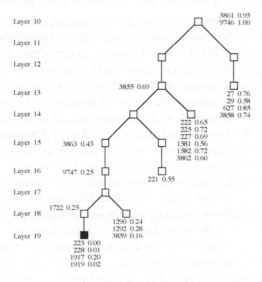

FIGURE 12.6

Classification produced by the DCL net, shows the input (left) and the Euclidean distance (right) from a given test input. For the test input 223 procedure 228 is the closest.

The quality of the classification was similar to the SOM, but this net has some advantages. The training and recall time is significantly better than the SOM, and the tree structure aids in the search for clones more distant than those in the immediate neighborhood.

12.3.11   Architecture of the Clone Detector Tool

The finished system can be viewed as consisting of two logical components, that support administrative tasks and user tasks. The administration task is to produce the classification data for a set of source code and the user task is to support the interrogation of the resulting classification in the search for clones.

12.3.11.1   The System Administrator

This main function of this part of the system is to produce a database of clone information. Initially a collection of source code procedures, which will be used to populate the database, is downloaded and preprocessed to produce a set of feature vectors, as described above. This is a nontrivial task and involves a partial source code parser built using the UNIX tools byacc, flex and newyacc, to extract the structure of the code, and then a C program to put together the feature vectors. These vectors are then used as a training set.

12.3.11.2   The User View and the Graphical User Interface (GUI)

Once a database has been generated, the navigation and interrogation of the results produced by the neural network is controlled through a GUI. The GUI was developed using GNU C++ and Neuron Data's Open Interface.

The main window of the GUI is shown in the Figure 12.7. There are five different areas to the main window. The first area contains the source procedure (top left). The second contains all the possible clones of that procedure (top right), and the three areas below display the actual code and where the differences between the two procedures occur. Once a potential clone has been identified, the user is then able to view the code not only of the source procedure, but also that of the potential clone. The two sections of code are displayed along side each other in the bottom half of the main window. The user is then able to view the code and see exactly how the sections differ and verify if that clone identified is actually a clone of the source procedure. In Figure 12.6, the only difference between the source procedure and the clone, which has been identified, is that the procedure name and one of the identifiers has been changed. The cloniness value is shown as 100% because the differences are so minor as to make the vector representation of the two procedures identical.

12.3.12   Using the Tool to Find Cloned Procedures

Once a database has been selected and loaded into the GUI the users must then identify a source procedure, this is done by using area 1 of the main window. The tool then automatically generates a list of possible clones. The criteria to which this list must conform are identified by the attributes in the top right-hand corner of the main window.

The attributes which determine the search criteria are contained in the top right-hand side of the window. They are

- cloniness—A simple similarity value. It uses the cosine measure of similarity of the feature vectors.
- threshold%—The minimum similarity that the reported clones have to match. This acts as a tolerance measure, a high value indicates that a high degree of accuracy is required.
- threshold (lines)—The minimum number of lines that the potential clones must have.

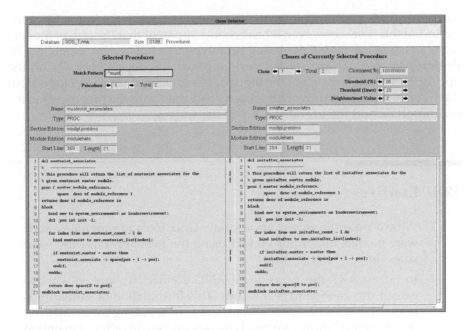

FIGURE 12.7
The user interface of the clone detector product.

- neighborhood value—The size of the area of the neural network where clones might be found. A high proportion of all clones for a particular procedure will be within the immediately surrounding area.

"Neighborhood" defines the search area, in the neural network, with the selected procedure belonging to the neural network class at the center of the search area.

Closely matching neural network classes are situated close together in the network. If the user wants to increase the search area to bring in more potential clones then the neighborhood search size can be increased. Increasing the search area will introduce more potential clones, but these will have a lower cloniness value. Rather than searching for clones for each procedure independently the tool allows for a list of all the clones in any particular database to be produced. However, as the size of the database grows the time to perform the operation grows exponentially.

12.3.13 Conclusions

Cloned software is prevalent in large software systems and this project has shown how it can be managed without complete reengineering of the code. It has also shown how neural computation can be used and integrated into fully functional tools. In fact, one of appeals of using neural nets is their relative ease of use when compared with other sophisticated techniques. Over the period of this work, however, it has become apparent that the task of moving from a successful, neural net-based prototype to a full system should not be underestimated. All neural network applications depend heavily on the appropriate preprocessing of the input data and postprocessing of the output data. A major part of our work has been concerned with the preprocessor, the user interface and the overall quality of the system.

12.4 Exercises

1. Write a program or use appropriate Unix tools to preprocess software models into input vectors using the twelve complexity measures detailed in the text.

2. In Fuzzy ART the fuzzy logic AND connective, *min*, is used to extend the method to real values. This formulation of similarity is unusual and a little inconsistent, since the resulting measure is not the canonical Norm 1 metric. It has arisen here in order to maintain the biological inspiration of the original ART model. It would be interesting to investigate the implication of using the Fuzzy Art similarity measure in a conventional clusterer. Write a cluster program to implement this similarity measure.

3. Using the source code program from 1, develop and investigate the results from a clone program that uses these complexity measures as its representation.

References

1. J.W. Sheppard and W.R. Simpson (1990), "Using a competitive learning neural network to evaluate software complexity", *ACM*, pp. 262–267.
2. K.H. Möller and D.J. Paulish (1993), *Software Metrics*, London: Chapman & Hall.
3. B.R. Schendler (1989), "How to Break the Software Logjam", *Fortune*, pp. 72–76, 25 September 1989.
4. C.M. Lott and H.D. Rombach (1993), "Measurement-based guidance of software projects using explicit project plans", *Information and Software Technology*, Vol. 35, No. 6/7, pp. 407–419.
5. T.J. McCabe (1976), "A complexity measure", *IEEE Transactions on Software Engineering*, Vol. 2, No. 4, pp. 3–15.
6. M.H. Halstead (1977), *Elements of Software Science*, New York: Elsevier Science.
7. P.G. Hamer and G.D. Frewin (1982), "M.H. Halstead's software science—a critical examination", in *Proceedings of the 6th International Conference on Software Engineering*, pp. 197–206.

8. T.C. Jones (1991), *Applied Software Metrics*, New York: McGraw-Hill.
9. B. Moore (1988), "ART1 and pattern clustering", in *Proceedings of the 1988 Connectionist Models Summer School*, eds. D. Touretsky, G. Hinton and T. Sejnowski, San Mateo: Morgan Kauffman, pp. 174–185.
10. T. Kohonen (1982), "Self-organized formation of topologically correct feature maps", *Biological, Cybernetics*, Vol. 43, pp. 56–69.
11. G.A. Carpenter and S. Grossberg (1987), "Adaptive Resonance Theory: stable self-organization neural recognition codes in response to arbitrary lists of input patterns", Eighth Annual Conference of the Cognitive Science Society, Hilldale, N.J., U.S.A.: Lawrence Erlbaum Associates.
12. G.A. Carpenter and S. Grossberg (1987), "A massively parallel architecture for a self-organizing neural pattern recognition machine", *Computer Vision, Graphics and Image Processing*, Vol. 37, pp. 54–115.
13. G.A. Carpenter, S. Grossberg, and D.B. Rosen (1991), "Fuzzy ART: fast stable learning and categorization of analog patterns by an adaptive resonance system", *Neural Networks*, Vol. 4, pp. 759–771.
14. D. DeSieno (1988), "Adding a conscience to competitive learning", *IEEE International Conference on Neural Networks*, Vol. 1, pp. 117–124.
15. P.G. Schyn (1991), "A modular neural network model of concept acquisition", *Cognitive Science*, Vol. 15, pp. 461–508.
16. N. Davey, P.C. Barson, S.D.H. Field, R. Frank, and D.S.W. Tansley (1995), "The Development of a Software Clone Detector", *International Journal of Applied Software Technology*, Vol. 1, No. 3/4, pp. 219–236.
17. W., Wayt Gibbs (1994), "Software's chronic crisis", *Scientific American*, pp. 72–81, September 1994.
18. R. Rannal, G. Toulouse, M.A. Virasoro (1986), "Ultrametrics for physicists", *Review Modern Physics A*, Vol. 58.
19. K. Butchart, N. Davey, R.G. Adams (1995), "A comparative study of three neural networks that use soft competition", in *Proceedings of International Workshop on the Applications of Neural Networks*, 1995.

13

Elements of Computational Intelligence for Network Management

P. Venkataram

CONTENTS

0-8493-1075-X/01/$0.00+$.50
© 2001 by CRC Press LLC

ABSTRACT   Communication networks are becoming exceedingly difficult to control because of increased traffic volatility, coupled with demands for higher efficiency, resilience, and economy. Management of such a network requires real-time information on the status of the network, exchanges, routes, common channel signalling system, traffic, etc., in order to provide an effective, efficient, high quality service, and efficient utilization of network resources. Conventional automation techniques for the management offer only a partial solution, since they do not address the full diagnosis/planning process of humans which is required for efficient management.

Network management problems are fertile fields for Computational Intelligence (CI) technology because of their complexity, reliance on experience, and the highly procedural approach used by human experts in their solution.

This chapter briefly discusses the network management and its complexity along with categories of network management being done presently. Later, it addresses some of the elements of CI used to solve the network management problem. The CI-assisted network management techniques described here combine the use of intelligent technologies like expert systems, knowledge-based systems, distributed AI, reasoning systems, neural networks, the machine learning capability, etc.

KEY WORDS: *Network management, fault management, fault detection, proactive anomaly detection, neural networks, feedforward network, backpropagation, Hidden Markov Model (HMM), feature extraction, segmentation, Bayesian network, router, Management Information Base (MIB), IP, UDP.*

13.1 What is Network Management

The communication networks are not only growing in large scale but are becoming more complex and heterogeneous as they support multi-vendor applications on a variety of underlying transmission facilities. Such expansion and complexity have induced growing challenges to manage the network elements according to the network service providers' and users' objectives and expectations.[1-3] As a result, a lot of research effort has gone into solving problems arising in this area and establishing standards that could be used across a broad spectrum of product types (e.g., hosts, routers, bridges, switches, telecommunication equipment) in a multi-vendor environment.[4]

The current telecommunications environment is based on international standards, such as the Intelligent Networks (IN)[5] and Telecommunication Management Networks (TMN),[6] representing the foundations for the uniform and efficient creation, provision, and management of advanced telecommunication services. Current IN and TMN architectures are based on highly centralized approaches for the location of service control and network (and service) management "intelligence", where related protocols (i.e., SNMP (Simple Network Management Protocol)[7] and CMIP (Common Management Information Protocol))[8] rely on the traditional client/server paradigm, service, and management programs, respectively.

Basically, a general network management system (NMS) contains four types of components:

Network Manager (or simply Manager), *agents* running on managed nodes, *Management protocols,* and *management information.* A manager uses the management protocol to communicate with agents running on the managed nodes.[2, 3]

13.1.1 Network Management Problems

Most of the problems that arise in today's network management systems are directly related to the end-to-end communication. The problems, generally, are Router problems, Bandwidth bottlenecks, Quantity, and Quality of data transferred on the network.

Traditionally, the primary objective of network management has focused on fault detection and restoration. While this is an admirable objective, network management encompasses a great deal more on operations and maintenance of the networks. Hardware is increasing in reliability, although this advantage is offset by the rapid growth in size and complexity of today's networks. Further, the complex distributed applications such as client/server systems has shifted this focus away from merely detecting a power or chip failure in a device to a statistical analysis of traffic necessary to meet stringent user's performance criteria.

The main network management requirements are to:

1. Have a nearly real-time identification of the network operational status, especially in overload and congestion situations on an origin-to-destination basis.
2. Diagnose and locate the degradation of performances and services by segmenting the network, thus either targeting better and more economical corrective actions, or sharpening the focus of planning.
3. Monitor and qualify the accessibility and availability of the network from a subscriber point of view, aiming at developing the procedures for an efficient management and planning of the available resources.

13.1.2   Network Management Models

The ISO/ANSI standards committee[9] has classified the sophisticated functionality required of network management systems into the well-known six categories. They are: *configuration management, fault management, performance management, security management, accounting management,* and *directory management.* The functional architecture defined by these six categories clearly identifies the different facets of network management and control, and enables a modular approach to be taken towards designing network management tools. However, there is considerable overlap and interaction between the various management systems.

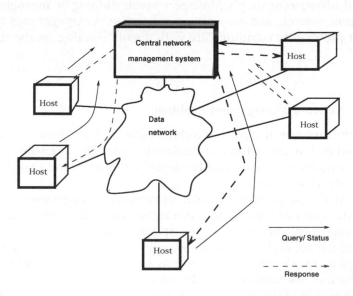

FIGURE 13.1
Centralized network management.

Contemporary network management systems, based on the platform-centered paradigm, hinder users from realizing the full potential of the network infrastructure on which their applications run. The paradigm needs to be augmented to allow for decentralized control and intelligence, distributed processing, and local interpretation of data semantics. This means that centralized/decentralized nodes, known as service central points (SCPs) and operating systems (OSs) agents are used for hosting services and managing programs, respectively. The network management models are categorized as follows:

13.1.2.1 Centralized Network Management

The Internet-standard network management framework uses SNMP as a reasonably general and extensible data-moving protocol. For example, the centralized SNMP paradigm[10] evolved for several reasons (see Figure 13.1). First, the most essential functions of network are well realized in this paradigm.

Agents running of the hosts are not capable of performing self-management when global knowledge is required. Second, all network entities need to be managed through a common interface. When many of these entities have limited computation power, it is necessary to pursue the "least common denominator" strategy. Unfortunately, in many cases this strategy does not allow for data to be processed where and when it is most efficient to do so.

13.1.2.2 Decentralized Management

Decentralized Network Management (DNM) or Management by delegation (MBD)[11] utilizes a decentralized paradigm that takes advantage of the increased computational power in network agents and decreases pressure on centralized NMS and spatial distribution (distribution over different network devices). In this paradigm agents that are capable of performing sophisticated management functions locally can take computing pressure off the centralized NMSs, and reduce the network overhead of management messages (see Figure 13.2).

13.1.2.3 Distributed Network Management

The two paradigms of network management presented above might be viewed as contrasting, competing, possibly even incompatible models. The reality is that the centralized paradigm and the MBD (or decentralized) paradigms are really just two points on a variety of continuous scales. An ideal network management system should be able to handle a full range of network management functions.

13.1.3 Network Management Information

A network may be interpreted as a collection of devices, and provides a facility for transferring data from one device to another. A typical network may

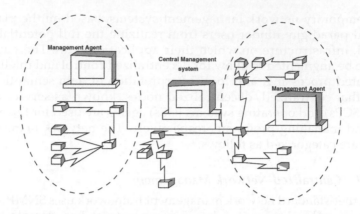

FIGURE 13.2
Decentralized network management.

include devices such as central processing units, monitors, printers, modems, etc. The information relating to network devices is called *network management information*.[12]

The network management protocols unify the syntax of managed data access, but leave semantic interpretation to applications. Since semantic heterogeneity of managed data has grown explosively in recent years, the task of developing management applications has grown more onerous.[13] The management information can be classified into three types (see Figure 13.3):

- **Sensor data:** these data represent the current network health.
- **Control data:** these capture the current setting of the networks. The control data is the higher level abstraction of the management and adopts two broad classes of objects: *Quasi-static* which describe the current network configuration, and *Dynamic objects* which relate network events (e.g., the transmission pockets).
- **Structured data:** these data describe the physical and logical construction of the network.

13.1.3.1　The Semantics of Network Management Information

The purpose of semantics of management information is to clarify the meanings of some real world entities. This kind of clarification and its usefulness, however, are very closely related to the application context and to the specific human community. The basic philosophy of the approach is to characterize the semantics of the management information by specifying the contexts in which the information can occur. In this way we can avoid the complexity associated to the traditional, behavior-oriented semantic frameworks, both in the semantic specification itself and in deriving useful results from these specifications for the purposes of management decisions.

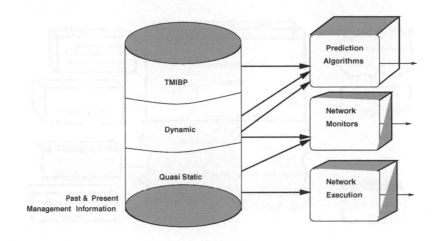

FIGURE 13.3
Types of network management information and their use.

13.2   Standard CI Techniques Used in Network Management

There are various CI methods used in network management of heterogeneous, hybrid, multi-vendor, high-traffic, and complex networks.[14, 15] At the highest levels in the management hierarchy, the CI methods are used to process the enormous amount of raw data about network health, status, etc. that is coming continuously from the network elements. CI techniques will also be needed to solve large resource allocation problems. At the lowest level, CI will be used online in order to accelerate the control process and reduce human intervention. Some of the CI systems[16, 17, 18] are used for representing and dealing with uncertainty and inconsistencies. Other methods are dealing with nonmonotonic reasoning including default reasoning. CI techniques are efficacious because:

- the inter-netted data communications networks are extremely complex with dynamic topologies and requirements for real-time response and critical reliability and availability requirements.
- network management, monitoring, and control functions must be distributed to ensure survivability of the network.
- skilled personnel for managing network resources, and controlling and maintaining the networks are scarce and mobile, therefore CI-based systems that capture their expertise are required.

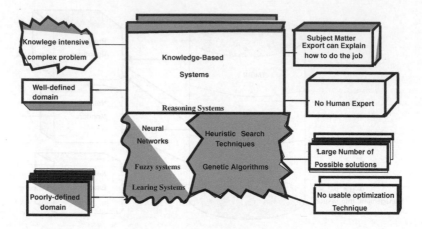

FIGURE 13.4
Elements of CI and their applications areas in network management.

- CI tools and techniques have demonstrated that they can provide improved network reliability (service, line quality, up-time) and can speed up the repair process.
- Accelerate the control process, try to automate that process, and if possible remove any human intervention from the control loop.

Different CI techniques are required to perform different network management tasks.[19] Many CI systems model uncertainty based with probabilities or level of confidence and probability inference computations. Some CI-based systems operate online, and they can be hardware based (like fuzzy logic and neural networks) rather than software-based.

Nowadays we can come across many network management techniques based on elements of CI-technology such as expert systems, knowledge-based systems, neural networks, distributed CI, reasoning systems, etc. Figure 13.4 illustrates some of the CI techniques used in different parts of the network management. In the following sections, we discuss some of the existing/proposed network management methods based on some of the elements of CI-Technology which are employed to perform different network management tasks either individually or collectively. For example, when a large number of possible solution situations are imminent in the network, heuristic search techniques or genetic algorithms may provide better network management techniques.

13.3   Expert System Based Methods

The future for expert systems in telecommunications looks very promising. With increasing competition, staff limitations or reductions, and the need for increased productivity and efficiency, expert systems is an extremely viable technology for addressing network management issues.

Expert systems are changing the programming landscape, particularly in network management, with currently operational systems demonstrating the ability to anticipate trouble and learn from experience. Expert systems applied in the telecommunications network management domain are being used mainly in the following functional areas: fault isolation and diagnosis, monitoring, scheduling, planning, configuring, interpretation, classification, design, network administration, and help desk.[20, 21]

13.3.1   Fault Isolation and Diagnosis

Fault isolation and diagnosis problem starts with the observation of some behavior which is recognized as a deviation from the expected or desirable, that is, a malfunction behavior is observed. Expert system based methods have been extensively used in this area.[22] Some of the systems are explained below.

13.3.1.1   Integrated Testing Expert System for Trunks (I-TEST)

I-Test[23] demonstrates the usefulness of integrated computational intelligence techniques into the network operations environment. I-Test was developed with a windowing system, graphical kernel system (GKS), and a set of high performance portable C programming language based on CI tools. I-TEST adequately builds to test and maintain trunks remotely from a switching control center. It has the following subsystems (see Figure 13.5):

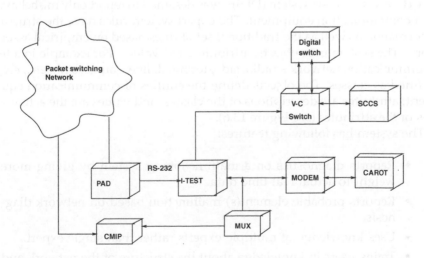

FIGURE 13.5
I-TEST diagnosis for telecommunication trunks.

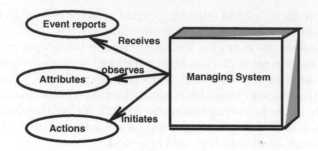

FIGURE 13.6
Interaction among the classes of Network Management in DES diagnostics concepts.

(a) The centralized automatic reporting on trunks (CAROT) system is used for transmission testing in a demand mode.

(b) The circuit installation and maintenance assistance package (CIMAP) is an administrative system that provides circuit order information and trouble-ticket generation.

(c) The switching control center system (SCCS) is primarily used for switch maintenance.

I-TEST combines traditional structural procedural knowledge and a typical rule-based production system into one inference mechanism to manage a complex trunks testing.

13.3.1.2 Diagnostic Expert System

The diagnostic expert system (DES)[24] was designed to report only misbehaving of a communication equipment. The expert system rules have the structural information as well as the traditional set of rules based on empirical associations. The system uses objects, attributes, and values. For example in a tele-communication network's radio, side, terminal, line, source, sink, etc., are all defined as classes. The objects define the entities of communication equipment, attributes are descriptions of the classes, and values are the actual values of the attributes (see Figure 13.6).

The system has following features:

- Reports diagnostics on a malfunctioning system by giving more weight to actual real-time data.
- Reports probable element(s) malfunction based on network diagnosis.
- Uses knowledge of multiple experts rather than single expert.
- Trains users in knowledge about the structure of the network and system.

This expert system effectively and efficiently diagnosis malfunctions in a large telecommunications network in a manner similar to an experienced field engineer. It operates using a real-time database created by the network monitoring system and network structure information. As a result, repairs are easier since the correct remote location and often exact cause and remedy of the malfunction are known.

13.3.1.3   An Automated Cable Expert System

Automated Cable Expert (ACE) system[23] is used to identify cable routes needing rehabilitation and preventive maintenance to anticipate and thereby avoid customer report trouble. ACE is a pioneering batch expert system running analysis software overnight. Some parts of the expert system work online and meet the reasonable timing constraints. It obtains data from the conventional cable repair administrative system (CRAS) and finds trouble spots diagnoses. ACE helps the technicians who are doing the analysis by providing user-friendly messages and a filtering function.

13.3.2   Monitoring Systems

The network monitoring systems are concerned with the observation of the status and behavior of network components' parameters. In other words, it does tracking of proper functioning at network's critical hardware and software resources.

13.3.2.1   DAD: Monitoring Subsystem

Monitoring and maintaining a modern communications network can be a formidable task even with automated support.[20] These problems are described in Datapac Advisor (DAD), which has successfully demonstrated the value of expert systems in managing the Canadian National Datapac network. The system architecture is based on loose coupling of four subsystems, which deal with real-time alarm monitoring, repairs assistance, trouble ticket creation, and database access.

The aim of DAD's monitoring subsystem is to respond to events and alarms as they occur on the network. It works directly in the real-time environment, analyzing the situation as each new alarm comes in. Each alarm is processed in six subsequent phases:

(a) Reading and formatting alarms
(b) Filtering out irrelevant information and redundant alarms
(c) Clustering together all alarms pertaining to a single problem
(d) Analyzing each problem to determine its nature, state, and progression
(e) Assigning priority to each problem
(f) Displaying the problem information to the operator

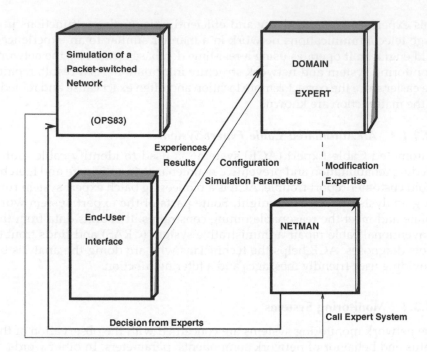

FIGURE 13.7
SIMNETMAN network management environment.

13.3.2.2   *SIMNETMAN*

A rule-based expert-system prototype, *SIMNETMAN,*[25] along with its companion system, NETMAN for network management, serves as a testbed for exploring the use of expert systems for network management by addressing the problems of traffic routing, congestion, and flow control for packet networks.

SIMNETMAN drives the simulator and invokes the NETMAN experts using an expert system approach (see Figure 13.7). This expert system is implemented by using language, OPS83.[26] Each rule in the system is made sensitive to exactly one kind of goal. Rules are invoked automatically by the production system interpreter when the invocation is appropriate. The control is data-driven; that is, to invoke a rule, a data pattern must match the goal on the left-hand side of the rule. The SIMNETMAN runs on IBM PC.

13.3.3   Network Maintenance and Troubleshooting Systems

In general, a typical expert system-based troubleshooter saves the user's symptoms and the actual problem resolution in a trouble history database. The next time a user reports a similar problem, troubleshooters will use these data and a combination of statistical and expert systems techniques to identify the most likely cause of the new problem.

13.3.3.1 Expert Tester

Expert Tester[27] was developed as an expert system for troubleshooting faulty special service analog telephone circuits. The implementation is as follows:

- Description of the Task
- Representation of knowledge in Tester
- Network models in Tester
- Diagnosis process

This is a production knowledge-based diagnostic expert system used to illustrate an implementation of the theory and the concepts that are used in the network diagnosis.

13.3.3.2 ATM Network Topology Tuner

The management system, LEN (Learning Expert for Network), performs the management tasks by monitoring the network simulator, learning traffic patterns, and triggering actions to tune topology.[28] In an ATM network the traffic is classified into two types: first, without expressing pipes, all traffic has to be stored and switched at each intermediate node; and second, with express pipes, traffic going through the pipes can bypass the intermediate switches. This expert tuner demonstrates how to learn the traffic patterns and tune the topology to the discovered patterns and will show the performance improvement.

13.3.3.3 Rule-Based Expert System (RBES)

The RBES is a real-time data driven expert system[16] to determine a near-optimal set of routing and flow controls that redirect calls around failed or congested elements and throttle the flow of traffic into the network under certain overload conditions. The RBES can handle the following anomalies: *node failure, link failure, link degradation, link congestion, general traffic overload.*

13.3.3.4 A Multimedia-Based Expert Network Troubleshooting System

The multimedia framework provided by Milan[29] has been applied to build a natural language front-end to an expert system for troubleshooting digital data services (DDS) in the circuits. The expert system's knowledge base contains description of the different circuit components and tests as well as layout specific circuits. The system's interactive interface allows the user to specify and monitor the desired network configuration. In the multimedia scenario, the interaction can be in terms of text, speech, or graphics.

The graphics mode is exercised via the mouse. The interaction might involve displaying a selected circuit and executing specified test sequences.

13.4　Knowledge-Based Systems

Knowledge-based systems are developed in wide varieties of tasks in telecommunication.[30] The knowledge-based techniques are more suitable candidates for network management.

13.4.1　Fault Detection Systems

　　1.　A knowledge-based approach for network trouble analysis

This knowledge-based approach is to analyze the multiple faults in a network.[31] In this work, knowledge systems of the various network elements have to interact with each other to satisfy service requests. Therefore, when a problem occurs in one network element, failure messages may be generated not only by network element but also by interacting with other network elements. There are three major steps in the process: *(1) relating messages to call events, (2) determining the decision point,* and *(3) identifying the defective network element.*

　　2.　Knowledge-based techniques for fault detection
　　　　in microwave communication

The technique is aimed at the diagnosis of faults in digital communication equipment.[32] The rule-based system and the machine learning system, both parts of this technique, provide solutions to the problem of fault diagnosis for digital microwave radio. The rule-based approach uses empirical rules of thumb to make the combinational effects of the manageable, whereas the machine learning techniques treats the variables as being of a continuous nature and uses adaptive pattern algorithms to provide solutions.

　　3.　MIDA

MIDA (Multi-function ISDN Diagnostic Advisor)[33] is a knowledge-based system that assists operators in the treatment of ISDN troubles. MIDA is a multifunction environment providing main functions that are the system core, and auxiliary features, supporting the diagnostical activities (such as diagnostic session suspend/resume, a context-sensitive help for operators that need an "easy to be used" system but also self-explaining and appropriate to their kind of work).

　　The system supports the operator during the phone dialogue with customers suggesting network element tests, interpreting answers, data, measures, performs trouble diagnosis, and suggests recovery actions. The tool guides to the whole sequence of operations needed to correctly locate the customer

ISDN trouble (possibly caused by wrong service configuration, exchange and subscriber loop anomaly, wrong terminal usage, etc.). It provides a correct interpretation of data and measures resulting from network remote tests, administrative information systems, and customer's answers and distinguishs actual faults from problems caused by a wrong customer's exploitation of service features.

4. Knowledge-based GUI for network surveillance
   and fault analysis

Building efficient graphic user interfaces (GUIs) to a network management system[34, 35] is one of the prerequisites to its overall success. Complex network management system loses much of its value if:

(a) network operations staff do not feel comfortable using the system, and

(b) the user interface is ignorant of the application domain and does not provide protection and guidance to the staff.

Some general requirements may be formulated for the new generation of user interface, which should be more knowledgeable about network management functions:

(a) The user interface should be a high level interface idiosyncratic to the domain. The end-user should be able to change the interface behavior within the defined semantic borders.

(b) The interface should be knowledge-based, which means two things: *first*, during the application development process the interface should support the validation of the consistency, correctness, and completeness of end-user specifications; and *second*, during the network management process the interface should provide intelligent event interpretation and analysis power.

13.4.2 Monitoring Systems

1. A Knowledge-based System of ATM Network Management

To achieve a high degree of performance in ATM network management, an intelligent network management system including SNMP-based network monitor and knowledge-based network control is built.[36] The key components of the system (see Figure 13.8) are following:

- **Network Monitor:** It is a SNMP-based network monitor. Upon receiving a large amount of data from SNMP-agent(s), the original data has to be analyzed and the valuable information extracted from them according to the "Filter" component. For example, to achieve the performance management and fault management, the

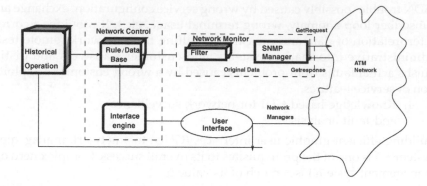

FIGURE 13.8
A KBS for ATM network management.

object values of the group are evaluated by the formula. If the error rates are bigger than these values of "alarm raising threshold," then an event will be generated and will drive the knowledge-based network control module.

- **Network Control:** The network control module is built by using the knowledge-based technology. The knowledge bases: includes a database and a rule base, are the most important components in the proposed technology. The functionality of database is to store these events from the "Filter" block. And the rule-base possess a lot of heuristic rules of diagnostic action.

13.4.3   Network Maintenance and Troubleshooting Systems

Some of the knowledge-based network maintenance and troubleshooting systems are:

Telecommunications network maintenance

Network-maintenance imposes additional demands on the operator, for example:

- the need for learning about, and interacting with, multiple types of equipment and network entities,
- coordination of symptoms and actions from multiple sources for problem isolation and sectionalization,
- coping with large amounts of data because of the increased network sizes and complexity,
- ability to recognize, prioritize, and react to the more significant problems quickly, and
- interworking with other operators and maintenance organizations, including the practice of effective problem escalation procedures.

A prototype knowledge-based Maintenance Advisor for DMS-100 digital switches (MAD)[37] has been designed for monitoring and maintenance in packet-switched networks. The knowledge of maintenance that is capturing in MAD is spread across many human experts. It includes a large set of troubleshooting flowcharts (more than 2300 pages) called the performance-oriented practices, and like the advisor, these are designed to aid craftspeople in clearing alarms.

13.5 Reasoning Systems

A number of different model logics have formalized, and inference rules comparable to propositional and predicate logics are available to permit different forms of nonmonotonic reasoning. Like modal logics, fuzzy logic was introduced to generalize and extended the expressiveness of traditional logics.

1. Model-based reasoning for sectionalization

In this model-based sectionalization, the network is represented by the components: a near-end switch, near-end element, carrier element, far-end element, and far-end switch.[38] The general strategy refers to the problem solving by reasoning the component-knowledge and the guidelines for fault detection. It includes performing tests at various test points based on symptoms using knowledge of the interconnectivity of the components, using the test results and knowledge of behavior of the components to reason about faults. The method works as follows:

(a) The general structure of the network and the general strategy for diagnosing the trouble in the network are used as reference points to start the problem-solving process.

(b) For the given symptoms, the general structure of the network is mapped to the specialized structure, and the general strategy.

(c) The knowledge of mapping is stored in the form of rules.

(d) The procedures in the specialized strategy are based on the symptoms and the knowledge of structure and behavior of the interconnected components.

2. Causal reasoning-based diagnosis

The task of diagnostic reasoning as a process of causal reasoning[28] from misbehavior to structural defects. The model of the domain, or a system, under consideration that is reasoned with, in order to generate information about the expected behavior of the system, has been variously called *causal* or *deep*. Given a representation of the behavior of the components of a device or a network system and a representation of its structure (i.e., the interconnection of the components), the ability to generate the behavioral description of the system as a whole is an important part of causal reasoning. This knowledge may need to be further organized to explicitly represent the hierarchical structure of the device or the system for the diagnostic process.

13.6 Distributed Artificial Intelligence-Based Methods

Distributed Artificial Intelligence (DAI)[39] is concerned with problems that arise when a group of loosely coupled problem-solving agents work together to solve a problem. Usually, each of these agents has a limited perspective about problem-solving activity and can only gain knowledge about events outside itself through communication and coordination with other agents. The three major aspects of network management that have been identified as being amenable to solution through distributed artificial intelligence are:

1. A distributed situation assessment task,
2. A distributed diagnosis problem, and
3. A distributed planning task.

13.6.1 Network Management by DAI

DAI is useful in domains in which action, perception, and/or control are naturally distributed. It is also a way to reproduce the human behavior when a group of experts works together to realize a task. For these reasons multi-agent systems are envisaged to control and manage future networks (see Figure 13.9).

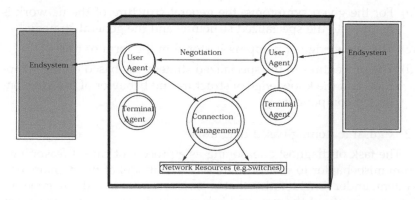

FIGURE 13.9
DAI based multi-agent system for network management.

The DAI field is often divided into two sub-areas[39]: *Distributed Problem Solving* (DPS) and *Multi-agent systems*. DPS considers the work of solving a number of modules that cooperate by dividing and sharing knowledge about the problem and the developing solution. Multi-agent systems include intelligent behavior among a collection of possibly preexisting, autonomous, intelligent "agents," when they can coordinate their knowledge, goals, skills, and plans to take action or solve problems. The agents in a multi-agent system may be working towards a single global goal or separate individual but interacting goals.

The basic goal is to bring the management intelligence, i.e., the management services, as close as possible to the managed source, i.e., its logical representation in the form of managed objects. This increases the intelligence and autonomy of a management agent by delegating management activities from the manager to the management agent(s). The management agent(s) is(are) acting as (a) proxy manager(s), which help to reduce the amount of communication between manager and management agent(s). The management agent has to be regarded as a specific agent execution environment supporting the remote execution of management scripts.

13.7 Heuristic Methods

Some emerging ideas in an area of CI technology are evolving as *heuristic search algorithms* for network management. The term *search* refers to the identification of a desired solution from a large collection of possible solutions. Heuristics are the knowledge used to make good judgments, or the strategies, tricks, or "rules of thumb" used to simplify the solution of problems. Heuristics play an important role in search strategies because of the exponential nature of most problems. Some of the desirable properties of heuristic search algorithms are: *admissibility, completeness, dominance,* and *optimality* properties.

1. Induction and deduction in network management

In this proposed framework,[29] knowledge-related underlying network is learned to capture network traffic pattern and refine the prespecified domain knowledge. The learning systems have more advanced abilities than nonlearning systems in performance and fault management, which require understanding of traffic patterns and knowledge of causality.

The proposed scheme is designed to operate on the standard platform of MIBs and CMIP. Two main contributions are the global abstraction, and the integration of learning and inference for autonomous management applications.

2. The temporal network management model

The temporal management information model represents the past, current, and future network state in a Temporal Management Information Base, (TMIB).[40] The TMIB is the core of a network management system as it provides the interface between the network administrator and all the functions of network management. One important parameter in TMIB design is to have the network administrator interact solely with the database, by using a temporal query language, so that, from the user's point of view, the TMIB embodies the network diachronically. As a result of this TMIB design, whenever the network administrator wishes to make changes in the network objects, such as changing the routing scheme, the administrator updates the appropriate variable in the database by using a temporal replace statement.

3. Management of distributed intrusion detection

Intrusion detection refers to the ability of a computer system to automatically determine that a security breach is in the process of occurring, or occurred at some time in the past. It is built upon the premise that an attack consists of some number of detectable security-relevant system events, such as attempted log-ons, file accesses, and so forth, and that these events can be collected and analyzed to reach meaningful conclusions.

Intrusion detection is an excellent candidate application for decentralized management.[41] There is a high motivation for decentralized intelligence and processing because it is very clear that centralized processing won't scale, and that network bandwidth won't accommodate all audit data being sent to a centralized point.

13.8 Neural Network-Based Methods

Neural Networks can contribute to this emerging new telecommunication management by providing fast, flexible, adaptive, and intelligent control.[17] The basic idea behind neural networks is to use modern parallel computation techniques to structure certain CI problems similar to the structure which underlines human brain functions. A typical neural network architecture might consist of three layers as shown in Figure 13.10. The *output layer* gives the results, i.e., a classification of some input pattern. The *input layer* reads in some features or attributes of that input pattern. Each input cell is potentially connected to all the *hidden layer* cells which are potentially connected to other hidden layer cells and to the output layer cells.

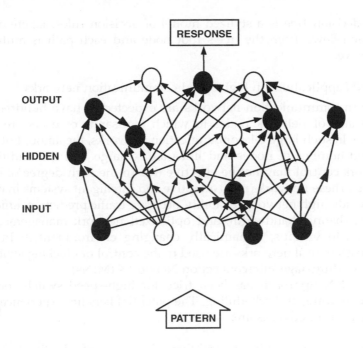

FIGURE 13.10
A typical neural network.

1. NN-based machine learning approach for network diagnosis

The machine-learning approach involves a series of steps for network diagnosis. The first step is to define and extract features from time-varying diagnostics data and transform that data into a standard classification format.

Once we have sample data, we can use several computer-based techniques to predict patterns of future faults. We can divide these computer-based learning methods roughly into three categories:

- weighted methods, such as linear discriminants and neural nets;
- symbolic methods, such as decision trees or decision rules; and
- case-based (nearest neighborhood) methods.

Both neural nets and linear discriminants make predictions on the basis of weighted functions. This linear discriminant uses a simple additive scoring function.

The neural net, on the other hand, can model more complex decision functions, typically nonlinear functions. Decision tree and decision rules pose solutions in the form of true or false conditions.

A decision tree is a stylized model of decision rules, where every decision flows from the tree's root node and each path is mutually exclusive.

2. NN applications in high-speed communication networks

Modern communication networks are expected to have hundreds or thousands of network nodes to which thousands of users are connected. In such large telecommunication networks, the amount of traffic and number of nodes and links are so large that the traditional network control may not be effective due to the high degree of complexity. These networks need adaptive and intelligent systems in order to provide high network reliability, accurate traffic prediction, efficient use of channel bandwidth, and optimized network management in relation to various, dynamically changing environments.[42] In this approach neural networks are used to the control of blocking switches such as Multistage Interconnection Networks (MINs).

This NN approach has been tried for high-speed switch control, packet routing, and scheduling. The potential benefits experienced by using neural networks are:

- an efficient adaptive control through the use of adaptive learning capacity of neural networks;
- sub-microsecond solutions to optimization problems due to massive parallelism; and
- a high degree of robustness.

3. An NN-based expert system for network topology error identification

An NN-based expert system is designed for network status identification with two major components: one is the rule-based part and the other is NN-based part. The rule-based expert system can be set up without many samples and the system has the ability to learn from new cases by itself. So, the expert system can take the advantages of these two components. In this artificial neural network (ANN),[43] where input neurons represent logical conditions, and output neurons represent results, an ANN-based logical procedure is constructed. The main feature of an ANN-based expert system is that it is a self-learning system. It is well known that a rule in an ANN is trained from samples, not predefined by humans. That means, the beginning of building an ANN-based expert system can include no rule at all. After inputting enough amount of samples, the expert system can form rules and logic by itself. At the same time, the most disadvantage is that to train the expert system a lot of samples are needed which are not easy to acquire in some subjects.

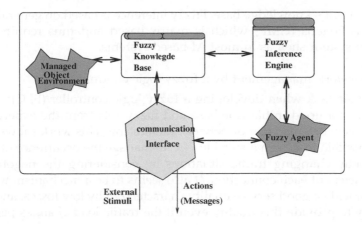

FIGURE 13.11
Fuzzy logic based network monitoring system.

13.9 Fuzzy Logic-Based Methods

Fuzzy logic is a suitable instrument to manage the highly growing networks. Fuzzy logic does not need a mathematical modeling of the controlled process, it needs no information about the structure and dynamic behavior of the system. Fuzzy logic is amenable to silicon implementation and provides an efficient mechanism for knowledge representation and inference. It is also suitable for handling vagueness and approximate reasoning prevasive in complex systems. Fuzzy systems are easy to implement. These are additional motivations for the present investigations on the usefulness of fuzzy logic controllers in communications.[44]

13.9.1 Monitoring Systems

1. Fuzzy agents for intelligent network monitoring

Fuzzy logic based multi-agent approach for network management[45] is shown in Figure 13.11. These multi-agents are self-executable software entities (daemons) designed to carry out a collection of prescribed tasks. These agents are nonautonomous because their actions are completely determined by their experience; rather they derive expertise from a collection of rules called knowledge-based which encode expert work monitoring strategies.

The fuzzy knowledge base is a collection of fuzzy rules which encodes experts' problem-solving strategies. These rules are activated by sensory device attached to the managed objects and onto the fuzzy agent. The *fuzzy inference engine* is a computational scheme for processing rules in the agent's knowledge base. Fuzzy inference is based on generalized Modus Ponens (GMP) which is matrix based and thus requires no expert system shell as in most AI-based systems.

2. Network management by a fuzzy logic controller

The main task when developing a fuzzy logic controller (FLC)[44] is to construct a good initial rule-base and fuzzy sets from the experience of the network manager or network monitoring. This work shows that it is possible to design a suitable FLC to manage the occurrence of permanently changing traffic situations by considering the negotiated parameters of each connection. It also seems to be a mechanism which can provide a good service quality characterized by low losses, and the ability to provide this quality even if the traffic load changes permanently.

13.10 Deductive Database Technology to Network   Management

Deductive databases offer the prospect of a more intelligent way of handling data by applying deductive rules to capture the complexity of network management information.[46] The deductive database system is suitable for developing a network management system because of two reasons,

1. It provides more expressive power than relational databases, and
2. It can describe recursive relationships between entities.

The development of a network management system by using deductive database begins with the description of the network entities as a set of facts and rules. Once the system is defined, a functional specification is developed that lists the type of queries a user wishes to be answered by the system. The use of logic in deductive database allows an elegant way for expressing facts, rules, and queries. In addition, it is also possible to add or remove rules without changing the overall structure of the program.

13.11 Integrated Network Management

Present day network technology demands for the network management functionalities, like fault management, performance management, etc., to interact among themselves toward an integrated model to achieve an efficient network management.[47] If each functionality deals with an intelligent

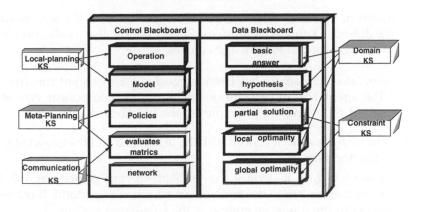

FIGURE 13.12
Typical blackboard architecture for network management.

agent/expert system, these expert systems solve the network management problem incrementally, cooperatively and run parallel, whenever possible. The advantages of the integrated models are:

- parallelism in executing the individual expert systems
- incremental problem solving by several expert systems for a complicated problem
- to enable different inference strategies by different expert systems.

Many Blackboard architecture based integrated network management systems have been proposed.

1. Black Board Architecture

This Blackboard Architecture (BBA)-based system[48] is generic even if some specific problems (diagnosis, performance, or security management) correspond better to the proposed system. Each task of the network management and control is considered as distributed problem solving (DPS). A global architecture has been proposed with the principle of blackboard architecture (see Figure 13.12) which has the following comments:

- a structured knowledge base called "blackboard," which contains the current state of the solution for a given problem. This shared data is analogous to the working memory accessed by many production rules and is a rule-based system. But this area is divided into separate areas of semantic abstraction called "levels."
- a collection of independent agents (or knowledge sources) which may lead and write one or more levels. They can be seen as a collection of independent processes able to cooperate to solve a problem. Each agent is specialized in one particular subtask.

- a control system, which insures the supervision of the actions of the different knowledge sources. This system is also a collection of integrated control-knowledge sources.

A separate architecture has been suggested for the agent construction. The agent has three components and each component can be reorganized following the principles of the blackboard architecture:

- *a knowledge module*: it concerns all the usual data and the knowledge of the domain;
- *a control module*: it defines the problem-solving strategy and the control mechanism to determine the next action to perform. It corresponds to the inference engine in the rule-based system;
- *a communication module*: it gives a model of the other agents in the environment. It also provides an interpreter of messages and a logic for communications with the other agents.

Each agent may be seen as an independent software processor with its own resources. The agents communicate by sharing the information if we look at a high level, i.e., at the entire multi-agent system level. They communicate through the blackboard which contains, at the beginning, the facts of the given problem. The agents read the information in the blackboard and write in it to reach a solution step by step.

2. Integrated Network Management by Extended BBA

An attempt is made to design Blackboard Architecture (BBA) based integrated network management.[49, 50] In it are five functional components of network management, namely: *fault diagnosis and performance management* by adductive reasoning, *configuration management and security management* by neural networks and *learning*. It is a centralized network management architecture, where each of the network management functions turns out to be an individual expert systems called knowledge bases (KSs) and share a global information base called global blackboard (GBB). Each KS may use a different inference mechanism depending on the domain. GBB represents the global network information collected by any of the monitoring protocols, using the object-oriented approach, and makes it available for all the expert systems. Each of the expert systems has a local blackboard (LBB) for local computations. The entire operations is coordinated by using a scheduler. These experts solve the network management problem incrementally and whenever possible, run parallel (see Figure 13.13).

3. Two-Layer Blackboard-Based ATM Network Management System

The network management is done in two layers:[51] *network layer* and *element layer.* In the network layer the management is more concerned

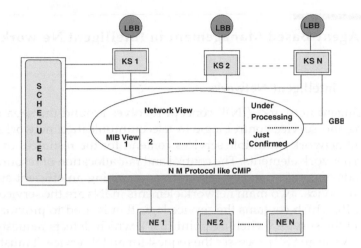

FIGURE 13.13
BBA-based integrated network management.

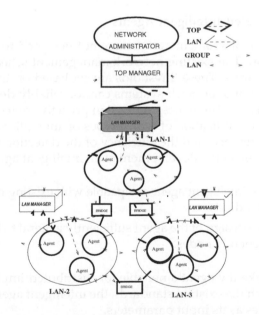

FIGURE 13.14
Intelligent agent-based integrated network management.

with the monitoring of network status and traffic whereas the element layer deals with the resource management. The system uses SNMP, CORBA, and TCP/IP protocols for exchange of management information and it is based on a blackboard system (see Figure 13.14) which is a group of knowledge modules collaborating with each other by a shared database to reach a satisfactory solution for network management.

13.12   Agent-Based Management in Intelligent Networks

13.12.1   Intelligent Networks

The intelligent networks (IN)[15] concept revolves around the separation of switching and service control logic in telecommunication networks. In the intelligent networks, multiple services contend for the resources of various common network elements. The control and fair allocation of resources during periods of congestion is essential to maintaining an efficient and well-perceived service. Two main network elements in INs are the service control point (SCP), which contains the service logic that is used to provide IN services; and the service switching point (SSP), which detects requests for IN services. While an SSP processes the request for an IN service, it must contact an SCP several times.

13.12.2   Distributed Intelligent Agents

Current network management systems are not designed to do provocative anomaly detection. The existing network management schemes use threshold to generate alarms. These thresholds are set based on the expertise of a human network manager. Such systems cannot reliably detect impending network problems.[52, 53] The issues involved in proactive detection of network problems are the identification of which types of anomalies can be detected proactively, performance, and the efficiency of the detection scheme.

The primary goals in the development of an intelligent agent are twofold:

- To make the intelligent agent compatible with existing network management standards.
- To make the design of the agent sufficiently general to be applicable to heterogeneous nodes.

These goals make the agent amenable for distributed implementation. To be compatible with the existing standards, the intelligent agent uses the standard MIB variables as its input parameters.

13.12.3   Agent-Based Network Management

The basic goal is to bring the management intelligence as close as possible to the managed resources, i.e., its logical representation in form of managed objects. This increases the intelligence and autonomy of a management agent by delegating management activities from the manager to the management agent(s).

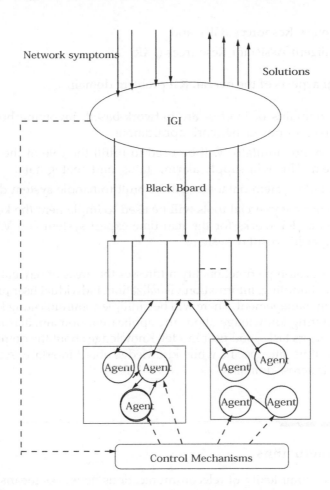

Network symptoms

Solutions

IGI

Black Board

Agent Agent   Agent Agent

Agent   Agent Agent

Control Mechanisms

FIGURE 13.15
A user interface-based network management.

13.12.4   Real-Time Intelligent User Interface Technology

The basic aim in such an interface is to use intelligence to enhance the user's interaction rather than perform domain diagnosis. Using human factors knowledge to combine advanced graphics with real-time expert system capabilities, the operator will be able to interact with an "intelligent assistant" capable of supporting the operator tasks. The focus in the provision of intelligence is on its use within the interface to assist users in locating, determining, and resolving system problems (see Figure 13.15). In the high-level architecture, we have adopted the notion of IGI (Intelligent Graphic Interface) having the following components:

1. User Interface (UI)
2. User Interface Resource (UIR)

3. Graphics Resources (GR), and

4. Intelligent Assistant Resources (IAR)

Particular aspects of the overall IGI problem domain:

1. The domains of interest are network-based, involving both telephone and power network applications.

2. A domain simulator will be used to fulfill the role of the control system. This will supply alarm, status, and analog data.

3. A simple system database will be built to handle system data.

4. Existing commercial tools will be used to implement the key components: RT works for the real-time expert system and VAPS for the graphic display system.

The target prototype functionality addresses the areas of navigation, monitoring, alarm handling, information visualization, individual user preferences, and situation management. There will be multiple communicating knowledge bases, separating knowledge about the application domain, alarms, display techniques, users tasks, and user model. Knowledge about the domain alarms and tasks will be site-specific while knowledge about interface issues will be more generic across domains.

13.13 Conclusions

The growing complexity of telecommunications networks means that their management is also becoming more and more complicated. Due to several factors, for example, amount of data, inconsistent and incoherent data, and time-constraints Artificial Intelligence techniques will be essential in order to carry out some of the management functions. Different AI techniques will be required to perform different tasks. At the highest levels in the management hierarchy, the key problem will be to process the enormous amount of raw data coming continuously from the network elements. AI techniques will be needed to solve large resource allocation problems. At the lowest level, AI will be used online in order to accelerate the control process and reduce human intervention. This chapter attempted to classify the network management problems and touched upon some of the existing AI techniques, which have proved their importance in the field of network management. A lot of work still needs to be done so that we can:

1. gain a better understanding of these intelligent techniques, and

2. determine how they can be applied to telecommunications problems.

References

1. W. Stallings, *Data and Computer Communications*, 5th ed., Prentice-Hall, Englewood Cliff, NJ, 1997.
2. A. Leinwand and K. Fang, *Network Management: A Practical Perspective*, Addison-Wesley, Reading, MA, 1993.
3. G.H. Floris, V. Broek, and M. Looijen, "Cost-effective management of international networks," *Intl. Jl. of Network Management*, vol. 7, pp. 137–146, 1997.
4. M.T. Rose, *The Simple Book: An Introduction to Internet Management*, Prentice-Hall, New York, 1996.
5. Lauer et al., "Broadband intelligent network architecture," *IEEE Proc. on Intelligent Networks Workshop (IN'95)*, Ottawa, May 1995.
6. S. Erfani and V. Sahin (Guest Editors), *Jl. of Network and Systems Management, Special Issue on TMN*, vol. 3, no. 1, 1996.
7. J. Case, M. Fedor, M. Schoffstall, and J. Davin, "A simple network management protocol", *RFC 1157*, 1990.
8. ISO/IEC IS 9596, *Information Technology-Open Systems Interconnection-Common Management Information Protocol*, 1990.
9. M. Chernick, K. Mill, R. Arnoff, and J. Strauch, "A survey of OSI network management standard activities", *Tech. Rep. NMSIG87/16 ICST-SNA-87*, Nat. Bureau of Stand., 1987.
10. K. McCloghrie and M. Rose, "Management information base for network management of TCP/IP-based internets: MIB II," *RFC 1213*, March 1991.
11. Y. Yemini, G. Goldshmidt, and S. Yemini, "Network management by delegation," *Proc. of Intl. Symp. on Integrated Network Management*, pp. 95–107, Washington, DC, April 1991.
12. T. Zhang and S. Covaci, "The semantics of network management information", *IEEE Proc. of INFOCOM*, pp. 489–496, 1996.
13. D. Wu, "Information exchange protocol: a new approach for future network management", *IEEE Proc. Of INFOCOM*, pp. 489–496, April 1995.
14. C. Muller, P. Veitch, E.H. Magill, and D.G. Smith, "Emerging AI Techniques for Network Management", *IEEE Proc. of GLOBECOM*, pp. 116–120, 1995.
15. G. Premkumar and P. Venkataram, "Artificial intelligence approaches to network management: recent advances and a survey", *Computer Commun.*, vol. 20, no. 4, 1313–1322, 1997.
16. A.A. Covo, T.M. Moruzzi, and E.D. Peterson, "AI-assisted telecommunications network management", *IEEE Proc. of GLOBCOM*, vol. 2, pp. 14.1.1–14.1.5, 1989.
17. K.D. Cebulka, M.J. Muller, and C.A. Riley, "Applications of artificial intelligence for meeting network management challenges in the 1990's," *IEEE Proc. of GLOBECOM*, vol. 2, pp. 14.4.1–14.4.6, 1989.
18. D. O'Leary, "AI and navigation on the Internet and intranets," *IEEE Expert*, vol. 11, No. 2, pp. 8–10, Apr. 1996.
19. J.F. Coates, "Artificial intelligence: observations on applications and control", *Computer Security Jl.*, vol. V, no. 1, pp. 39–47, 1988.
20. E.C. Ericson, L.T. Ericson, and Minoli (Eds), *Expert Systems Applications in Integrated Network Management*, Artech House, New York, 1989.
21. H.J. Watson and R.I. Mann, "Expert Systems: Past, Present, and Future", *Jl. of Info. Sys. Management*, vol. 5, no. 4, pp. 39–46, 1988.

22. L. Bernstein and C.M. Yuhas, "Expert systems in network management—the second revolution," *IEEE Jl. on Special Areas in Commun.*, vol. 6, no. 5, pp. 784–787, June 1988.
23. D.D. Liu and D.A. Pelz, "I-TEST: integrated testing expert system for trunks", *IEEE Jl. on Selected Areas in Commun.*, vol. 6, no. 5, pp. 800–804, June 1988.
24. V. Nuckolls, "Telecommunications diagnostic expert system," *IEEE Proc. Of GLOBECOM*, vol. 2, pp. 14.5.1–14.5.5, 1989.
25. W.D. Zhan, S. Thanawastien, and L.M. Delcambre, "SIMNETMAN: An expert workstation for designing rule-based network management systems," *IEEE Network; Mag.*, pp. 35–42, Sept. 1988.
26. C.L. Forgy, "The OPS83 user's manual," *Production Systems Technologies Inc.*, Dec. 1985.
27. R.O. Yudkin, "On testing communication networks", *IEEE Jl. on Select. Areas in Commun.*, vol. 6, no. 5, pp. 805–812, June 1988.
28. Y.D. Lin and M. Gerla, "Induction and deduction for autonomous networks," *IEEE Jl. on Select. Areas in Commun.*, vol. 11, no. 9, pp. 1415–1525, Dec. 1993.
29. K. Ganeshan and M. Ganti, "A multimedia front-end for an expert network management system," *IEEE Jl. on Select. Areas in Commun.*, vol. 6, no. 5, pp. 788–791, June 1988.
30. K.J. Macleish, "Mapping the integration of artificial intelligence into telecommunications," *IEEE Jl. on Selected Areas in Commun.*, vol. 6, no. 5, pp. 892–898, June 1988.
31. E.Y. Lai, "A knowledge-based approach to intelligent network trouble analysis," *IEEE Proc. of GLOBECOM*, vol. 2, pp. 14.6.1–14.6.5, 1989.
32. K.E. Brown, C.F.N. Cowan, T.M. Crawford, and P.M. Grant, "Knowledge-based techniques for fault detection in digital microwave radio communication equipment," *IEEE Jl. on Select. Areas on Commun.*, vol. 6, no. 5, pp. 819–827, June 1988.
33. S. Barra, R. Simeoni, and P. Vailati, "MIDA: a knowledge-based system supporting ISDN customer trouble management," *IEEE Proc. GLOBECOM*, pp. 600–603, 1996.
34. T. Saydam, "A generic multimedia service and management architecture," *Intl. Jl. of Network Management and Architecture*, vol. 2, no. 4, pp. 233–255, 1997.
35. G. Jakobson, A. Lemmon, and M. Weissman, "Knowledge-based GUI for network surveillance and fault analysis," *IEEE Proc. of Network Operations and Management Symp.*, vol. 3, pp. 846–855, 1994.
36. J.L. Chen, H.F. Sun, C.H. Tsai, and C.Y. Wu, "A knowledge-based system for ATM network management," *IEEE Proc. of GLOBECOM*, pp. 548–552, 1994.
37. D. Peacocke and S. Rabie, "Knowledge-based maintenance in networks," *IEEE Jl. on Select. Areas in Commun.*, vol. 6, no. 5, pp. 813–818, June 1988.
38. N.A. Khan, P.H. Callahan, R. Dube, J.L. Tsay, and W.V. Dusen, "An engineering approach to model-based troubleshooting in communication networks," *IEEE Jl. on Select. Areas on Commun.*, vol. 6, no. 5, pp. 792–799, June 1988.
39. T. Magedanz, K. Rothermel, and S. Krause, "Intelligent agents: an emerging technology for next generation telecommunications?," *IEEE Proc. of INFOCOM*, pp. 464–472, 1996.
40. T.K. Apostolopoulos and V.C. Daskalou, "Temporal network management model concepts and Implementation Issues," *Computer Commun.*, vol. 20, no. 25, pp. 94–708, 1997.

41. K. Meyer, M. Erlinger, J. Netser, and C. Sunshine, "Decentralizing control and intelligence in network management," *IEEE Proc. of the 4th Intl. Symp. on Integrated Network Management*, May 1995.
42. Young-Keun Park and Gyungho Lee, "Applications of Neural Networks in High-Speed Communication Networks," *IEEE Communication Mag.*, pp. 68–74, Oct. 1995.
43. T. Tain, M. Zhu, and B. Zhang, "An artificial neural network-based expert system for network topological error identification," *IEEE Proc. of Intl. Conf. on Neural Networks*, vol. 2, pp. 882–886, 1995.
44. D. Jensen, "B-ISDN network management by a fuzzy logic controller," *IEEE Proc. of GLOBCOM*, vol. 2, pp. 799–804, 1994.
45. T.D. Ndousse, "Distributed fuzzy agents: a framework for intelligent network monitoring," *IEEE Proc. of GLOBCOM*, pp. 867–871, 1997.
46. N. Sharda, R. Fatri, and M. Abid, "Applying deductive database technology to network management," *Computer Commun. Review*, 42–54, 1996.
47. I. Krishnan and W. Zimmer, *Integrated Network Management, II*, North-Holland, 1991.
48. D. Gaiti, "A proposal for integrating intelligent management in the intelligent network conceptual model," *Computer Networks and ISDN Systems*, vol. 28, pp. 689–699, 1996.
49. G. Premkumar and P. Venkataram, "Extended blackboard architecture for network management", *IEEE Proc. of Symp. of Intelligent Systems*, pp. 35–41, Bangalore, Nov. 1993.
50. G. Premkumar, "Integrated network management using extended blackboard architecture," *Doctoral Thesis*, Dept. of ECE, Indian Institute of Science, July 1996.
51. J.L. Chen, "Two-layer ATM network management system," *IEEE Proc. of Intl. Conf. on System, Man and Cybernautics*, vol. 2, pp. 1233–1238, 1995.
52. P. Maes, "Agents that reduce work and information overload," *Commun. ACM*, vol. 37, no. 7, pp. 8–10, 1994.
53. P. Maes, "Intelligent software," *Scientific American*, vol. 273, no. 3, pp. 84–86, Sept. 1995.

14

Intelligent Monitoring for Network Fault Management

Cynthia S. Hood and Chuanyi Ji

CONTENTS

14.1 Introduction

As communication networks continue to increase in size and complexity to support new applications and large numbers of new users, understanding and managing network behavior becomes increasingly difficult. Many applications, such as video, require networks to maintain a higher standard of network availability and reliability, thereby making effective network fault management critical. The dynamic nature and heterogeneity of current networks makes this more difficult. Fundamental changes to the network occur much more frequently due to the growing demands on the network and the

0-8493-1075-X/01/$0.00+$.50
© 2001 by CRC Press LLC

availability of new, improved components and applications. With network components developed in an open environment, a network can be configured by mixing and matching several vendors' hardware and software. While this allows the network to utilize the latest technologies and be customized to the needs of the users, it also increases the risk of faults or problems.[18]

Network management is a broad term that has been defined according to the Open Systems Interconnection (OSI) systems management specification developed by the International Organization for Standardization (ISO). It includes the following five functional areas: fault management, performance management, accounting management, configuration and name management, and security management. These functional areas are covered by a Network Manager (NM) that may be an automated system, a human, or a combination of both. A NM may be responsible for managing an entire network or just part of the network, depending on the size of the network. The NM does its job by processing information received from the network nodes (servers, routers, bridges, terminals, etc.). How well the NM performs its job is highly dependent on the quality and completeness of the information received from the nodes.

The collection and processing of data at each node is referred to as network monitoring. There is a tremendous amount of data collected at each node that must be processed to extract pertinent information to be sent to the NM. The definition of pertinent varies, but in general the NM expects each node to notify it of any abnormal or anomalous behavior. To accomplish this, the monitoring function at the node must be able to process the collected data and detect the anomalous behavior. Given the complexities and dynamics of current networks, this is a very difficult task. Given a variety of network measurements, what constitutes normal behavior? When network behavior can potentially change on an hourly basis, how is it possible to distinguish between normal and anomalous behavior? Lastly, how are the collected measurements related?

The goal of intelligent network monitoring is to identify problems at a stage when they can be addressed and catastrophic failures can be avoided. Ideally this can be done with minimal performance degradation, thereby making it transparent to the network users. The network monitoring process can be broken into three stages: (1) Information gathering, (2) Information processing, and (3) Information fusion. The following sections describe each of these stages in detail and discuss how the processing in each stage relates back to the questions posed above, as well as several different techniques for stages (2) and (3).

14.2 Network Management Background

The network management system architecture consists of a centralized network manager along with many agents. The agents reside in various network nodes and collect data.[18] They communicate with the central network manager through a network management protocol. There are two sets of standardized network management protocols: (1) for TCP/IP networks and (2) for OSI networks. The TCP/IP management protocol is Simple Network Management Protocol (SNMP) and the OSI protocol is Common Management Information Protocol (CMIP). Although there are distinct features of each of these protocols, they both provide basic network management tools. For the purposes of fault management, they both provide methods for collecting, storing, and processing measurements at a single node as well as means for communicating the information to a centralized network manager. In this work, data was collected from an Internet using SNMP. Although the specifics of SNMP are used, the methods can be generalized to OSI networks as well.

SNMP provides a structure for organizing the information collected and maintained by the agents. This structure is implemented in the Management Information Base (MIB) at each agent. The MIB contains a set of measurement variables pertinent to that particular network node. There are a standard set of measurement variables specified in Reference 10 but the equipment manufacturer can add measurements considered important or delete ones deemed unimportant. For this reason, the specific measurements collected on different nodes can vary. Currently, SNMP agents are included in most pieces of equipment for the Internet, making this a convenient way to gather information without requiring specialized hardware.

The SNMP MIB is organized into groups. Data was collected from the interface, IP, and UDP groups. The measurement variables within each group characterize the activity of the group (i.e., interface, IP, or UDP). Along with data collection, SNMP also provides a protocol for communication between the agents and the network manager. This protocol allows the manager to query the MIB for current information, change information in the MIB, and receive notification of certain events occurring at the agent. Detailed information on SNMP can be found in Reference 10. The basic SNMP functionality is extended with the Remote Network Monitoring MIB (RMON).[14] RMON provides the agent with the opportunity to do some processing on the measurements. The specific type of processing must be specified when the agent is installed by the human network manager.

14.3 Information Gathering

Information gathering is the part of the monitoring process where information is collected from the network. The type of information collected and the frequency of the collection depends on the network. Network management agents (e.g., SNMP) or specialized software and hardware can be used to collect the measurements. Existing collection tools range from simple SNMP

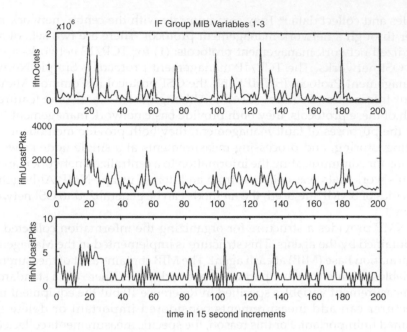

FIGURE 14.1
Examples of MIB variable data.

agents to commercial products that integrate SNMP agents and provide a graphical user interface as well as a set of analysis tools. The two key issues for information gathering are: 1) identifying the measurements to collect and 2) establishing the collection frequency for each.

The types of measurements studied are numerical. The assumption is that these measurements reflect some aspect of the system, thereby providing a means to learn about the system's behavior. The measurement variables used in this work are MIB variables collected from a TCP/IP network. The MIB variables studied are all integer counters. Some of the MIB variables were representative of network traffic and others were representative of unusual or error conditions. Examples of MIB variables are shown in Figure 14.1. The plots indicate the magnitude of the variables over a 50-minute period of time. The time unit is 15 seconds (the polling interval). The polling interval was chosen to maximize the information collected without placing undue burden on the network. This compromise was necessary to collect the data on an operational network.

A pragmatic view of the information-gathering stage of the monitoring process was taken. There was no attempt to find an optimal set of variables to measure. The MIB variables collected and studied were those that were available on the university network. A polling frequency was chosen that was acceptable to the network administrator. An acceptable polling frequency is one whereby the data collection overhead for the network nodes being polled does not degrade performance noticeably, the percentage of additional network traffic due to the data collection is minimal (only a small fraction of the total traffic), and the storage space required was reasonable. Therefore no attempt was made to address the open issues with information gathering.

14.4 Information Processing

The goal of the information processing part of the monitoring system is to transform the data collected into meaningful information. The measurements need to be put into some context. Recall that the goal of intelligent network monitoring is to identify problems at an early stage. This implies identifying when one or more of the measurements is indicating problematic behavior in the network. How can this be accomplished? There is no simple answer to that question. There are several complexities that must be addressed. The first is that for most variables that are measured there is no clear definition of problematic behavior. Most of the measurement variables studied are closely related to network traffic and to date, the modeling and characterization of network traffic signals is an active research area.[17] In Reference 16 the research in this area is related to network management measurement variables. Since *a priori* models of the normal or abnormal behavior don't exist, both must be estimated from the data.

14.4.1 Hybrid Neural Network—Hidden Markov Model

The first attempt at information processing involved using a hybrid neural network—Hidden Markov Model (HMM) system[6] as described in Reference 12. Two states (normal and abnormal) are defined for each measurement. The state of each measurement is estimated from the data. The neural network is used for its learning and generalization capabilities. It is assumed that the underlying process is Markov in nature, but the states cannot be observed (i.e., they are hidden); therefore an HMM is chosen. This model provides the ability to consider temporal context which is very important.

Each MIB has a hybrid neural network—HMM to determine the MIB's state. For the neural network portion, a feedforward network is trained using the standard backpropagation algorithm to provide the *a posteriori* state probabilities $p(s_i | \theta(t))$, where s_i is the state, $\theta(t)$ is the time-series data of the MIB variables at time t and m is the number of states. Given these probabilities, the HMM then provides temporal context by estimating the state probabilities as

$$p(s_i | \theta(t)) = \frac{\alpha_i(t)}{\sum_{j=1}^{m} \alpha_j(t)},$$

where

$$\alpha_i(t) = p(\theta(t)|s_i) \sum_{j=1}^{m} a_{ij}\alpha_j(t-1).$$

It is assumed that a_{ij}, the state transition probabilities are known ahead of time. By Bayes' rule

$$p(\theta(t)|s_i) = \frac{p(s_i|\theta(t))p(\theta(t))}{p(s_i)},$$

where $p(s_i)$ are the assumed prior state probabilities. Since the probability of the measured data $p(\theta(t))$ is independent of the state, this can be omitted from the calculation without loss of generality. The state probabilities are then used to estimate the state of the MIB variable. The estimation is done by selecting the state with the highest posterior state probability.

Data was collected from a single node on the Internet over a period of 14 days. An abnormal situation was created by congesting the network with broadcast packets. MIBs from the TCP and IP layers were collected. The neural networks used had one input, three hidden nodes, and two outputs. There was a single neural network for each MIB variable (hence the single input). The number of hidden nodes was varied and it was observed that three was the minimum number of hidden nodes that performed well in all of the neural networks. For simplicity the same configuration was adopted for each neural network. The neural networks were trained using 6000 training samples. Artificially generated abnormal data (approximately 10%) was added to the training set to compensate for the lack of abnormal samples in the network data (approx. 0.5%). The abnormal samples were drawn from a uniform distribution. This means that the additional abnormal samples provided no additional information about the abnornal class (i.e., no new examples, just more of the same examples to train by). The HMM transition probabilities were chosen based on discussions with network managers and it was assumed that the prior probabilities of both states were equally likely.

The neural network in this system learns both the normal and abnormal behavior of the network through training samples of both types. For each data sample, the neural network estimates the *a posteriori* state probabilities. The quality of these estimates depends on the training set. Here the limitation is the abnormal samples. The amount of data collected during known abnormal times is always going to be small relative to the entire data set. Along with the problems due to the small quantity data, there are also issues with the quality of the data. For the neural network to do its job, it must be trained with different examples that cover a wide range of abnormal behavior. This type of data is extremely difficult to get. Although the neural network will have some generalization capability (i.e., the ability to correctly classify abnormal data that differs from the abnormal data in the training set), it will not perform well on unknown faults. Since networks change over time and faults may look different depending on load, configuration, etc., it is concluded that this approach is limited.

To address the shortcomings of the above approach a different direction is tried. Instead of trying to learn both the normal and abnormal behavior of the MIBs, only the normal behavior is learned. Good quality "normal" behavior data can be collected relatively easily. The approach taken is one of change detection. Changes from the normal behavior are identified instead of explicitly detecting abnormal behavior. The contention is that the majority of changes detected will in fact be abnormalities. One of the most common techniques used for network fault management (thresholding) can be considered as a change detection approach.

14.4.2 Change Detection—Bayesian Network

Thresholds are the primary method currently used in both practice[9,14] and research[2,3] for detecting abnormal behavior. The feature is not the value of the threshold itself, but the information on whether or not the threshold has been exceeded by a particular measurement variable. There can be both upper and lower thresholds, in which case the feature is the information on whether the variable is within the thresholds. One of the difficulties with thresholds is properly setting the threshold level, since thresholds strongly depend on the traffic level. If the thresholds are set too high, they might never be exceeded, thereby letting problems go undetected. If the thresholds are set too low, they might be exceeded too often, thereby flooding the network manager with false alarms. While properly set thresholds do a good job of detecting large rises and falls in a measurement variable, more subtle behavioral changes are missed.

In Figure 14.2(a) the level of the signal, or measured variable, increases and in Figure 14.2(b), the variance of the signal changes. These changes can be symptoms of something problematic occurring in the network. Detection of the more subtle signs of problems can allow corrective action to be taken to avoid a bigger problem. In addition, identification of the problem also becomes easier with a more complete description of the symptoms rather than just the extreme cases.

To detect the more subtle changes in the nature of the measured variables, the parameters of a second order autoregressive process $AR(2)$ are used as features:[7]

$$y(t) = a_1 y(t - 1) + a_2 y(t - 2) + \epsilon(t),$$

where $y(t)$ is the value of the signal at time t, a_1, and a_2 are the AR parameters that are used as features, and $\epsilon(t)$ is a white noise process. These features, like threshold features, are monitored for each measurement variable.

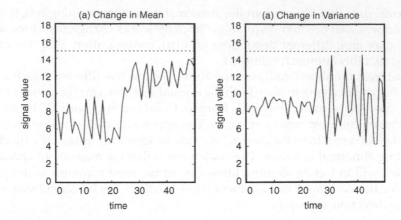

FIGURE 14.2
Examples of subtle behavior not detected using thresholds (dotted lines).

Each MIB variable is sampled every 15 seconds. From the time series data, a 20-sample window is used to estimate the $AR(2)$ parameters. This window includes the current samples along with the 19 previous samples. This sliding window scheme allows the current samples to be temporally correlated with some past samples. The parameters are completely reestimated each time using least-squares. Using the $AR(2)$ parameters as features, the more subtle changes in Figure 14.2 can be detected. AR parameters were chosen for simplicity. If an underlying stochastic process that governs the data is composed of piecewise Gaussian processes, the extracted AR parameters would be able to characterize the process well (assuming the order of the AR model is properly chosen). Otherwise, the AR parameters would correspond to the best linear approximation in terms of mean-square-error. Higher order AR processes have also been tried, but no significant gain has been observed.

These features are used to learn the probability distributions of the MIB variables. Through the features, the distribution of normal behavior for each of the MIBs can be estimated. As before, there is difficulty with the distribution of abnormal behavior. Since this distribution cannot be learned from the data, this distribution is treated as unknown and a uniform distribution over a range of the $AR(2)$ parameters is assumed. Any sample that falls outside this range is considered abnormal.

Although the distribution of normal behavior for each of the MIBs can be learned, it is important to recognize that the behavior changes over time. On most networks, the traffic in the middle of the day is very different than the traffic during the night. Since the changes can be drastic, it is important to adjust the model of normal as behavior changes. A learning window is defined as a sliding window where the baseline normal behavior is learned. The size of the learning window can be adjusted to fit the dynamics of the network. The assumption here is that since abnormal behavior happens infrequently, this behavior will not be learned (i.e., no special action is taken to keep abnormal samples out of the data).

This method can be improved upon by segmenting the data before the features are calculated.[8] Using segmentation, statistically similar data is grouped into variable length segments. The improvement is twofold: (1) the statistics calculated from each segment are more representative, and (2) signal processing techniques requiring a stationary signal can be used within each segment. Although the assumption of a stationary signal was being used with the sliding window method, it was discovered that this assumption was frequently violated.

In terms of monitoring, segmentation provides the benefit of temporally correlating the observations. Since many of the network signals are bursty, the temporal correlation can help distinguish between a burst and a change in the nature of the signal. To do the segmentation, each MIB variable is modeled as a quasistationary autoregressive (AR) process. Within each segment, the statistical properties of the time series are assumed to remain constant, and are described by a set of AR parameters. The segmentation algorithm is described in Reference 1.

14.5 Information Fusion

The last part of this process is to put together all of the processed information in a meaningful way. This is a very difficult problem because the relationships between the information are not well known. In fact, these relationships may change over time as the network changes. To tackle this problem, some simplifying assumptions are made.

To combine the information, the hierarchical nature of network protocols is exploited. The measurements collected were all related to network protocols so it was natural to organize the information into a hierarchy. The measurements are at the lowest level of the hierarchy. The next level of the hierarchy is the specific protocol (e.g., TCP, IP). For example, in the hierarchy TCP is the parent of all of the TCP-related measurements. The TCP measurements are combined to estimate the state (normal or abnormal) of TCP. The highest level of the hierarchy is where the information from each of the protocols is combined to estimate the state of the network. This model does not explicitly deal with the organization of the protocol stack. It treats each protocol as an independent part of the network.

14.5.1 Hybrid Neural Network—Hidden Markov Model

From a probabilistic perspective, fusing the information is extremely complicated. A joint distribution on all the pieces of the network (or protocol) is

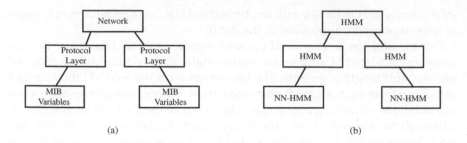

FIGURE 14.3
(a) Logical structure of hierarchy, and (b) implementation of hierarchy.

desirable, but to do this requires significant knowledge of the relationships between the various parts of the network (e.g., joint distributions of TCP, IP and TCP, UDP). This information is not known, so some assumptions must be made. In the first attempt at this problem, it was assumed that all of the variables collected were independent. This greatly simplifies things and allows the joint probabilities to be easily calculated. A linear combination of the lower-level states is used to estimate higher-level state probabilities.[6] The hierarchy is pictured in Figure 14.3.

At each of the middle levels (see Figure 14.3(b)), the outputs from the previous level are combined linearly to estimate the higher-level state probabilities $p(S_i | \Phi(t))$, where S_i is defined as a higher-level state probability and $\Phi(t)$ is $\theta(t)$ and the outputs from previous level HMM(s). Recall that two states (normal and abnormal) have been defined where state S1 is normal and state S2 is abnormal. These states are defined for each of the higher-level entities (i.e., IP, TCP and Network).

$$p(S_1|\Phi(t)) = p(s_{11}, s_{12}|\Phi(t)) + 0.5[p(s_{11}, s_{22}|\Phi(t)) + p(s_{21}, s_{12}|\Phi(t))],$$

$$p(S_2|\Phi(t)) = p(s_{21}, s_{22}|\Phi(t)) + 0.5[p(s_{11}, s_{22}|\Phi(t)) + p(s_{21}, s_{12}|\Phi(t))],$$

where s_{ij} represents the ith state of the jth output from the previous level. In the cases where one of the inputs is normal and the other is abnormal, the probability mass is split evenly. As mentioned above, the variables monitored were assumed to be independent, simplifying the calculation of the joint probabilities.

The results obtained using this method are shown in Figure 14.4.

Each figure indicates the posterior probability of being in an abnormal state. A simple Maximum A Posteriori (MAP) scheme is used to estimate the state, so any time that $p(abnormal)$ is greater than 0.5, the state is deemed abnormal. Two periods of time were observed where the network was so severely congested that traffic was stopped. Both of these times were detected at all layers of the hierarchy, correctly indicating a network problem. Starting from the lowest level, Figures 14.4(d)–(g) show the alarms generated after each MIB variable is temporally correlated through the hybrid neural network-HMM.

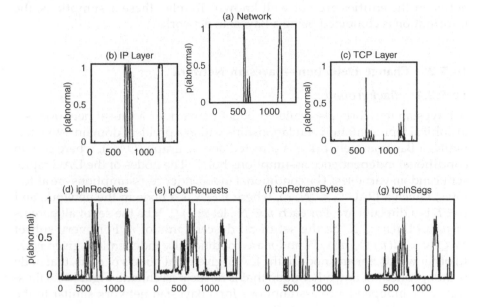

FIGURE 14.4
Hierarchy output at each level for hybrid NN-HMM system.

At the next level, the amount of time in the abnormal state decreases as expected. There are only four intervals in which the IP layer is identified as abnormal (Figure 14.4 b), and only one where the TCP layer is identified as abnormal (Figure 14.4c). At the highest level of the hierarchy, there are two intervals where the entire network is in an abnormal state (Figure 14.4a) corresponding to the situation observed.

These results are promising. This system is capable of spatial and temporal correlation. The limitations of this system are (1) the need for abnormal training data and (2) the restrictive assumptions regarding independence of the measurements. The first limitation is most problematic. It is difficult to obtain data that includes labeled examples of network faults. When training a neural network it is important that the training set provides good coverage of the sample space to be learned. With so few data points one cannot infer much about the abnormal space and without good coverage the neural network generalization capability is lost. The goal is to design a system that can detect unknown faults so other methods for learning the behavior must be investigated.

It is also recognized that the independence assumptions are not realistic in many cases. Looking at the information collected, one quickly can see the strong correlation between many of the MIBs. It is also true that the states of the protocol layers are not independent of each other. For example, a problem in the IP layer can impact the operation of the TCP layer. Once again there is the difficulty of calculating the joint probabilities when the relationships between the entities are not well known. To relax these assumptions, the information is combined using a Bayesian network.[7]

14.5.2   Change Detection—Bayesian Network

14.5.2.1   *Background*

A Bayesian network, also called a belief network or a causal network, is a graphical representation of relationships within a problem domain. More formally, a Bayesian network is a directed acyclic graph (DAG), where certain conditional independence assumptions hold.[11] The nodes of the DAG represent random variables. The conditional independence assumptions are as follows: Given a DAG $G = (N, E)$, where $n \in N$ is a node in the network, and $e \in E$ is a directed arc. For each $n \in N$, let $p(n) \subseteq N$ be the set of all parents of n, and $d(n) \subseteq N$ be the set of all descendants of n. For every subset $W \subseteq N - (d(n) \cup \{n\})$, W and n are conditionally independent given $p(n)$. In other words, for any node in the DAG, given that node's parents, that node is independent of any other node that is not its descendant. Figure 14.5 illustrates the independence assumptions for a Bayesian network similar to the one used for the monitoring system.

Given the observed information or evidence, these assumptions facilitate the estimation of the conditional probabilities of any of the nodes (or random variables) in the Bayesian network. Algorithms for estimating these probabilities are given in Reference 11. The efficiency of the algorithm depends on the structure of the Bayesian network. The strength of Bayesian networks is that they provide a theoretical framework for combining statistical data with

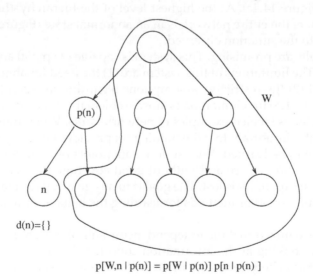

$$p[W,n \mid p(n)] = p[W \mid p(n)] \, p[n \mid p(n)]$$

FIGURE 14.5
Example of independence assumptions for a particular Bayesian network.

prior knowledge about the problem domain. Therefore, they are particularly useful in practical applications.

Bayesian networks have been widely used for medical diagnosis[13, 4] and troubleshooting.[5] In the communication network field they have been proposed to diagnose faults in Linear Lightwave Networks.[2] In Reference 2, standard methods have been used for detection and the Bayesian networks are used for diagnosis only. In this work, Bayesian networks are proposed as a mechanism for combining the information for detecting network anomalies.

14.5.2.2 The Bayesian Network for Monitoring

To begin with, the random variables or nodes in the Bayesian network are defined. There are two types of variables: those that are observed, and those that are not observed and thus need to be estimated. The variables that are not observed are called the internal variables. The observed variables directly correspond to the MIB variables. The internal variables are defined to be *Network, interface* (IF), IP, and UDP. The IF, IP, and UDP variables correspond to the MIB groups. Logically they represent different types of network functionality. The MIB variables within a group are the measurement variables for that network function. The Network variable is defined to correspond to all of the network functionality. It is assumed that the network is comprised only of the functions corresponding to the MIB groups monitored—IF, IP, and UDP. The structure of the Bayesian network is shown in Figure 14.6. The arrows between the nodes in Figure 14.6 go from cause to effect.

Analogous to medical diagnosis examples, the observed MIB variables are considered to be symptoms of larger problems. In this model, the health of the network is the most general information estimated and can be considered to be an underlying influence on the rest of the nodes in the Bayesian network.

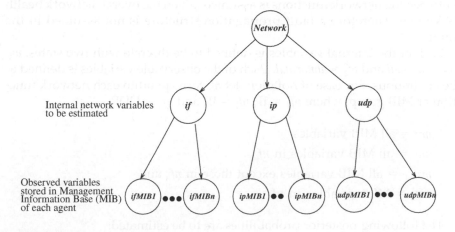

FIGURE 14.6
Bayesian network for fault detection.

The overall health of the network directly influences the health of the three functions of the network (i.e., IF, IP, and UDP). This is indicated in the model by the arrows from the network random variable to the IF, IP, and UDP random variables. For each network function, the health of that function directly influences the values of the individual measurement or MIB variables for that function. One way of interpreting this is that a problem in the network will cause symptoms to occur within the network functions, and likewise, problems within a network function will cause symptoms to show up in the MIB variables for that function.

The model has been designed based on these intuitive relationships from the structure of the MIB. The Bayesian network model requires the conditional independence assumptions described in the previous section. It is assumed that given knowledge of the health of the network, the health of the network functions are independent. In addition, given the health of a network function, the measurement variables for that function are independent and the measurement variables for other network functions do not contribute any additional information. These assumptions are reasonable since each of the network functions represent independent functional components of the network. These components may fail independently, although there is a relationship between the functions, and serious problems in one component can eventually impact the other components. Since propagation of a fault through the functional components depends on the type and location of the fault (i.e., faults may propagate from low-level functions to high-level functions, or vice versa),[15] it is difficult to incorporate a propagation structure into the model that will accommodate all (or at least most) types of propagation. In addition, the relationships between the nodes of the Bayesian network (network functions and MIB variables) are complex and not well understood, so even if the propagation structure was fixed, this would still be a difficult problem. The proposed is a simple model where no *a priori* relationship between the network functions is assumed, given the overall network health is known. Therefore a fault propagation structure is not assumed in the model.

Each of the internal variables is defined to be discrete with two states, w_k: $w_1 = normal$ and $w_2 = abnormal$. Each of the observable variables is defined to be continuous. For ease of notation, let nf_i correspond to each network function or MIB group, where $nf_1 = IF$, $nf_2 = IP$, and $nf_3 = UDP$,

mv = all MIB variables,

mv_i = all MIB variables in nf_i,

$mv_{\neg i}$ = all MIB variables except those in nf_i, and

mv_{ij} = MIB variable j in nf_i.

The following posterior probabilities are to be estimated:

$$p(Network = w_k | mv); \quad k = 1, 2, \tag{14.1}$$

and

$$p(nf_i = w_k|mv); \quad i = 1, 2, 3; \quad k = 1, 2, \tag{14.2}$$

Due to the tree or singly connected structure of the Bayesian network, these probabilities can be calculated efficiently either using Pearl's algorithm[11] or directly. The equations for the direct calculation can easily be derived using the conditional independence assumption stated above. In particular, the probability in (14.1) can be factored as

$$p(Network = w_k|mv) = \frac{(\Pi_{i=1}^3 p(mv_i|Network = w_k))p(Network = w_k)}{p(mv)};$$

$$k = 1, 2 \tag{14.3}$$

using the definition of conditional probability and the conditional independence assumption. Likewise, the probability in (14.2) can be factored as

$$p(nf_i = w_k|mv) = \frac{p(mv_i|nf_i = w_k)p(nf_i = w_k|mv_{-i})p(mv_{-i})}{p(mv)};$$

$$i = 1, 2, 3; \quad k = 1, 2 \tag{14.4}$$

because the variables in network function i are determined by the health of function i; the other variables do not contribute any additional information. The joint probability of all the MIB variables in the denominator of both Equations (14.3) and (14.4) need not be specified, since it is not dependent on the observations and can be considered a normalization factor. The joint probability of the MIB variables not related to the network function currently being estimated in Equation (14.4), $p(mv_{-i})$, will be canceled out.

Using the conditional independence assumption again, the remaining quantities can be calculated as:

$$p(mv_i|Network = w_k) = \sum_{l=1}^{2} p(mv_i|nf_i = w_l)p(nf_i = w_l|Network = w_k);$$

$$i = 1, 2, 3; \quad k = 1, 2, \tag{14.5}$$

$$p(nf_i = w_k|mv_{-i}) = \sum_{l=1}^{2} p(nf_i = w_k|Network = w_l)$$

$$p(Network = w_l | mv_{-i}); \quad i = 1, 2, 3; \quad k = 1, 2. \tag{14.6}$$

and

$$p(Network = w_k | mv_{-i}) =$$

$$\frac{\left(\prod_{i=1}^{3, \, l \ne i} p(mv_l | Network = w_k) \right) p(Network = w_k)}{p(mv_{-i})};$$

$$i = 1, 2, 3; \quad k = 1, 2. \tag{14.7}$$

Finally,

$$p(mv_i | nf_i = w_k) = \prod_{mv_i} p(mv_i | nf_i = w_k); \quad i = 1, 2, 3; \quad k = 1, 2 \tag{14.8}$$

The following quantities are necessary to calculate the probabilities in (14.1) and (14.2):

$$p(Network = w_k); \quad k = 1, 2, \tag{14.9}$$

$$p(nf_i = w_l | Network = w_k); \quad i = 1, 2, 3; \quad k = 1, 2; l = 1, 2, \tag{14.10}$$

and

$$p(mv_{ij} | nf_i = w_k), \forall \, mv_{ij} \in nf_i; i = 1, 2, 3; \quad k = 1, 2. \tag{14.11}$$

The prior probabilities in (14.9) and the conditional probabilities in (14.10) are estimated using prior knowledge of the network behavior gained from observations and conversations with the network managers. These probabilities remain constant throughout the monitoring process. The conditional probabilities in (14.11) are estimated using the observed MIB variables.

Since the monitoring is done locally, all of the evidence or probabilities estimated from the observed MIB variables are available to the system. This enables the system to calculate the desired posterior probabilities using a complete and current set of observations.

14.5.2.3   Results

The CD-Bayesian network system was tested on a set of ten faults observed on the university network over a period of six months. Most of the faults (9 out of 10) were recorded as *server not responding*. The remaining fault is a report of excess Ethernet collisions on one of the subnets. Although nine of the faults were the same, the types of changes observed in the data from fault to fault (i.e., no fault signatures) were not the same. This can be traced to the fact that the faults were causes by different sets of circumstances or root causes.

One of the faults the system was tested on was a fault that was reported as *server not responding* between 6:33 am and 6:36 am on December 23. The measurements are collected from a router. The faulty server was on one of the subnets connected to the router. The results for the posterior probabilities estimated:

$$p(Network = abnormal \mid MIBs),$$
$$p(IF = abnormal \mid MIBs),$$
$$p(IP = abnormal \mid MIBs),$$
$$p(UDP = abnormal \mid MIBs).$$

are shown in Figure 14.7. The asterisks denote the fileserver downtime period. Abnormal behavior is detected in the network approximately 12 minutes before the server is reported unreachable. Anomalies are present before the problem in all three network functions, but only IP detects an anomaly during the crash. Simular results were obtained when a 4-hour learning window was used.

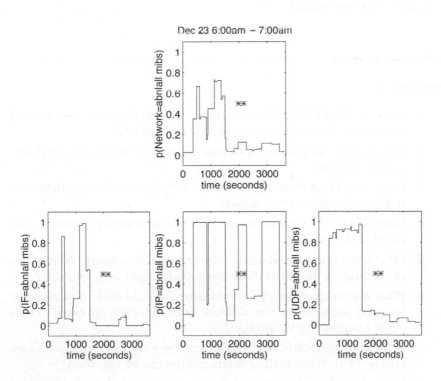

FIGURE 14.7
Results using 1-hour learning window.

The results are encouraging. By studying the statistical characteristics of network measurements it is possible to learn the normal behavior of the network and detect deviations from this normal. A learning window is used to continually adjust the view or baseline of normal behavior. By detecting subtle changes from normal and then correlating the information in a Bayesian network it is possible to do proactive fault detection. The detection system can warn the network manager of impending problems so catastrophic failures can be avoided.

This work is viewed as a first step in the direction of an automated fault management system that can generalize from network to network with minimal network specific information required *a priori*. The fault management problem is very complex and the nature of the problem evolves as networks evolve. Future work involves expanding the scope of the experiments as well as further investigation of the methods.

Both the information processing and information fusion areas need more work. For the information processing piece, the search for better features to extract is ongoing. In the information fusion the relationships between the variables is an area of ongoing work. Current work focuses on trying to find common threads across different networks. The goal is to model the relationships as accurately as possible. With better understanding of the problem the approaches can be refined.

Problems

1. How can the Hidden Markov Model (HMM) approach be extended to account for nonstationary network behavior?

2. Why is it so difficult for a neural network to learn abnormal network behavior?

3. Is the length of training an issue for the network fault detection problem? Explain your answer.

4. Discuss the differences between the change detection approach and the HMM approach.

5. Why is temporal correlation of the samples important? What types of methods (other than those described) might be applicable?

6. What are the qualities of a good feature? Would mean and variance be good features for the network fault detection problem? What other features can be studied?

7. A hierarchical relationship between the variables is defined (see Figure 14.6). What further relationships can be specified?

8. How does the estimation of the posterior probabilities change if the Bayesian network structure is changed to incorporate additional relationships?

References

1. U. Appel, A. V. Brandt, "Adaptive Sequential Segmentation of Piecewise Stationary Time Series," *Information Sciences*, vol. 29, 1983, pp. 27–56.
2. R. H. Deng, A. A. Lazar, W. Wang, "A Probabilistic Approach to Fault Diagnosis in Linear Lightwave Networks," *IEEE Journal on Selected Areas in Communications*, vol. 11, no. 9, Dec. 1993, pp. 1438–1448.
3. G. Goldszmidt, Y. Yemini, "Distributed Management by Delegation," *Proceedings of the 15th International Conference on Distributed Computing Systems*, June 1995.
4. D. Heckerman, "A tractable algorithm for diagnosing multiple diseases," *Proceedings of the Fifth Workshop on Uncertainty in Artificial Intelligence*, Windsor, ON, 1989, pp. 174–181.
5. D. Heckerman, J. S. Breese, K. Rommelse, "Decision-Theoretic Troubleshooting," *Communications of the ACM*, vol. 38, March 1995, pp. 49–57.
6. C. S. Hood, C. Ji, "Intelligent Network Monitoring," *Proceedings of the IEEE Neural Networks for Signal Processing*, Boston, 1995, pp. 521-529.
7. C. S. Hood, C. Ji, "Proactive Network Fault Detection," *IEEE Transactions on Reliability*, vol. 46, no. 3, September 1997, pp. 333–341.
8. C. S. Hood, C. Ji, "Intelligent Agents for Proactive Fault Detection," *IEEE Internet Computing*, March/April 1998, pp. 65–72.
9. E. L. Madruga, L. M. R. Tarouco, "Fault Management Tools for a Cooperative and Decentralized Network Operations Environment," *IEEE Journal on Selected Areas in Communications*, vol. 12, no. 6, Aug. 1994, pp. 1121–1130.
10. K. McCloghrie, M. Rose, "Management Information Base for Network Management of TCP/IP-based internets: MIB-II," RFC 1213, March 1991.
11. J. Pearl, *Probabilistic Reasoning in Intelligent Systems: Networks of Plausible Inference*. San Mateo, CA, Morgan Kaufman, 1988.
12. P. Smyth, "Markov Monitoring with Unknown States," *IEEE Journal on Selected Areas in Communications*, vol. 12, 1994, pp. 1600–1612.
13. D. J. Spiegelhalter, A. P. Dawid, S. L. Lauritzen, R. G. Cowell, "Bayesian Analysis in Expert Systems," *Statistical Science*, vol. 8, no. 3, 1993, pp. 219–288.
14. S. Waldbusseer, "Remote network monitoring management information base," RFC 1271, Nov. 1991.
15. Z. Wang, "Model of network faults," *Integrated Network Management I*, B. Meandzija and J. Westcott (eds.), New York, NY, Elsevier Science Publishing Company, 1989.
16. M. Wei, "Computer Communication Network Traffic Modeling, Analysis and Test," Masters Thesis, Rensselaer Polytechnic Institute, 1996.
17. W. Willinger, M. S. Taqqu, W. E. Leland, D. V. Wilson, "Self-Similarity in High-Speed Packet Traffic: Analysis and Modeling of Ethernet Traffic Measurements," *Statistical Science*, vol. 10, no. 1, pp. 67–85.
18. Y. Yemini, "A Critical Survey of Network Management Protocol Standards," *Telecommunications Network Management into the 21st Century*, S. Aidarous and T. Plevyak (eds.), New York, NY, IEEE Press, 1994.

15

A Hybrid Genetic Approach for Channel Reuse in Multiple Access Telecommunication Networks

Ioannis E. Kassotakis, Maria E. Markaki, and Athanasios V. Vasilakos

CONTENTS

0-8493-1075-X/01/$0.00+$.50
© 2001 by CRC Press LLC

ABSTRACT   The evolving Broadband Integrated Services Digital Network (B-ISDN) is reinforcing the demand for high-speed and high-performance multiple access networks. The number of channels available to support the isochronous traffic in these networks is limited by technology, due to implementation costs. We introduce a method using channel sharing/reusing in an effort to provide efficient management of isochronous traffic under this limitation. The proposed method is based on a hybrid genetic algorithm and aims to accomplish the establishment of a maximal number of connections with the minimal number of isochronous channels. Experimental results are provided and they are compared with those of a deterministic graph coloring algorithm. The performance of the proposed algorithm in all simulation runs reveals the robustness, the flexibility, and the efficiency of using evolutionary approaches to complex real-world problems.

KEY WORDS: *Hybrid Genetic Algorithm, NP-Complete, Isochronous Service, Channel Reuse.*

15.1   Introduction

A common design issue to all modern telecommunication networks is the efficient use of the available bandwidth (b/w). In all cases, the network stations have access to a multiple access physical medium of finite b/w required to support all the communication services. The physical medium can either be the air (e.g., wireless, radio, mobile networks) or a hardwired common medium (e.g., cable, optical fiber). In both cases, the concept of *channel* is used to describe and model a finite amount of b/w, which is adequate to satisfy the needs of the basic service provided to the network users. The b/w of a channel may be a frequency band for Frequency Division Multiple Access (FDMA) networks or a *time slot* for the Time Division Multiple Access (TDMA) networks.

The network services are divided into two major categories: The connectionless services that do not have strict time constraints (file transfer, e-mail, etc.) and the time critical connection-oriented services (video conferencing, telephony etc.) that require the continuous availability of guarantee b/w to achieve the desired quality of service (QoS). The connection-oriented services, when implemented at high speed multiple access networks are usually referred as isochronous services and the b/w shares (Virtual Channels) that they use when they employ TDMA schemes, are called Isochronous Channels (usually 64 Kbit/sec).

Channel management is an important aspect for high-performance networks to guarantee the QoS of critical and real-time traffic. It is evident that if channel assignment is improperly implemented, it can be b/w abusive. Hence, *Channel sharing/reusing* is one approach toward efficient management of isochronous traffic, since it increases the network's throughput and

decreases the fraction of blocked messages. Typical applications where channel sharing/reusing is critical are the following:

- file transfers in companies, organizations with hierarchical structure, where multi-person connections are established between
  neighboring bureaus;
- video telephony and conferencing;
- collaborative systems with electronic libraries, databases, etc.,
  located in stations between the active stations;
- LAN interconnection of heterogeneous types, i.e., Ethernet, Token
  ring, FDDI, Fast Ethernet by a high speed, distributed dual bus, where
  traffic is concentrated mainly between the stations of each LAN.

In this study, we are interested in improving the provision of isochronous
service and particularly in the problem of reusing isochronous channels,
referred as *Isochronous Channel Reuse Problem (ICRP)*, in topologies that
involve a medium of multiple access (DQDB, FDDI, Ethernet, etc.). The task
for the solution of ICRP is to assign isochronous channels to newcoming
requests with the twofold issue of serving the maximum number of isochronous connections and employing the least number of channels.

Beyond its application to isochronous traffic in ISDN networks, the Isochronous Channel Reuse Problem (ICRP) applies to all types of communication networks. In satellite communication systems, the cochannel interference has
become a major factor of determining system designs and operations to secure
the communications quality.[1] Moreover, with the dramatic increase of geosta-
tionary satellites, two or more adjacent satellites can often cause the intersystem
interference by using the same frequency. In order to cope with the cochannel
interference problem, the rearrangement of frequency assignments has been considered as an effective countermeasure in practical situations. Similar channel
assignment considerations apply for the TDMA and FDMA cellular networks.[2, 3]

Mathematically speaking, channel assignment appears as a combinatorial
optimization problem that is closely related to graph coloring, and as such
known to be NP-complete.[4] Thus, an exact search for the best solution in real-
time applications is infeasible, due to an exponentially growing calculation
time. The fact that ICRP can be modeled as a graph coloring problem has led
researchers to use a wide variety of algorithms for its solution. These algorithms range from deterministic graph coloring algorithms to more intelligent schemes involving neural networks, simulated annealing, etc.

The use of intelligent techniques is becoming an increasingly attractive
practice in an effort to overcome the complex combinational problems that
arise during the design and development of High Speed Networks. The classic combinational and analytical methods fail to produce solutions within the
strict time constraints that these networks are bounded to operate.[5, 6, 7]

In this chapter, we propose a hybrid genetic algorithm (HGA) for the ICRP
solution. A hybrid genetic algorithm is a genetic algorithm coupled with a

local improvement operator. The reason for combining a genetic algorithm with a local search algorithm is that they complement each other and therefore the reliability properties of the genetic algorithms are combined with the accuracy of hill-climbing methods. Genetic algorithms and hybrid genetic algorithms were found to be well suited for similar NP-complete problems.[8, 9, 10, 11] since they adopt a global strategy, rely on intelligent randomization, and explore a number of possible alternatives simultaneously. The performance of the HGA is compared via simulation to that of a graph coloring algorithm (GCA) proposed in Reference 12. A Dual Bus Distributed Queue (DQDB) network has been used as a testbed for the evaluation of the algorithms' performance.

The rest of the paper is organized as follows: In Section 15.2, we provide the definition of ICRP and a brief description of the DQDB testbed network. Section 15.3 presents the deterministic graph coloring algorithm (GCA) and discusses the genetic algorithms. The customization of hybrid genetic algorithm (HGA) for the ICRP solution is addressed in Section 15.4. Section 15.5 introduces the simulation models and assumptions and Section 15.6 presents and comments on the simulation results. Section 15.7 concludes the paper.

15.2 Problem Definition

In this section, we overview the DQDB network, used as a test-bed in our simulation experiments, we state the ICRP, and formulate the ICRP as a graph coloring problem.

15.2.1 Isochronous Traffic and DQDB Topology Definitions

Consider a DQDB network that interconnects N nodes labeled 1, 2, . . . , N (station index) from left to right as shown in Figure 15.1.[13] The underlying network of the DQDB consists of two unidirectional slotted buses operating in opposite directions. This allows full duplex communications between each pair of nodes, providing isochronous (circuit switched) and nonisochronous (packet switched) services. Every node receives and transmits on both buses, via read and write connections, so bus selection is based on the destination. Nodes use F-Bus to transmit to stations of higher indexes and R-Bus to transmit to stations of lower indexes. Transmissions on the two buses are independent; thus the effective data rate of a DQDB network is twice the data rate of the bus.

The data on each bus is formatted by the node at the head of each bus, called slot generator, into either DQDB layer management information octets or fixed-length units called *slots* with a length of 53 octets. Each slot contains an Access Control Field (ACF) that contains the protocol control information and a segment that forms the payload of the slot. The management

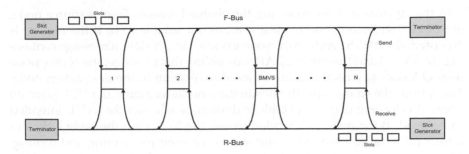

FIGURE 15.1
15.1 The DQDB network topology.

information octets are used to maintain the operational integrity of the network. The slots are used to carry data between the nodes. All data flow terminates at the end of each bus, at the slot sink.

The DQDB layer provides two modes of access control to the dual bus. These are Queued Arbitrated and Pre-Arbitrated, which use QA and PA slots for data transfer, respectively. Queued arbitrated access is controlled by the distributed queuing protocol[14, 15] and would be used typically to provide nonisochronous services, while pre-arbitrated access would be used typically to provide isochronous services. Next, we describe in detail only the access to PA slots since the PA slots are used to provide isochronous service.[14]

Access to PA slots: For each PA slot, there are 48 octets of payload, each of which can be occupied by a single station to form a 64 kbps isochronous channel. A sequence of coherent PA slots can either provide 48 distinct 64 kbps isochronous channels or a number of c_i*64 kbps isochronous channels with $\Sigma c_i = 48$. To maintain this b/w, the slot generator located at the head of each bus must generate at least one PA slot every 125 μsec for channels with a total b/w requirement of (48*64) kbps. To support larger b/w requirements, the slot generator may generate several PA slots in every 125 μsec. The header of a PA slot consists of a Virtual Channel Identifier (VCI), which is used to identify the PA slots that belong to the same isochronous channel. In order to identify which octet in a PA slot is occupied by a specific isochronous channel, an offset value is used. Thus, a VCI/Offset pair uniquely identifies a 64 kbps isochronous channel.

The source node that wants to establish an isochronous connection sends a channel request to the Bandwidth Manager and VCI Server (BMVS) via QA slots. The BMVS is an administrator node usually located at the center of the dual bus (Figure 15.1). The designation of slots to PA access and the marking of the PA slot header is controlled by BMVS. In particular, the BMVS must write the Virtual Channel Identifier (VCI) field into PA slots. The BMVS also ensures that the PA slots for each VCI are provided in a periodic manner on the bus to guarantee that sufficient b/w is available for isochronous service users.

In this approach, after receiving the channel request from a source node, the BMVS will accept the request if there is available b/w. If the request is accepted, the BMVS sends to the source node, via QA slots, a message containing the VCI(s) and the offset(s). After receiving this message, the source node uses QA slots to pass this information downstream to the destination node. Then, both the source and the destination nodes examine the VCI value on every PA slot, and use a local table to determine whether the VCI is intended to be used. If the node is intended to use the VCI, then the table indicates which offset octets within the slot should be used for reading and writing. The source node writes the octets in positions marked for writing and the destination node reads the octets in positions marked for reading. If the VCI is not in use by the node, then the PA slot is ignored. At the end of the session, the source node sends a message to BMVS, via QA slots, to indicate the end of the connection. Then, BMVS marks the particular offset(s) of the VCI(s) as available and is free to allocate them for another connection request.

15.2.2　Statement of ICRP

The terminology used in this paper is as in Reference 12. Let $X = \{x_1, x_2, \ldots, x_{n_c}\}$ denote a set of established isochronous connections, in which $x_i = \{s_i, d_i, O_i\}$ stands for an established isochronous connection between stations s_i and d_i with a set O_i of consecutive offsets (O_i represents the b/w in use). Let $Y = \{y_1, y_2, \ldots, y_{n_r}\}$, $y_i = (s_i, d_i, w_i)$ denote a set of isochronous requests; where y_i is the isochronous request issued from station s_i, and d_i, w_i are the destination station and the required b/w (number of 64 kbps isochronous channels), respectively. It is assumed that a station can issue several isochronous requests simultaneously to the same or different destinations. This means that for a pair of stations, the establishment of multiple isochronous connections is allowed. We say that a request y_i is *intersected* with another request y_j (or an established isochronous connection x_k) if $s_i = s_j$ or $s_i < s_j < d_i$ ($s_i = s_k$ or $s_i < s_k < d_i$). It is evident that two intersected isochronous requests/connections can not share the same isochronous channel(s).

Let the consecutive offsets to be used by request y_i be called range R_i. We say that $R(X, Y) = \{R_1, R_2, \ldots, R_{n_r}\}$ is a *ranges arrangement* of Y; where $R_i = (s_i, d_i, \{o_i\})$ and o_i is the arranged starting offset of request y_i, if for each pair of intersected requests in Y and/or established isochronous connections in X, their arranged ranges do not occupy an overlapped offset. Let $w(R(X, Y))$ denote the total number of assigned offsets of $R(X, Y)$. Let $R^*(X, Y)$ denote the optimal ranges arrangement of (X, Y) which has the minimum number of assigned offsets $w(R^*(X, Y))$. The problem considered can now be defined:

Isochronous Channel Reuse Problem (ICRP): *Given a set X of established isochronous connections a set Y of isochronous requests, find an optimal ranges arrangement $R^*(X, Y)$.*

In this chapter, in order to simplify the implementation, we assume that all the established isochronous connections as well as the isochronous requests have the same b/w, equal to that of a single isochronous channel (64 kbps).

15.2.3 A Graph Coloring Modeling of ICRP

In this section we briefly state the graph coloring problem[16] and show how the ICRP can be formulated as a graph coloring problem. A *graph* is a set of points called *vertices* with connections called *edges* between pairs of vertices. Given a graph $G = (V, E)$ with vertex set V and edge set E and given a positive integer k, one has to find a *k-coloring* of G, i.e., a partition of V into k independent sets $V_1, \ldots, V_k$ (a set V_i of vertices is called independent if no two vertices in V_i are linked by an edge in G). If such a partition exists, G is called *k-colorable*.

This problem has received much attention for several reasons. First it belongs to the class of problems known as NP-complete. In addition, a large number of practical problems can be formulated in terms of coloring a graph, including clustering and scheduling problems. Additionally, the ICRP can be restated as Graph Coloring Problem if we assume the following:

- The concept of an isochronous channel corresponds to that of a color.
- The established connections are represented as colored vertices.
- The isochronous requests are represented as uncolored vertices.
- Two vertices are connected with an edge if they are intersected.

Let us consider a set X of connections ($x_1 = (1, 7, \{1\})$, $x_2 = (4, 6, \{2\})$, $x_3 = (5, 6, \{3\})$, $x_4 = (7, 8, \{3\})$) and a set Y of isochronous requests ($y_1 = (7, 8, 1)$, $y_2 = (6, 8, 1)$, $y_3 = (3, 8, 1)$, $y_4 = (4, 8, 1)$, $y_5 = (5, 6, 1)$). The graph G with $V = X \cup Y$ is depicted in Figure 15.2, where the intersected pairs of vertices can be easily distinguished. For instance, it can be easily seen that y_1 is intersected with y_2, y_3, y_4, and x_4.

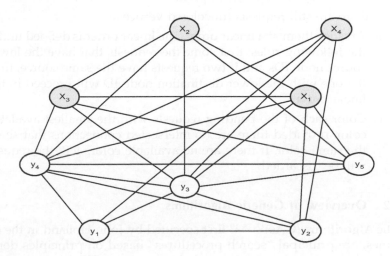

FIGURE 15.2
Example graph.

The ICRP can be restated as follows: Given a graph $G = (V, E)$ with vertex set $V = X \cup Y$ and edge set E, assign any of the uncolored vertices (requests) a color from the set of available colors, with the condition that no pair of uncolored (requests) and colored vertices (connections) connected with an edge shares the same color.

15.3 Algorithms for the ICRP Solution

15.3.1 Review of Graph Coloring Algorithms

In Reference 12 the authors propose an $O(n^3)$ graph coloring algorithm (GCA) to solve the ICRP, where n is the sum of the number of established isochronous connections and the number of isochronous requests. The following steps can describe the channel assignment method of GCA:

1. For every isochronous request find the number of intersected requests and impose a descending order on them.
2. Traversing the requests in this order, perform the following actions for every request:
   a. Find the colors used by the connections that are intersected with the request (unpreferred colors). From the set of available colors remove the unpreferred colors.
   b. If at least one color is found, assign the color (channel) with the lower VCI value to the request.
3. If there are still requests (uncolored vertices):
   a. Arrange them at a linear order. The linear order is defined under the following rules: first come the requests that have the lower source node IDs and if two requests have the same source, then the one with the lower destination node ID will proceed in the linear order.
   b. Color each of the pending requests with the smallest available color, provided there are no intersected connections that share the same color. If there are no available colors left, the request will be discarded.

15.3.2 Overview of Genetic Algorithms

Genetic Algorithms (GAs)[17, 18, 19] first specified by John Holland in the early seventies, are principal "search procedures" based on principles derived from the dynamics of natural population genetics. They have been successfully applied to numerous large space problems where no efficient polynomial-time algorithm is known, such as NP-complete problems.

The problem to be solved in Genetic Algorithms should be represented as one-dimensional or multi-dimensional structures, which represent a search point in the search space. This means that the problem should be encoded, that is, to find a pattern to represent each solution, called *chromosome*, as a chain of characters taken in a finite alphabet. The Simple Genetic Algorithm (SGA)[20] relies on a binary alphabet, and thus solutions appear as chains of "0" and "1". The GAs operate on chromosomes grouped into a set called *population*. Successive populations are called *generations*. Each chromosome is evaluated by the *fitness function*, which reflects its merit and its chances to survive in the next generation.

The operation of the Simple Genetic Algorithm (SGA) can be summarized as follows. The initial solutions, chosen randomly, are encoded as binary strings (chromosomes) and they form the initial population. Each chromosome is associated with a fitness value. The current population is evolved creating a new one with higher fitness value, by using three operators:

a. Selection—This operator selects a chromosome from the current population to survive in the next one. The probability of a chromosome to be selected is proportional to its fitness value. This operator is an artificial version of nature's Darwinian survival of the fittest chromosomes.

b. Crossover—This operator exchanges portions between selected chromosomes, as in nature it recombines the genetic material of two parent chromosomes to produce two children. The chromosomes are broken at the same (random) place and combined together creating two new children chromosomes.

c. Mutation—This operator causes sporadic and random alteration of the binary bits (changing a 0 to 1 and vice versa) of the chromosomes, as in nature it plays the role of regenerating lost genetic material.

The fitness values of the new chromosomes are evaluated by the fitness function, before the next selection process. From that point the algorithm will be repeated. The algorithm stops after a fixed number of generations, or when a chosen criterion reaches a predefined threshold. When optimization heuristics that can provide some domain-based guidance to the search process are incorporated into the GA, the resulting algorithm is called hybrid genetic algorithm. It is expected that if knowledge of the problem is embodied into the GA, then the convergence speed of the hybrid GA will increase.

15.4 Proposed Hybrid GA for the ICRP Solution

The proposed algorithm[21] has the typical components of a hybrid genetic algorithm and along with its specific characteristics it is going to be described next in detail. Figure 15.3 shows the algorithm's layout.

1. Create initial population of *pop_size* chromosomes

2. Evaluate (*pop_size*)

   repeat *pop_size*/2 times

3.   Select 2 parents P_1, P_2 from the population *pop_size*

4.   Offsprings $C_1, C_2 \leftarrow$ Crossover (P_1, P_2)

5.   Offsprings $C_1', C_2' \leftarrow$ Mutate (C_1, C_2)

6.   Replace if better $P_1 \leftarrow C_1'$

$$P_2 \leftarrow C_2'$$

   end;

7. Find the best chromosome of the population, λ

8. Optimize (λ)

9. Output (λ)

10. Update stopping condition

11. Repeat steps 2 to 10 until the stopping condition is met

FIGURE 15.3
The outline of the proposed algorithm.

15.4.1   Representation and Population Initialization

Denote *chan* and n_r^j as the isochronous channels and the requests of chromosome *j*, respectively. Consider w_i^j as the isochronous channel assigned to *i* request of *j* chromosome. For solving the ICRP a binary encoding is used to represent the *j* chromosome, i.e., $V^j = (w_1^j, w_2^j, \ldots, w_{nr}^j)$. The binary matrix representation of *j* chromosome is formed as follows: The matrix is [*chan* $\times$ n_r^j] and if the channel w^j is assigned to the request *i*, the entry (w^j, *i*) of the matrix is set to 1. The remaining entries of the matrix for request *n* are set to 0. The binary encoding of $V^j = (2, 2, 1, 4, 3)$ with *chan* = 4 and $n^j = 5$ and a more general chromosome representation are given below:

$$
\text{chrom}^j =
\begin{bmatrix}
0 & 0 & 1 & 0 & 0 \\
1 & 1 & 0 & 0 & 0 \\
0 & 0 & 0 & 0 & 1 \\
0 & 0 & 0 & 1 & 0
\end{bmatrix}
, \quad
\begin{matrix}
 & & & 1 & \cdots & i & \cdots & n_r^j \\
1 & & & & & & & \\
\vdots & & & & & & & \\
w^j & & & & \cdots & & & \\
\vdots & & & & & & & \\
chan & & & & & & &
\end{matrix}
\begin{bmatrix}
 \\
 \\
 \\
 \\
\end{bmatrix}
$$

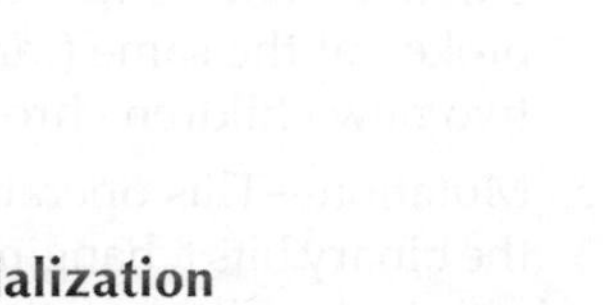

We define as *pop_size* the number of chromosomes. The *pop_size* chromosomes of the initial population are randomly initialized, by assigning to each of the n^j chromosome's requests a randomly generated channel such that

$$1 \leq w_i^j \leq chan, \quad \forall(i = 1, 2, \ldots, n_r^j, j = 1, 2, \ldots, pop\_size) \quad (15.1)$$

15.4.2 Evaluation

The evaluation of a chromosome takes place in order to test its fitness as a solution and is achieved by making use of a fitness function. Let n_c^j denote the set of established connections of j chromosome. Consider z_i^j as the isochronous channel assigned to i connection of j chromosome. In our approach, the fitness function of j chromosome, F^j, is defined as the number of intersected pairs of n^j requests and n^j connections that employ the same isochronous channel. Hence,

$$F^j = \sum_{i=1}^{n_r} f_i^j \quad (15.2)$$

where f_i^j is defined as the fitness function for the i request of j chromosome and is calculated as shown in the following code.

```
for (i=1; i ≤ n^j; i++)
{
    f_i^j: = 0;
    for (k = 1; k ≤ n^j; k++)
    {
        if ((i request intersected with k connection) and (w_i^j = z_k^j)) f_i^j: = f_i^j + 1;
    }
}
```

15.4.3 Selection

A variety of selection strategies are used, some of which use the fitness value directly (tournament with tournament size of two, roulette-wheel, FPS) and others order the population and allocate parents by rank with a scaling factor of two (linear scaling, sigma truncation). With the exception of tournament and roulette-wheel selections, which are commonly used in GAs, the rest of the selection strategies are briefly described. More details can be found in References 20, 22. FPS (Fitness Proportionate Selection) selects the first parent in proportion to its fitness, while the second parent is picked at random. Linear scaling requires a linear relationship between the scaled fitness, F_s^j, and the raw fitness, F^j, of j chromosome as follows:

$$F_s^j = a F^j + b \quad (15.3)$$

where the coefficients a and b are chosen to ensure that the average population member will receive one offspring on average and the best will receive copies equal to the scaling factor. In sigma truncation the scaled fitness, F_s^j, is given by:

$$F_s^j = F^j - (\text{aver } \{F^j, j = 1, 2, \ldots, pop\_size\} - c\sigma) \qquad (15.4)$$

where c is the scaling factor taken equal to two and σ is the standard deviation of the fitness values of the pop_size chromosomes.

In the phase of selection, a new fitness function is adopted for each of the pop_size chromosomes. It is apparent that during the GA's operation the hard constraint is the requirement that the requests intersected with established connections cannot share the same channel. However, in the phase of selection we are also interested in the number of currently active connections. Hence, due to the soft constraint of maximizing the active connections, the new evaluation function for the j chromosome, F^{j*} becomes:

$$F^{j*} = \max\{F^i, i = 1, 2, \ldots, pop\_size\} - F^j + n_c^j \qquad (15.5)$$

Considering the new fitness function, the fittest chromosome is the one that has the highest fitness value of Equation (15.5) determined by the number of intersected pairs of requests and connections that share the same channel and the number of active isochronous connections.

15.4.4   Crossover

The crossover probability $pcross$ recombines the genetic material of two parent chromosomes to generate two new ones. We use a one-point crossover; thus the part of the two binary chromosomes i and j after the crossover point is swapped to generate two children. The crossover point stands for column position and is a random number between 1 and min (n^i, n^j).

15.4.5   Mutation

The mutation operator, applied after the crossover operator on each of the offspring chromosomes independently, offers the opportunity for new genetic material to be introduced into the population. If no crossover is to be performed, the mutation operator is applied on the parent chromosomes. In both cases, each of the n^j isochronous requests of j chromosome is assigned a random channel from [1, $chan$]. The operator's probability, $pmutation$ is proportional to the contribution the operator has made to the chromosome's fitness. If the use of the randomly selected channel improves the fitness value of the chromosome, we set $pmutation$ to 0.4, or else to 0.1.

15.4.6 Replacement

The children chromosomes replace their parent chromosomes in the population to maintain a fixed-population size *pop_size*, if their fitness values are lower than the fitness values of their parents. Since the steady-state GA does not guarantee that the best member in the current population will be present in the next, we include the elitist strategy.[17] The elitism ensures that the chromosome with the best performance always survives intact into the next generation (replaces the chromosome with the worst performance) and we introduce the constraint that the successive replacements with the same best member of the population will not be more than [N/5], in order to avoid premature convergence.

15.4.7 Local Improvement Operator

The motivation for combining the genetic algorithm with a local search algorithm to form a hybrid algorithm is that the genetic algorithm will try to optimize globally, while the local optimizer will try to optimize locally.[23] We introduce a new local improvement operator, which is called Heuristic Algorithm and it works as follows. The algorithm is described in Figure 15.4.

```
for (i=1; i≤ nᵣ^λ ; i++)

    {

            init_ch := wᵢ^λ ;   {store the initial channel}

            init_fit := fᵢ^λ ;   {store the fitness value of init_ch channel}

            vch :=0;

            repeat

                    vch :=vch+1;

                    wᵢ^λ :=vch;       {assign vch channel to the i request}

                    next_fit := fᵢ^λ ;   {compute the fitness value of vch channel}

            until ((4*next_fit<=init_fit) or (vch>=chan));

            if (4*next_fit>init_fit) wᵢ^λ :=init_ch;   {assign the initial channel}

    }
```

FIGURE 15.4
A pseudo-code of the heuristic algorithm.

Our goal is to improve the channel assignment of the pending requests of the best chromosome $\lambda(F^\lambda \leq F^i, \forall_i 1, 2, ..., pop\_size)$, by minimizing the value of its fitness function. We examine separately every n_r^λ request of the λ chromosome. For each request we assign initially the channel 1 and compute the value of the fitness function. If the computed fitness value is less than 25% of the initial fitness value, the newly assigned channel replaces the initial one. Otherwise, we try the second channel and so on until we find a channel that satisfies the above requirement. In case none of the *chan* channels fulfills the condition, the initial channel is reassigned and we proceed with the next request until all n_r^λ requests are inspected. The output of the algorithm will be the solution λ optimized according to the previously described procedure.

15.4.8   Stop Criterion

The algorithm is halted when the generation counter exceeds the maximum number of generations *maxgen* or when convergence is indicated. A good indication of convergence is that a performance index has not changed in a given number of generations, which in our experiments is set to maxgen/2 generations. The performance index is the throughput, which is defined as the total number of connections made by all nodes divided by the total number of requests issued by all nodes. Additionally, the algorithm terminates in case we find the "optimal" solution, i.e., a chromosome with no pending requests, during the procedures of crossover, mutation or local improvement.

15.4.9   Time Complexity

It is easy to check that the time complexity of the parent selection is $O(pop\_size)$. For j chromosome, the crossover operator takes n^j *chan*. For $n = n^j + n^j \, \forall_j = 1, 2, ..., pop\_size$, since $n^j*chan \leq n*chan$ the crossover operator takes $O(n)$ time in the worst case. Similarly, we find that the mutation operator takes $O(n^2)$ time and the time complexity of the local improvement is $O(n^2)$. Therefore, the time complexity for each generation (select 2 parents, apply the crossover and the mutation operator *pop_size*/2 times, next find the best of the population and finally locally improve it) is:

$$\text{Time Complexity} < O(n^2*pop\_size^2) \tag{15.6}$$

The GA's setup time (steps 1 and 2 in Figure 15.3) is bounded by $O(n^2 * pop\_size)$.[24]

15.5   Simulation Models and Assumptions

In order to evaluate the performance of the HGA introduced in the previous section, two experimental scenarios were considered. In the first scenario referred to as *static load*, HGA was tested under static conditions, i.e., a pre-defined number of connections and requests is generated. In the second referred to as *dynamic load*, a dynamic environment is simulated, where both isochronous requests issued by the DQDB nodes and releases of the established connections are of Poisson distribution. The experimental performance of HGA is also compared with that of GCA.

15.5.1 Static Load Simulations

For the first experimental scenario the ICRP was formulated as follows:

A number of nodes (10 in our experiments) are arranged on a unidirectional DQDB bus of (chan) isochronous channel capacity. At the initial state there are (Con) = (chan) randomly established connections among the nodes and a pool of (Req) requests awaiting channel allocation. Both algorithms are tasked to serve the maximum number of (Req) requests by reusing the (chan) available channels.

The number of (*Req*) requests is normalized and expressed as a function of the available channels (*Req/chan*), thus representing the network's load. The higher the load, the harder the allocation problem becomes.

Two performance measures are used for the evaluation of the ICRP solution:

a. The established connections per available channel (*Conn/chan*) *<Efficiency>*.

b. The percentage of the served requests (*Conn/Req*) *<Throughput>*.

c. The time that each algorithm requires to reach its optimal allocation solution was also monitored. Both algorithms were simulated on the same (630 MIPS) machine using the same resources at all simulation experiments (runs). The allocation time measured in msec corresponds to the time complexity of each algorithm.

15.5.2 Dynamic Load Simulations

Both algorithms were tested on a DQDB network of ten nodes. The isochronous requests generated by each node are transmitted to the BMVS administration node. The BMVS node concentrates the requests from all nodes and periodically employs the allocation algorithm to select the isochronous channel that will serve each of the accumulated requests. The number of requests that are served at every allocation is set to 100. The traffic model that is considered for the generation of the channel requests is a uniform distribution. Thus, every node has equal probability to request a connection with any of the other nodes.

The following assumptions have been made for the DQDB network simulation:

- Network size is set to *Nodes* = 10;
- The stations are equally spaced along the dual bus;
- The load of the network is equally distributed among the stations;
- The arrival rate of the isochronous requests (r_i) in each station is a Poisson process;
- The departure rate of the established connections in each station (r_o) is a Poisson process.

The network load is expressed by the pair of the parameters (r_i, r_o). The quotient r_i/r_o, named *Load Quotient (LQ)*, is a measure of the load intensity. Higher *LQ* values indicate heavier network load.

For the evaluation of the algorithm performance in the DQDB network environment, the following two performance measures have been considered:

- The total number of established *connections* after all stations have issued a number of 10000 requests.
- The number of *active connections* maintained by all stations during the simulation run.

15.5.3   HGA Parameters

HGA requires a number of tuning parameters to be preset. After many experimental runs for tuning the genetic algorithm the following parameter values were determined:

- Population size is set to 2* *Nodes* = 20;
- Number of generations is set to 10 for the dynamic load simulation runs and to 1 for the static load simulation runs. It was concluded that for the static load runs, no more than one generation was required to converge to the optimal solution.
- Crossover probability is set to 0.8 (see Reference 19);
- Mutation probability is set to either 0.1 or 0.4 as discussed in previous section.
- Several different selection schemes (linear, tournament, FPS, sigma, roulette) were tested and their results are presented.

The HGA operators have been extensively presented in Section 15.5. The different selection schemes have been simulated in an effort to study their effect on the HGA performance. The selection of the proper GA parameters is a process based both on extensive experimentation and the experience gained by the authors during previous work at the same field. At a real application, there should be a predefined set of parameters that will automatically be applied according to the traffic model of the user requirements.

15.6 Simulation Results

The simulation results presented in this section are derived from 20 simulation runs for each experiment. Experience has showed that at least 20 runs are
needed to achieve statistically reasonable results (confidence interval of at
least 0.9), regardless of the network size, since the variation of the on-line
node number is automatically compensated by the proportional adjustment
of the GA population size (*pop_size* = 2* *Nodes*).

15.6.1 Static Load Simulations

In Figure 15.5, we investigate the efficiency of each algorithm at different
loads (*Req/chan*). At the lower side of the load curve (2.8 & 3), the GCA and
the HGA(s) appear to have similar performance. However, as the number of
the requests per channel increases, the HGA(s) continue to accommodate an
increasing number of requests using the limited number of network
resources (isochronous channels), while GCA diverts to lower number of
connections per channel. It is evident that among the HGA(s) using different
selection schemes, no significant difference can be noted. For the particular
case of *chan* = *Con* = 50, Table 15.1 presents the number of established connections for different request pool size (*Req*). In Figure 15.6 we investigate the
throughput of the algorithms by measuring the percentage of the served
requests at different load. It is evident that as the algorithm load becomes
increasingly heavier, the HGA(s) successfully maintain their QoS by preserving the percentage of the served requests. On the other hand, GCA fails to
maintain a constant percentage of served requests.

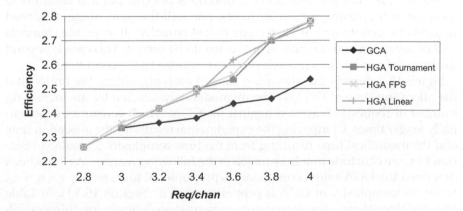

FIGURE 15.5
Channel allocation efficiency at static load.

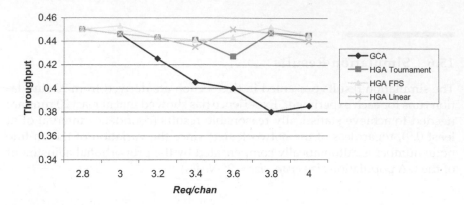

FIGURE 15.6
Channel allocation throughput at static load.

TABLE 15.1

Established Connections for Different Values of *Req* with *Chan* = *Con* = 50

Req	GCA	GA Tournament	GA Roulette	HGA FPS	HGA Sigma	HGA Linear
90	63	63	63	63	63	63
100	67	67	63	68	68	67
110	68	71	72	71	72	71
120	69	75	75	75	73	74
130	72	77	79	78	80	81
140	73	85	86	86	84	85
150	77	89	87	89	87	88
Average	69.857	75.285	75	75.714	75.285	75.571

At both experiments we must note that if excessive numbers of requests are considered, neither the number of connections per channel will continue to increase at increasingly heavier loads, nor will the percentage of served requests be maintained. It is obvious that eventually all available channels will be saturated and no more requests would be served. This case is beyond the scope of our work as it corresponds to extreme traffic conditions.

Figure 15.7 depicts the allocation time of each algorithm. We notice that the allocation time of HGA(s) is only moderately affected by the increasing number of requests compared against the GCA, which consumes exponentially longer times. Comparing the experimental results for the allocation time and the theoretical time resulting from the time complexity formulae of Section 15.4, we conclude that they match in the following manner. As it has been described the HGA's time complexity is proportional to n^2, where $n = n_r + n_c$, while the complexity of GCA is proportional to n^3 (Section 15.3.1). In Table 15.2, the algorithms' allocation time is presented analytically for different values of n. It is evident that the increase of run time curve is steeper for GCA corresponding to the $O(n^3)$ time complexity against the $O(n^2 * pop\_size^2)$ of HGA.

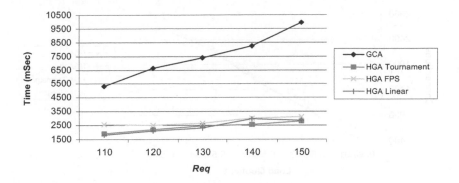

FIGURE 15.7
Channel allocation time at static load.

TABLE 15.2

Channel Allocation Time (mSec) for different values of *Req* with
Chan = *Con* = 50

REQs	GCA	HGA Tournament	HGA Roulette	HGA FPS	HGA Sigma	HGA Linear
110	5330	1920	2310	2580	2030	**1820**
120	6650	2200	3070	2520	2360	**2090**
130	7360	2480	2420	2640	2530	**2310**
140	8240	**2580**	2800	3020	2800	2970
150	9950	**2800**	3190	3130	3180	2850
Average	7506	**2396**	2758	2778	2580	2408

15.6.2 Dynamic Load Simulations

In Figure 15.8, we present the total number of established connections at different values of LQ. At the heavier load (LQ = 0.5/0.01) the HGA(s) achieve considerably higher number of connections compared to GCA. As the load becomes lighter, i.e., the release rate r_o increases, the total number of connections grows. At these conditions, all algorithms tend to achieve similar numbers of connections with HGA(s) always in favor.

In Figure 15.9, we monitor the number of active connections along the network simulation run. It is obvious that when all nodes have generated approximately 10,000 requests, the network has reached its steady state, since the number of *active connections* is stabilized. The HGA(s) achieve a considerably higher number of active connections than GCA given the same load. Considering the convergence to the steady state, the GAs appear to seize faster the available b/w (channels) and accommodate at any given instant the maximum number of requests. No safe conclusion can be derived that a specific selection scheme performs better than the others in all cases. We can only note that the linear scaling scheme achieves slightly better steady state performance, since it converges to the higher number of active connections.

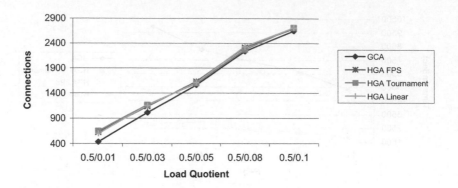

FIGURE 15.8
Established connections at dynamic load.

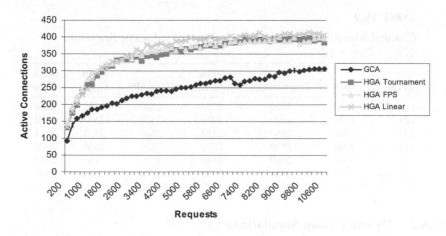

FIGURE 15.9
Active connections at dynamic load.

15.7 Concluding Remarks

Motivated by the limitation on the number of channels that can be supported and the increasing demand for giving access to a huge number of users, we have considered the use of channel sharing/reusing in multiple access networks. A hybrid genetic algorithm (HGA) has been presented for solving the NP-complete problem in-hand (ICRP) and is compared to a representative deterministic graph coloring algorithm (GCA), via simulation on a DQDB testbed network.

The results have shown that HGA is an interesting novel approach to ICRP. We concluded that HGA maintains its capability to reach optimal solutions for the ICRP at a wide range of network loads. A study of the simulation results reveals that the GCA performance is comparable to that of HGA at light/medium load, while HGA solutions massively outperform the GCA ones at heavy network load. This happens due to the fact that deterministic algorithms are more efficient when applied in small-size problems. On the contrary, as the problem size increases, robust and general procedures are favored. Deterministic algorithms may be trapped in local optimum, in contrast to genetic algorithms. In hybrid genetic algorithms, the combination of a local optimization algorithm with mechanisms as selection, crossover, mutation, the population-wide search, the capability of moving from one solution to another significantly different, permit to scan a bigger neighborhood of solutions and therefore to move towards increasingly better optima.

The CPU time that HGA consumes in order to reach its solution is proved to be considerably lower than the time required by the GCA. This conclusion matches with the theoretically computed time complexity in terms of magnitude. The nature of genetic algorithms to intuitively search for optimal solutions against the combinational methods employed by the deterministic graph coloring algorithms, ensures considerably lower search time for the HGA especially at heavy traffic conditions.

The application of HGA can be extended to a variety of resource allocation problems at multiple access networks (e.g., satellite, mobile networks), where channel sharing/reusing can be used to enhance the channel management.

Problems

1. Implement the classical GA algorithm for the ICRP problem. Keeping constant all but one of the GA parameters to be tuned, analyze the behavior of the simulation results by varying one parameter at a time. Compare the results with those found by the HGA approach.

2. Evaluate the advantages and disadvantages of using the HGA approach for the ICRP problem.

3. Apply the HGA approach to Channel assignment problem in cellular mobile networks.

4. Compare the HGA approach with Neural Network approach and other heuristic approaches (i.e., Simulated Annealing) for the channel assignment problem in cellular mobile networks.
   Design and implement a common simulative framework (CSF) and a common testbed framework (CTF) to evaluate deterministic, HGA, Neural Networks, and heuristic approaches for channel management in mobile networks.

5. Is it possible to combine two different approaches to solve the ICRP problem? (which is NP-Complete).

References

1. N. Funabiki and S. Nishikawa, "A Gradual Neural-Network Approach for Frequency Assignment in Satellite Communication Systems," *IEEE Trans. on Neural Networks*, vol. 8, No. 6, 1359–1370, 1997.
2. M. Duque-Anton, D. Kunz, and B. Ruber, "Channel Assignment for Cellular Radio Using Simulated Annealing," *IEEE Trans. on Vehicular Technology*, vol. 42, No. 1, 14–21, 1993.
3. G. Chakraborty and Y. Hirano, "Genetic Algorithm for Broadcast Scheduling in Packet Radio Networks," in *Proc. IEEE Int. Conf. Evolutionary Computation*, 183–188, 1998.
4. S. Olariu, "An Optimal Greedy Heuristic to Color Interval Graphs," *Information Processing Letters*, vol. 37, 21–25, 1991.
5. H.G. Sandalidis, P.P. Stavroulakis, and J. Rodriguez-Tellez, "An Efficient Evolutionary Algorithm for Channel Resource Management in Cellular Mobile Systems," *IEEE Trans. on Evolutionary Computation*, vol. 2, No. 4, 125–137, 1998.
6. V. Schnecke and O. Vornberger, "Hybrid Genetic Algorithms for Constrained Placement Problems," *IEEE Trans. on Evolutionary Computation*, vol. 1, No. 4, 226–277, 1997.
7. L.D. Chou and J.C. Wu, "Bandwidth Allocation in ATM Networks Using Genetic Algorithms and Neural Networks," in *Proc. IEEE GLOBECOM'97*, 962–966, 1997.
8. J. M. Renders and S.P. Flasse, Hybrid Methods using Genetic Algorithms for Global Optimization," *IEEE Trans. Syst., Man, Cybern. B*, vol. 26, No. 2, 243–258, 1996.
9. T. Bäck and S. Khuri, "An Evolutionary Heuristic for the Maximum Independent Set Problem," in *Proc. IEEE INFOCOM'94*, 531–535, 1994.
10. R. Elbaum and M. Sidi, "Topological Design of Local-Area Networks Using Genetic Algorithms," *IEEE/ACM Trans. Networking*, vol. 4, No. 5, 766–778, 1996.
11. H. Esbensen, "Finding (Near-) Optimal Steiner Trees in Large Graphs," in *Proc. Int. Conf. on Genetic Algorithms and Their Applications*, 485–491, 1995.
12. N.-F. Huang and H.-I. Liu, "A Study of Isochronous Channel Reuse in DQDB Metropolitan Area Networks," *IEEE/ACM Trans. Networking*, vol. 6. No. 4, 475–484, 1998.
13. W. Stallings, *Local and Metropolitan Area Networks*, 4th ed., New York: Macmillan, 1993.
14. IEEE Standard: *Distributed Queue Dual Bus(DQDB) Metropolitan Area Network(MAN)*, Media Access Control and Physical Layer Protocol Documents, P802.6/D12, 1990.
15. R.M. Newman, Z.L. Budrikis, and J.L. Hullett, "The QPSX Man," *IEEE Communications Magazine*, vol. 26, 20–28, 1988.
16. M.C. Golumbic, *Algorithmic Graph Theory and Perfect Graphs*, New York: Academic Press, 1980.
17. L. Davis, *Handbook of Genetic Algorithms*, New York: Van Nostrand Reinhold, 1991.
18. G. J.E. Rawlings, *Foundation of Genetic Algorithms*, San Mateo, California: Morgan Kaufmann Publishers, 1991.
19. T. Bäck, *Evolutionary Algorithms in Theory and Practice*, New York: Oxford University Press, 1996.

20. D.E. Goldberg, *Genetic Algorithm in Search, Optimization, and Machine Learning*, Reading, Massachusetts: Addison-Wesley, 1989.
21. M. Markaki, A. Vasilakos, and I. Kassotakis, "A Hybrid Genetic Algorithm for the Provision of Isochronous Service in High Speed Networks," in *Proc. IEEE Int. Conf. Evolutionary Computation*, 133–137, 1997.
22. L. Chambers, *Practical Handbook of Genetic Algorithms Volume II*, Florida: CRC Press, 1995.
23. J.A. Miller, W.D. Potter, R.V. Gandham, and C.N. Lapena, "An Evaluation of Local Improvement Operator for Genetic Algorithms," *IEEE Trans. Syst, Man, Cybern.*, vol. 23, No. 5, 1340–1351, 1993.
24. M.R. Garey and D. S. Johnson, *Computers and Intractability*, San Francisco: W.H. Freeman and Company, 4–11, 1979.

16

*Use of Computational Intelligence Techniques for Designing Optical Networks**

Imrich Chlamtac, Andrea Fumagalli, and Luca Valcarenghi

CONTENTS

KEY WORDS: *Cellular systems, Hopfield neural networks, feedforward neural networks, winner-take-all neural network, fixed channel assignment, dynamic channel assignment, interface constraints.*

* This work was partially supported by the NSF under contract NCR-9628189.

0-8493-1075-X/01/$0.00+$.50
© 2001 by CRC Press LLC

16.1   Introduction

Today's communication networks are growing in size, complexity, and bandwidth. In parallel, users' applications are becoming increasingly demanding in terms of network latency and quality of service in general. It is, therefore, necessary that network designers and managers possess adequate fast techniques to optimally design and manage these networks in a way that can serve users' requirements.

This applies in particular to optical networks. Fiber optics is turning out to be essential for supplying the huge bandwidth required by today's and future data and real-time applications, and are expected to become the mainstay of the next generation Internet. Due to the complexity and unique constraints of optical networks, many of the design issues have not, or cannot be, dealt with by conventional optimization methods, such as Linear Programming (LP), Non-Linear Programming (NLP), and similar.

In this chapter we show that heuristic techniques, often applied in Computational Intelligence (CI), Artificial Intelligence (AI), and Operational Research (OR) optimization problems, can be used in optical networks to provide near-optimal and fast solutions for efficient network design and management. As a vehicle to demonstrate our approach, we show how Simulated Annealing (SA) can be used to solve the problem of optimally placing optical amplifiers in Wavelength Division Multiplexed (WDM) optical networks. The SA technique is shown to have two appealing fundamental properties: the necessary speed for user-interactive design and the flexibility to promptly adapt the cost function being optimized to changing system constraints and requirements.

16.2   Network Heuristic Methods Based on CI

Modern heuristic methods have evolved significantly during the last decade and have been applied to a number of optimization problems. To mention just a few examples, Simulated Annealing, Tabu Search, Genetic Algorithms, and Artificial Neural Networks are among the best known among these methods. In this section we briefly discuss the connection between CI and the above heuristic methods. We provide a high level view of the various methods, followed by a detailed description of Simulated Annealing, the technique whose use is demonstrated in this chapter.

16.2.1   CI and Heuristics

"Computational Intelligence (CI) has emerged few years ago" (1992)[1] and, "as a term, it has become a rubric for efforts in neural networks, fuzzy systems, and evolutionary computation."[2] The first attempt to define the term CI[1, 3] has been done by Bezdek[4] by asserting that " . . . A system is computationally intelligent when it: deals with only numerical (low-level) data, has a pattern recognition component, does not use knowledge in the AI sense . . . ". In fact in Reference 12 it is stated that "the central focus in traditional AI research has been on emulating human behavior by extracting rules and knowledge from human experts" while the CI technologies are considered to be "methods of computation that can be used to adapt solutions to new problems and do not rely on explicit human knowledge.[2]

However, "the boundary between CI and AI is not distinct."[3] Artificial Intelligence is defined as "the design and study of computer programs that behave intelligently."[5] The word "intelligently" indicates programs which are able to "respond flexibly in situations that were not specifically anticipated by the programmer."[5] Moreover, expert systems based on AI, also referred to as knowledge-based systems, are currently being applied to several classification problems such as interpretation, diagnosis, and monitoring as well as to problems of planning and design.[6]

The heuristic techniques described and utilized in this chapter are not limited to the "building blocks of CI"* but can be seen as general "intelligent artifacts."[7] Therefore, we refer to CI as "the study of the design of intelligent agents" (or artifacts).[7] In fact "the central scientific goal of CI is to understand the principles that make intelligent behavior possible in natural or artificial systems. The main hypothesis is that reasoning is computation. The central engineering goal is to specify methods for the design of useful, intelligent artifacts."[7]

Several combinatorial and search problems encountered in CI require exponential solution time; see, for instance, the Traveling Salesman Problem (TSP).[8] In such cases, heuristic techniques, which provide possibly only a suboptimal solution but which require polynomial time,[9] are generally applied. These heuristics are often derived from the observation that the search time can be reduced considerably when the structure of the search space is taken into account. Consequently, the number of attempts necessary to converge to a near-optimal solution can be reduced. To do so, an evaluation function must be defined that estimates the distance of a given solution from the optimum.

The design of telecommunication networks presents a number of problems related to resource optimization and allocation in which the computational complexity is exponential in the problem size, i.e., in the number of unknowns. The exhaustive search for the optimal solution in these problems requires the evaluation of an immense number of possible solutions and may involve real and/or integer variables, so that the time required to find an optimal solution becomes unacceptable.

In such cases heuristic techniques employed in the CI search problems can be used to help design the network suboptimally. The evaluation function is

---

* Neural Networks, Genetic Algorithms (GAs), Fuzzy Systems (FSs), Evolutionary Programming (EP) (see Reference 1).

defined as the cost function that returns the total cost of the various solutions, or states, in the solution space. The objective of the heuristic technique is to find the state, or its best approximation, that minimizes or maximizes, depending on the problem under consideration, the cost function value.

One of the most interesting and challenging features of heuristic techniques is the possibility for the user to trade, by means of a set of adjustable parameters, the solution time with the quality of the solution, depending on the problem under investigation. Another argument in favor of heuristics is their flexibility and capability "of coping with more complicated (and more realistic) objective functions and/or constraints than exact algorithms" do.[10] For example, when using algorithms based on Simulated Annealing, Genetic Algorithms, Tabu Search, or Artificial Neural Networks, approximation of the cost, or *objective function*, is not necessary. Contrary to this, other optimization techniques, such as LP solvers, are restricted to linear objective functions and may require an approximated formulation of the actual cost function. By directly optimizing the actual cost function, the heuristics yield an accurate and exact analysis of the network under design.

We next describe some of the more popular heuristic techniques generally used in communication networks design. For more details on these techniques, the interested reader is referred to Reference 10.

16.2.2   Genetic Algorithms

Genetic Algorithms (GAs) were originally introduced in 1975.[11] A survey on some of their applications to practical optimization problems is given in Reference 12. The name *genetic algorithm* takes its origin from the analogy between the representation of a complex structure by means of a vector of components, and the genetic structure of a chromosome. Generally speaking, a GA consists of three basic operations: *Selection, Genetic Operation*, and *Replacement*.[13] By using these operations, the algorithm modifies the values of the data structure, or chromosome, to generate a new solution to the problem, which is improved over the previously found solutions. The classic GA approach is described next.

1. The problem under optimization is mapped into a function that indicates the *fitness* of any given solution. A population of solutions is defined as a group of solutions from which candidates can be selected to generate new solutions or to be the final solution to the problem.

2. An initial population is randomly generated. Any solution is encoded in a vector called *chromosome* whose basic elements are called *genes*. The gene takes on a value from a set of *alleles*.

3. Every chromosome is evaluated by means of the fitness function that assigns to it a value according to the objective function of the problem.

4. Every chromosome is assigned a reproduction probability that is proportional to its fitness.

5. A particular group of chromosomes (parents) is selected from the population to generate the offspring using the defined genetic operations.

6. The fitness of the offspring is evaluated in a way similar to their parents and the chromosomes in the current population are then replaced by their offspring, based on a given replacement strategy.

7. The process ends when the established termination conditions are met.

If correctly performed, the simulated evolution from the initial population leads to a highly evolved solution of the problem represented by the best chromosome found in the final population, i.e., the best approximation of the optimal solution.

The encoding scheme adopted for the chromosomes is a key factor in GA, as it may considerably limit the window of information observed on the system. The bit-string encoding is the most commonly used due to its simplicity and tractability, but other encoding schemes are possible, such as real number representations. The GA performance also depends on the types of genetic operation that are used to generate the offspring from the parents. Two types are generally considered: *crossover* and *mutation*.

The crossover is a recombination operator that combines subparts of two parent chromosomes to produce an offspring chromosome that contains some parts of both parents' genetic material. Its operation rate is determined by a given probability, termed p_c.

The mutation introduces random variations in the chromosome according to a given probability, termed p_m. These variations can be global or local.

In addition to this general structure of the algorithm, there are various parametric choices. It is also possible for the initial population to have a significant impact on the final solution. Different implementations of the GA algorithm, which are best suited to solve specific problems, can be found in literature.[14] For example *Messy Genetic Algorithm* (mGA)[15] has demonstrated to be effective in a number of problems.

16.2.3 Tabu Search

Tabu Search (TS) has its origin in 1970[16] when it was used in combinatorial procedures applied to nonlinear covering problems. Since then it has been used in a variety of problems ranging from scheduling and computer channel balancing to cluster analysis and space planning.[17] The latest applications of this optimization algorithm include integrated circuit design and time tabling problems in which TS has been proven to obtain high-quality solutions with modest computational effort.[17]

The simplest form of TS presents two key elements:

- constraining the search by classifying certain moves in the search space as forbidden (i.e., tabu), and
- partially freeing the search with a short-term memory function that provides "strategic forgetting."

In its most general form, the objective function $f(x)$ in TS may be either linear or nonlinear and it may incorporate penalties that drive the search towards solutions that satisfy the problem constraints.

The elements of the TS algorithm are:

- a set of permitted solutions, X, defined by the problem-constraining conditions. The constraints are, generally, taken into consideration at any step of the search;
- a function, s, that identifies the moves in the set of the permitted moves, $S(x)$, from a given solution $x \in X$. This function defines the neighborhood of x;
- a tabu list, T, that indicates the set of prohibited moves that cannot be made from the given solution x. This peculiar element of TS keeps track of previously visited solutions to prevent that these solutions are revisited unless the trajectory followed by the search is a new one;
- a move evaluator function, *OPTIMUM*, that associates each move a *move_value*. This value represents the goodness of the move with respect to the problem-objective function.

In its general definition, the TS algorithm operates as follows.

1. An initial solution, $x \in X$, is chosen. Both the current solution, $x^* = x$, and the best found solution, $x^{best} = x$, are set equal to this initial solution. An iteration counter, k, is set to zero and the tabu list is set empty.

2. If set $S(x^*) - T$ is not empty, counter k is incremented and the best possible move (considering also nonimproving moves) applicable to x^* in set $S(x^*) - T$ is chosen. The choice is based on the evaluation function *OPTIMUM*. Tabu moves in T may be also chosen when they satisfy particular aspiration criteria (e.g., if the tabu move results in a solution better than any solution visited so far).

3. The newly chosen solution is set as the current solution, $x^* = x$. The best solution is updated if the cost of the new solution x is better than the cost associated with the best solution found so far.

4. The algorithm ends when either the maximum number of iterations has been reached or $S(x^* - T) = \phi$. When none of the former

termination conditions is met the algorithm updates the tabu list *T* and returns to point 2.

The best solution found during the execution of the algorithm represents the best approximation of the problem-optimum solution. To improve the TS performance, a number of alternative approaches has been derived from this simple version of the algorithm. These approaches are based on complex tabu list procedures and aspiration criteria whose scope is to avoid deadlock in local minima.

16.2.4   Artificial Neural Networks

The Artificial Neural Network (ANN) method emulates the structure of biological neural networks and their natural way to encode and solve a given problem.

The main elements of the biological Central Nervous System (CNS) are the neurons. These biological cells communicate with each other by means of electrical signals conveyed by *synapses*. The network formed by neurons and synapses excites or inhibits biological functions regulated by the CNS. The basic idea of the ANN method is to emulate some of the fundamental functions found in the CNS to construct mathematical models capable of performing intelligent data processing.

The basic computational entities of an ANN are the neurons, v_i, which take on either real values in the [0, 1] (or [−1,1]) range or binary values $s_i = \{0, 1\}$ (or $\{−1, 1\}$). In the chapter the term v_i refers to a general neuron with either binary or real values, while s_i is used to indicate a neuron with only binary values. Common to most neural models is the *local updating rule*:

$$v_i = g\left(\sum_j w_{ij}v_j - \theta_i\right) \tag{16.1}$$

where v_j represents all neurons feeding neuron v_i, w_{ij} represents the weights (or synapses) given to each respective feeding neuron, and θ_i is a threshold term. The weights can have either positive (excitatory) or negative (inhibitory) values. The *transfer function* $g(\cdot)$, that determines the state of the neuron is nonlinear and typically has the following form:

$$g(x) = 0.5[1 + \tanh(x / T)] \tag{16.2}$$

where temperature T controls the inverse gain: at low temperature function $g(\cdot)$ is steep, at high temperature it becomes practically constant. The limit $(T \to 0)$ corresponds to a binary neuron s_i. This artificially created neuron mimics the real biological functions in which the effect of several simultaneous signals arriving at the neuron is approximately linearly additive, whereas the resulting output is a strongly nonlinear all-or-none type of function.

Neural network models are based on one of two possible architectures: the feed-forward and the feed-back.[10] In the feed-forward architecture the synapses between the nodes are monodirectional and the electrical signals move from a set of input nodes to a set of output nodes (usually these nodes are respectively placed at the bottom and at the top of the graphical representation of the ANN). As opposed to the former architecture, feed-back networks have bidirectional synapses. Feed-forward networks are used in problems in which it is possible to distinguish between the input and the output layers. For example, in feature recognition problems input nodes constitute the input pattern and output nodes represent the recognized feature. Feed-back networks, on the other hand, have been shown to be capable of finding good solutions in difficult combinatorial optimization problems, such as the Graph Bisection Problem (GBP) and its generalization, the Graph Partitioning Problem (GPP), the Traveling Salesman Problem (TSP), scheduling problems, and so on.[10] The optimization of feed-back networks is obtained by minimizing the energy function associated with the network, which typically assumes the following form:

$$E = -\frac{1}{2}\sum_{i,j} w_{ij} s_i s_j \qquad (16.3)$$

where s_i are binary neurons. In its general form, the energy function makes use of real neurons v_i too.

Given a problem to optimize, the ANN model is constructed by associating one neuron, either v_i or s_i, with each of the N variables of the problem. The choice on the set of weights, w_{ij}, is critical to achieve satisfactory results. This choice is not trivial as the weights are fixed for each problem and cannot be dynamically adapted during the optimization process. Neurons are allowed to reach a stable state, $\vec{s} = (s_1, s_2, \ldots, s_i, \ldots, s_N)$ or $\vec{v} = (v_1, v_2, \ldots, v_i, \ldots, v_N)$, that yields minimum energy in the system. While minimization of equation 16.3 takes place, the values of the neurons are constantly updated according to the local updating rule in equation 16.1. The solution to the problem is directly derived from the reached stable state.

16.3 Simulated Annealing

In the previous section we reviewed some of the better known methods adopted in CI to solve optimization problems in communication networks. Most of the problems to which these methods have been applied belong to the class of combinatorial problems. However, in general, in the design of communication networks the variables under optimization may assume both real and integer values.

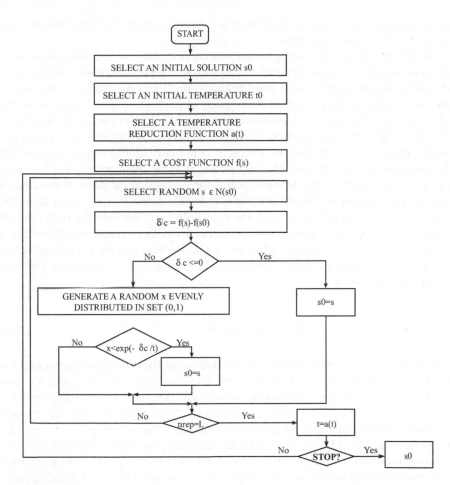

FIGURE 16.1
SA basic algorithm.

This section describes Simulated Annealing (SA), a heuristic technique that can efficiently deal with optimization problems that involve both integer and real variables. This section presents a general description of the SA algorithm, detailing the function of the various parameters utilized. A practical application of the heuristic to the design of optical networks is given in Section 16.4.

Among the heuristic algorithms available today for optimizing complex cost functions, SA is one of the most frequently used due to its versatility and effectiveness.[10] The concepts that form the basis of SA were first published by Metropolis et al.[18] as part of an algorithm that simulates the cooling of material in a heat bath (a process known as annealing). Thirty years later, Kirkpatric et al.[19] suggested that this type of simulation could be used to search the feasible solutions of a generic optimization problem with the objective to converge to an optimal solution.

The SA technique can be regarded as a variant of the well-known technique of local search. Contrary to traditional forms of local search which employ a strictly descent strategy in a minimization problem, SA tolerates ascending moves. SA thus offers a higher probability of reaching a good approximation of the global optimum by overcoming the risk of being trapped in one local minimum. A high level description of the SA algorithm is depicted in Figure 16.1, under the assumption that the problem being considered requires us to find the minimum cost solution. The algorithm, starting from an initial solution s_0, samples the neighboring feasible solutions $(N(s_0))$ at random. If the cost of the newly randomly chosen solution, $f(s)$ with $s \in N(s_0)$, is lower than the cost of the current solution, $f(s_0)$, the new solution becomes the new current solution. If the new solution cost is higher than the current solution cost, a random decision is made on whether or not the new solution is still accepted in place of the better current solution. The probability of accepting the new solution is given by

$$P\{\text{acceptance } \delta c\} = e^{-\frac{\delta c}{t}}, \tag{16.4}$$

where δc is the cost gap between the new and the current solution and t is a varying parameter called *temperature*. The probability of accepting a worse solution is inversely proportional to the ratio between the cost gap and the temperature. Therefore, with a sufficiently high temperature, the chosen solutions can freely move in the solution space without being constrained by local optima. In addition, by gradually reducing the temperature during the simulation from an initial value t_0 to a final value t_f it is possible to guarantee that the accepted solutions converge to a minimum which is a good approximation of the global optimum. In the algorithm version shown in Figure 16.1, SA convergence is obtained by decreasing the temperature with a reduction function $a(t)$. Between two consecutive reductions of temperature, *nrep* neighboring feasible solutions are drawn by the algorithm. An often-used temperature-reduction function (also adopted in this chapter) is the geometric-reduction function:

$$t^+ = a(t) = at \tag{16.5}$$

where t^+ is the next temperature value, $0 < a < 1$ is the cooling factor and t is the current temperature value. The total number of feasible solutions drawn during the entire SA run is, therefore,

$$iter_{\min} = nrep \cdot \left\lceil \frac{\ln(t_f / t_0)}{\ln a} \right\rceil \tag{16.6}$$

The value of $iter_{min}$ is a good indicator of the computational complexity of the simulation for a chosen set of parameters. The best solution found during the algorithm search, which often coincides with the current solution s_0 at the time the algorithm stops, represents the heuristic optimal solution.

16.4 SA for Optimal Amplifier Placement in Multi-Wavelength Optical Networks

To keep up with today's growing demand for bandwidth, novel architectures for designing optical networks are being explored. One of the most promising architecture is that of an all-optical network, in which end-to-end transmission is achieved without electronic conversion of the transmitted data at the intermediate nodes in the network.[20] All-optical networks are not limited by the relatively low bandwidth of the electronic processing occurring at the switching nodes (e.g., routers) associated with conventional networks, so that a better utilization of the optical fiber bandwidth, in the dozens of THz, can re-realized.

One of the fundamental problems arising in the design of these networks is the optimal placement of optical devices. For example, in Wavelength Division Multiplexed (WDM) networks, optical amplifiers need to be placed to regenerate the optical signals on the various WDM channels and compensate for the propagation and splitting losses encountered by the signals while traveling from source to destination. Optical amplifiers also make it possible to maintain optical transparency (i.e., avoid the need for electro/optical conversion of the optical signal at the regeneration point) in large networks, providing the necessary scalability.

In this section we present the problem of optimally placing optical amplifiers in broadcast-and-select WDM networks, termed "the Generalized optimal Placement of optical Amplifiers" or GPA problem. The following subsections present a formal description of the GPA problem and show how SA can be used to yield optimal or near-optimal solutions to the problem.

16.4.1 Rationale and Related Work

Currently available technology makes it possible to utilize a significant portion of the 30-THz optical fiber bandwidth by using Wavelength Division Multiplexing (WDM) multi-channel systems.[21] This technique allows the establishment of orthogonal communication channels in the same fiber on different wavelengths. A simple and practical solution to achieve all-optical transmission in a WDM network is the so-called broadcast-and-select approach.[22] In a broadcast-and-select network each source station is assigned a *distinct* wavelength for transmission, and the transmitted optical signal is broadcast to every destination station of the network. In order to receive the signal the intended destination station must tune its receiver to the source node's wavelength. To compensate for the propagation and splitting losses encountered by the optical signal, thus guaranteeing a satisfactory signal to noise ratio at every receiving station, optical amplification may be necessary in the network. The need for amplification depends upon the total number of wavelengths (stations), fiber layout, and network span. Erbium-Doped Fiber Amplifiers (EDFA)[23] and Semiconductor Optical Amplifiers (SOA)[24] can provide the necessary optical amplification. However, since the price of an Optical Amplifier (OA) is currently significant, on the order of tens of thousands of dollars, the total number of OAs required in the network significantly affects the overall system cost. Moreover, the location of an amplifier may affect installation and maintenance costs. In other words, alternative network designs that require the same number of amplifiers may not necessarily yield the same total network cost.

In the past, several studies dealt with the minimization of the number of amplifiers in star-based optical networks.[25, 26, 27] These approaches simplified the amplifier location problem by either adding or relaxing some problem constraints. For example, in Reference 25 the number of amplifiers is minimized under the assumption that the power of the signals across the wavelengths is uniform, i.e., using the so-called equally-powered wavelength method. In a more recent work, an Unequally-Powered Wavelength Method[26] has been proposed which leads to a Mixed-Integer Non-Linear Problem (MINLP) solved by relaxing some of the integral constraints. The latter approach may further reduce the number of amplifiers in the system due to the larger solution space originated by the allowed unequal distribution of power when compared to the equal-powered case.

The study presented in this chapter generalizes the Unequally-Powered Wavelength Method, by introducing the Generalized optimal Placement of optical Amplifiers in broadcast-and-select WDM networks (GPA) problem. In the GPA problem, the number of optical amplifiers in the network is minimized taking into account that:

1. the fiber layout can have star, tree and ring topologies,
2. installation and maintenance costs of the amplifier may depend on its location, i.e., the amplified link.

In particular, while the number of amplifiers is minimized in the system, an attempt is made to place the amplifiers on links that require the lowest installation and/or maintenance costs.

By going beyond the star/tree topology and by taking into account the amplifier location-dependent costs, the presented solution of the GPA problem provides a practical and versatile technique for designing cost-effective WDM networks.

16.4.2 The GPA Problem Formulation

In this subsection we summarize those parts of the network and amplifier models that are most related to the adaptation of the SA algorithm to solve the GPA problem. The interested reader is referred to Reference 28 for additional details of the problem formulation.

16.4.2.1 The Network Model

The network model is used to represent the signal power levels on each distinct wavelength in every link of the system, the transmission power at every source station, the total number of amplifiers, and their respective gain. By means of the network model, it is also possible to verify whether or not a chosen solution (number of amplifiers and their location in the network) satisfies the physical constraints imposed by the topology and the optical devices. As a result, two classes of solutions are defined: the feasible and the nonfeasible solutions.

The network model is derived from the star and tree model proposed in References 26 and 27. Our formulation adds two features:

1. the ring fiber layout, and
2. the location dependent cost of the amplifier.

Figure 16.2 shows an example of star-based topology. (The construction of a ring-based network can be found in Reference 28.) Recall that according to the broadcast-and-select concept, each source station has a signature wavelength, and any destination station can receive from the source station on that wavelength by properly tuning its receiver. Consequently, N wavelengths are necessary to connect N stations. The star topology comprises transmitting/receiving stations and nonreflective passive stars called *splitting nodes*. The latter nodes are the primary cause of power loss as they evenly split the power of any incoming signal among all the output ports except the output paired with the input on which the signal arrives.

The physical constraints due to the optical components are: P_{NLmax}, the maximum total power allowed in the fiber without generating substantial nonlinear effects; α, the fiber attenuation; P_{sen}, the minimum power level necessary in the fiber to guarantee a satisfactory signal to noise ratio; P_{max}, the transmitter maximum power*; and G_{max}, the maximum gain achievable by the amplifier. The unknowns of the problem are the power levels transmitted at the N source nodes, p_i^{xmit}, and the total signal gain in each of the L links of the network, sg_l.**

To be feasible, a solution must satisfy the following constraints on each link l:

---

* In our formulation this is also the maximum output power of the amplifier.
** The number of amplifiers necessary on a link is directly derived from the link total gain.

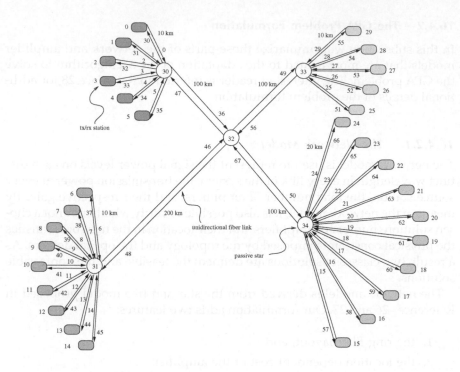

FIGURE 16.2
A star-based broadcast-and-select topology.

$$p_l^{\min,\text{beg}} \geq p_{\text{sen}} \tag{16.7}$$

$$p_l^{\min,\text{beg}} + sg_l - \alpha \cdot L_l \geq P_{\text{sen}} \tag{16.8}$$

$$P_l^{\text{beg}} \leq P_{\text{max}} \tag{16.9}$$

$$P_l^{\text{beg}}(dB) + sg_l - \alpha \cdot L_l \leq P_{\text{max}} \tag{16.10}$$

where $p_l^{\min,\text{beg}}$ is the power (in dBm) of the minimum power wavelength at the beginning of link l, L_l is the length of link l, and p_l^{beg} is the total power (on all wavelengths) at the beginning of link l in mW. A set of equations are used to derive quantities p_l^{beg} and $p_l^{\min,\text{beg}}$ from the problem unknowns, p_i^{xmit} and sg_l. Notice that variable $p_l^{\min,\text{beg}}$ is not linear in the unknowns as its value is the minimum among the power levels of the wavelengths at the start of the corresponding link.

For the sake of simplicity, the location-dependent costs of the amplifier is modeled by assuming that some links in the system have higher-than-average location costs. (More complex models of the location-dependent costs are possible, without requiring significant modifications of the proposed approach.) The set of *deprecated* links, $DL = \{d1, d2, \ldots, di, \ldots, dL\}$, is defined as the set of links for which the amplifier location-dependent costs are higher than average. Besides minimizing the total number of amplifiers in the system, the final, feasible solution must also minimize the number of amplifiers on these links.

16.4.2.2 The Amplifier Model

The saturation gain model described in Reference 25 is used in this chapter to characterize the amplifier gain. More complex gain models, such as the one proposed in Reference 26, can also be used without affecting the correctness or the efficiency of the proposed heuristic approach. The gain function is

$$G = \min\{G_{max}, (P_{max} - P_{in})\} \tag{16.11}$$

where P_{in} is the total power at the amplifier input, expressed in dBm, G_{max}, and P_{max} are characteristic parameters of the amplifier defined in Section 16.4.2.1, expressed in dB and dBm, respectively.* The maximum gain achievable by an amplifier on link $l(gmax_l)$ is derived from Equation 16.11, thus it depends on the power distribution across the wavelengths on link l that feed the amplifier. This distribution clearly determines the overall power in the fiber, and at the same time determines how soon amplification is required on a given link before the signal with the lowest power level reaches the allowed minimum power level, p_{sen}.

16.4.3 Solution of the GPA Problem by SA

This subsection describes the details of the SA algorithm deployed to solve the GPA problem, including the representation of the generic solution s, the neighborhood structure employed to choose the next solution $N(s)$ and the cost function $f(\cdot)$.

In solving the GPA problem we show that SA, which tolerates ascending moves in the case of minimization problems, can reach a good approximation of the global optimum. A second positive effect of the SA ascending moves is the independence of the final result from the starting solution. Finally, in order to cope with the dual objective of minimizing the number of amplifiers and, simultaneously, avoiding the placement of amplifiers on links where the installation and maintenance costs are higher than average, SA allows to define a hierarchical cost function.

The effectiveness of the SA approach is assessed by comparing its results with the lower bounds presented in References 26 and 27. The comparison is based on the number of amplifiers necessary in the network, under the assumption that all links are equally suitable for amplification, and reveals that the SA approach yields results that are optimal or near-optimal.

---

* G_{max} is the maximum amplifier gain in dB and P_{max} is its maximum total output power.

16.4.3.1 Solution Representation

A straightforward representation of the generic solution, s, is the set of real values $\{\{sg_l\}, \{p_i^{\text{xmit}}\}\}$. The total number of elements in the set equals $L + N$, where L is the number of links in the system and N is the number of receiving/transmitting stations.

A more concise representation of the solution, one that requires only L variables, can be obtained using one variable to represent both the transmitted power of a station and the gain of the corresponding unidirectional link connecting the station to the splitting node. Although this representation does not allow us to distinguish between the gain and the station's transmitted power, it provides the necessary information to correctly sample the solution space. Its advantage is that all the solutions characterized by the same value of the sum $sg_l + p_{sl}^{\text{xmit}}$ can be represented by a single solution in the simulation. Thus, each time a solution is sampled, the simulation is actually exploring an infinite number of solutions that are equivalent from the point of view of power budget at the splitting nodes.

16.4.3.2 Neighborhood Solution Structure

The strategy used to choose the next solution from the current solution is critical for the success of the SA technique.[29] Among the several strategies that were studied to solve the GPA problem, the most efficient and practical one is based on the idea of changing one link gain at a time. The link (l) is randomly chosen, and its gain sg_l is modified by adding a random value that is uniformly distributed between $[-\beta G_{\max}, +\beta G_{\max}]$, where β is a varying parameter chosen to achieve gain variations that are neither too small, nor too large.[29] If the resulting value for sg_l is negative, sg_l is set equal to 0. If the obtained solution is feasible, i.e., it satisfies the physical constraints given in Section 16.4.2.1, its cost is compared with the cost of the current solution. Otherwise, a new random attempt is made to find another neighboring solution that is feasible. If the number of solutions consecutively sampled without finding a feasible solution reaches a user-defined maximum value the search stops and returns the best found solution.

16.4.3.3 The Hierarchical Cost Function

The objective of our optimization is twofold: to minimize the overall number of amplifiers required in the system and, at the same time, minimize the number of amplifiers required on the deprecated links. The integer characteristic of the GPA problem originates from the nature of the cost function, i.e., the total number of amplifiers necessary in the network. This number is derived from the total supplied gain (sg_l) and the maximum gain achievable by an amplifier on each link ($gmax_l$). The *primary* cost function of the GPA problem is, therefore:

$$N_L = \sum_{l=0}^{L-1} \left\lceil \frac{sg_l}{gmax_l} \right\rceil \tag{16.12}$$

where N_L is the total number of amplifiers. We define ΔN_L as the difference in the number of amplifiers between the sampled solution and the current solution.

Between two solutions for which $\Delta N_L = 0$, the solution requiring fewer amplifiers on the deprecated links is preferred. A *secondary* cost function is thus introduced in the SA algorithm to be used only when the primary cost function is not sufficient to identify the better solution. The secondary cost function is obtained by computing the total number of amplifiers placed on the deprecated links (N_{DL}). When the two solutions under comparison have the same number of amplifiers on deprecated links, the secondary cost function returns the total gain, in dBm, required on the deprecated links. The secondary cost function is, therefore, defined as follows. Let s_0 be the current solution and s the sampled feasible solution. Define N_{DL}^s as the secondary cost function of solution s

$$N_{DL}^s = \sum_{l \in DL} \left[\frac{sg_l}{g \max_l} \right] \qquad \text{if } N_{DL}^s \neq N_{DL}^{s_0} \qquad (16.13)$$

$$N_{DL}^s = \sum_{l \in DL} sg_l \qquad \text{otherwise} \qquad (16.14)$$

We define δN_{DL} as

$$\delta N_{DL} = \frac{N_{DL}^s - N_{DL}^{s_0}}{N_{DL}^{s_0}}. \qquad (16.15)$$

A weight, K, is used to "magnify" the secondary cost difference. By varying K the designer can shift the emphasis of the optimization between the total number of amplifiers in the system and the number of amplifiers on the deprecated links.

When δN_{DL} equals zero, the sampled solution is always chosen as the new current solution.

In summary, the cost difference between the sampled feasible solution and the current solution, $\delta = f(s) - f(s_0)$, is:

$$\delta = \Delta N_L, \qquad \text{if } \Delta N_L \neq 0$$

$$\delta = K \cdot \delta N_{DL}, \qquad \text{if } \Delta N_L = 0$$

16.4.3.4 Setting of the SA Parameters and Results

The proposed SA based solution to the GPA problem relies on a number of parameters, including the cooling factor, a, the initial temperature, t_0, the final temperature, t_f, the number of solutions attempted at each temperature, *nrep*,

TABLE 16.1

Network Physical Constraints

Parameter	Description	Interval	Utilized value
P_{sen}	Minimum received signal	−30 dBm at 1 Gbps	−30 dBm
G_{max}	Small signal maximum gain	≥ 37 dB EDFA[23]	20 dB
P_{NLmax}	Maximum linearity power	10 mW	10 mW
P_{max}	Maximum output power		0 dBm
α	Fiber attenuation		0.2 dB/km

the gain variation factor, β, and the weight of the secondary cost function, K. The values for these parameters must be chosen to achieve a near-optimal solution with the minimum necessary computational time, e.g., with the minimum value of total iterations, $iter_{min}$. Clearly, there exists a trade-off between the goodness of the near-optimal solution and the required computational time. For example, reducing the cooling rate, i.e., increasing the cooling factor, a, may increase the probability to find a better approximation for the global optimum but, at the same time, it increases the computational time required by the simulation (see Equation 16.6).

We next quantify the trade-off for the most critical SA parameters. In particular, we focus our attention on the cooling schedule (characterized by parameters a and $nrep$) and the weight of the secondary function, K. For a complete analysis of the effect of the SA parameters on the suboptimal solution, the interested reader is referred to Reference 28.

Figure 16.2 depicts the star network used to carry out our analysis. The system parameter values are shown in Table 16.1.

For this network, originally proposed in References 26 and 27, a lower bound on the number of amplifiers is known and equals 14. This lower bound is used to determine the ability of the SA approach to find the minimum in the shortest possible time, while taking into account, at the same time, the location of the amplifiers in the network. Our study was carried out using 20 distinct simulations* for each set of chosen parameter values and collecting the statistics on the solutions found, e.g., how many simulations reached the lower bound, the average number of amplifiers found and the total simulation time on a Pentium 200 MHz-based machine.

Results obtained using various cooling schedules (i.e., different values of a and $nrep$) are shown in Table 16.2. The other SA parameters are constant for all simulations: $t_0 = 0.9$, $t_f = 0.1$, $\beta = 3.0$ and $K = 0$, i.e., the set of deprecated links is empty.

All simulations used the same initial conditions: 1) the power injected by all transmitters is $p_{init2} = P_{max}(dBm) − sl_{max}(dB) − 10 \cdot \log_{10}(N − 1)$, where sl_{max} is the maximum value among all splitting losses occurring at the splitting nodes, 2) amplifiers are placed using the "As Late As Possible" (ALAP)

---

* The 20 simulations are obtained using 20 distinct seeds for the random number generator.

TABLE 16.2

Statistics for the Results Obtained Varying a and $nrep$

Simulation results							
			Number of amplifiers				Avr. sim. time
Simulation constants						Avr. sim. time	per opt. sol.
n°	a	$nrep$	14	15	16	17	seconds	seconds
1	0.82	136	1	5	7	7	898	17,960
2	0.91	204	13	5	1	1	2891	4447
3	0.92	340	18	1	1	0	12,078	13,420
4	0.93	204	14	5	1	0	3782	5402

technique proposed in Reference 27. According to this technique, an amplifier is placed only when the weakest signal (wavelength) power in the link reaches p_{sen}.

The results shown in Table 16.2 indicate the trade-off between the solution time and the success rate in finding the optimal solution. In the first set of simulations, whose average duration was approximately 15 minutes, only one out of 20 solutions reached the minimum number of amplifiers (14) while seven solutions required as many as 17 amplifiers. In this case the relatively fast cooling factor, $a = 0.82$, and the small number of sampled solutions at each temperature, $nrep = 136$, do not provide the sufficient computational time to escape from local optima.

By gradually slowing down the cooling factor and increasing the number of sampled solutions at each temperature, the percentage of solutions reaching the minimum number of amplifiers was increased. For example, in the second set of simulations, using $a = 0.91$ and $nrep = 204$, the number of optimal solution grows to 13. Notice that the average simulation time in set 2 was about three times longer than the simulation time in set 1.

By further increasing the cooling factor, a, it becomes possible to obtain a larger number of optimal solutions as shown in simulation set 4. However, the increased success in finding a larger number of optimal solutions, approximately 8%, does not justify the increased average simulation time that is about 30% longer than the simulation time in set 2. In other words, increasing the cooling factor alone, past a certain value, is not advisable. In order to achieve higher success rates in finding the optimal solution both parameters a and $nrep$ must be increased as shown in simulation set 3. In this case 18 optimal solutions were found using a cooling factor that is less than the one used in set 4. The number of sampled solutions at each temperature is now 340, requiring an average simulation time of about 3.5 hours.

The previous observations underline the importance of the trade-off between the number of optimal solutions found by SA and the required average simulation time. If, for example, we compared the four simulation sets considering the average simulation time per optimal solution found, set 2 yields the best trade-off with an optimal solution found every 4447 seconds.

Generally, it is thus advisable to explore different combinations of values for a and $nrep$ until the desired trade-off between solution optimality and simulation time is found. The relationship between a and $nrep$ can be better clarified by modeling the algorithm as a sequence of homogeneous Markov chains of finite length, each chain being associated with one value of the temperature, t.[30] In this case $nrep$ determines the number of states of the homogeneous Markov chain sampled at every temperature and a determines the decrement of the control parameter (the temperature), t. To obtain a high probability to find the global optimum of the cost function one has to choose

- a value for $nrep$ such that at every temperature level the homogeneous Markov chain has sufficient time to reach the quasi-equilibrium condition and

- a value for a "such that small Markov chain lengths suffice to re-establish quasi equilibrium after the decrement of the control parameter.[30]"

The other parameter to focus on is the weight of the secondary cost function, K. By varying its value it is possible to affect the *placement* of the amplifiers in the network, possibly avoiding the deprecated links in set DL.

For the sake of demonstration, our experiment arbitrarily assumed that link 47 (see Figure 16.2) is not suitable for amplification, thus $DL = \{47\}$. Figure 16.3 shows three distinct amplifier placements obtained using the following SA

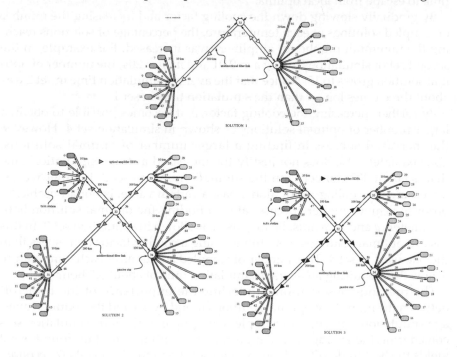

FIGURE 16.3
Three different amplifier placements obtained with $DL = \phi$ (top), $DL = \{47\}$ with $K = 20$ (bottom left) and $K = 50, 100$ (bottom right).

parameters: $a = 0.93$, $t_0 = 0.9$, $nrep = 204$, $\beta = 3.0$, and initial condition p_{init2}. The top placement is obtained assuming that there are no deprecated links, i.e., $DL = \phi$, or equivalently $K = 0$. The two bottom placements are obtained assuming $DL = \{47\}$, with $K = 20$ (left) and $K = 50, 100$ (right). The results obtained from the simulations are summarized in Table 16.3.

The figure shows that the 14 amplifiers are placed where the amplification is mostly needed, i.e., along the longest links and close to the splitting nodes with greater splitting loss. As a result, in the top solution with $DL = \phi$ and $K = 0$, 2 amplifiers are needed on link 47. The number of amplifiers on link 47 is decreased to 1 when $K = 50$ or $K = 100$, as shown in the bottom right solution. The total number of amplifiers in the network remains unchanged as now link 46 requires 3 amplifiers when compared to the 2 amplifiers on the same link needed in the top solution. The effect generated by parameter K is to "shift" one amplifier from deprecated link 47 to link 46. Notice that this result is not achieved in the bottom left solution obtained with $K = 20$.

Table 16.3 presents a more exhaustive description of how parameter K affects the placing of the amplifiers. As intuitively suggested, by increasing the weight of the secondary function, K, more solutions with fewer number of amplifiers in the deprecated link are found. For example, when $K = 20$ the solutions are distributed among the three placements shown in Figure 16.3. By further increasing the value of K ($K = 50, 100$) most of the found solutions have only one amplifier in the deprecated link (sol. #3). Interestingly, the total number of optimal solutions found by SA decreases from 14 (simulation set 5 with $K = 0$) to 7 (simulation set 7 with $K = 50$ and simulation set 8 with $K = 100$). This fact is due to the interaction between K and the other simulation parameters: a high value of K gives the secondary cost function 1.15 a greater importance with respect to the primary cost function, thus potentially compromising the simulation ability to find an optimum.

Lastly, an important property of SA is its ability to obtain solutions that are independent from the chosen initial solution. This property holds in the current study too, as demonstrated in Figure 16.4 where the slopes of two simulations with identical simulation parameters, but different initial (feasible) solutions, are plotted. The two solutions require the same initial number of amplifiers (48) but the power levels of the transmitters at the stations are dif-

TABLE 16.3

Statistics of the Results Obtained by Varying K with $DL = \{47\}$

Simulation constants		Simulation results					
		Number of amplifiers			Solution occurrences		
$n°$	K	14	15	16	Sol. #1	Sol. #2	Sol. #3
5	0	14	5	1	14	0	0
6	20	13	4	3	2	5	6
7	50	7	10	3	0	0	7
8	100	7	10	3	1	0	6

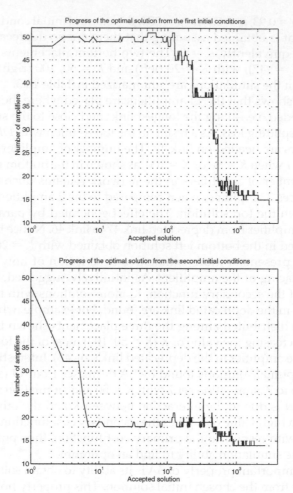

FIGURE 16.4
The number of amplifiers in the current solution as a function of the number of feasible solutions drawn in the simulation. p_{sen} is the initial solution used in the plot on the left, p_{init2} is the initial solution used in the plot on the right.

ferent. The plot on the left-hand side is obtained by initializing the transmitted power of every station to the minimum admissible value p_{sen} (this initial condition is referred to as p_{sen}). The plot on the right-hand side is obtained by initializing the transmitted power of every station to $p_{init2} = p_{max}(dBm) - sl_{max}(dB) - 10 \cdot \log_{10}(N - 1)$. The SA parameters used in either simulations are $a = 0.93$, $t_0 = 0.9$, $nrep = 204$, $\beta = 3$, $K = 0$. In each plot, the leftmost value indicates the cost of the initial solution, the rightmost value indicates the cost of the final solution found by SA. Despite the different initial solutions, both simulations reach the lower bound of 14 amplifiers. The slopes in Figure 16.4 also demonstrate the converging behavior of the cost function as temperature t decreases.

TABLE 16.4

Results for the Simulation Whose Slope is Shown on the Right-Hand Side
of Figure 16.2

Link ID	36	46	47	48	49	50	56	67
Num. of amplif.	1	2	2	3	2	2	1	1
Gtot (dB)	15.07	39.82	25.9	43.48	22.55	26.78	17.56	19.74
G_{ampli} (dB)	15.07	20	15.9	16.46	15.73	16.95	17.56	19.74
		19.82	10	16.46	6.82	9.83		
				10.56				
Placement (km)	58.08	70.42	45.35	45.31	45.31	45.31	54.15	72.33
		167.25	95.35	127.63	79.38	94.45		
				180.45				
CPU time				Total				
			1 h 2 min 46 s (Pentium 200 MHz)					

The numerical results obtained by the simulation on the right-hand side of
Figure 16.4 are summarized in Table 16.4. Links are identified using the link
identifier given in Figure 16.2. By comparing our results with those obtained
in previous work[26, 27] we conclude that on the considered star network, SA
not only requires shorter simulation time, when compared to the solution
time of other existing solutions,* but it also reaches the lower bound of 14
amplifiers, thus requiring two amplifiers less than the number of amplifiers
found in References 26 and 27.

In summary, having run a large number of simulations on various network
topologies[28, 31] it is possible to establish several empirical rules and identify
initial parameter values that can be used as a starting point when applying
SA to solve the GPA problem. A satisfactory set of values for the SA parame-
ters is for example $t_0 = 0.9$, $t_f = 0.1$, $a = 0.93$, $nrep = 630$, $\beta = 3.0$, and $K = 50$.
More in general, for small networks with few links some of these values, such
as a and $nrep$, can be relaxed without affecting the optimality of the solution
found. For large networks comprising several links parameters a and $nrep$
must be increased to maintain satisfactory results. A reasonable relationship
between $nrep$ and the size of the network is, $nrep \leq 10 \times L$, where L is the
number of links in the network.** A practical approach to determining the
value for a consists of using $t_0 = 0.9$, and gradually increasing the value for a
from 0.9 until simulations with different initial solutions obtain similar results.

16.5 Conclusion

In this chapter we reviewed some of the well-known optimization heuristic
techniques used in CI. Several of them have been applied to various optimization

---

* The solutions presented in Reference 27 requires more than ten hours on an Otherwise-
Unloaded DEC 5000/240 machine.
** This way the probability that, at each temperature, a link is never drawn for a gain modifica-
tion is negligible.

problems in traditional networks in the past. In this chapter we have shown that Simulated Annealing (SA), can be successfully applied to help resolve design problems arising in the emerging optical networks. Analyzing and solving the so-called Generalized Placement of optical Amplifiers (GPA) problem, allowed us to demonstrate that a solution provided through Simulated Annealing has the flexibility to cope with the hierarchical cost function necessary to minimize multiple parameters. Specifically, we have shown that in our case both the number of amplifiers in the system and the respective location-dependent costs can be simultaneously optimized. Furthermore, beyond showing the versatility of SA when applied to network design, we demonstrated the need for careful tuning of the SA parameters to obtain optimal or near-optimal solutions to the GPA problem with minimal computational time.

In conclusion, by optimally solving the GPA problem it was shown that CI-related methods can yield effective, practical, and versatile tools for designing cost-effective WDM networks. One can, therefore, expect similar heuristic techniques to be successfully applied to various aspects of the design of emerging network architectures with increasing success and frequency in the future.

16.6 Acknowledgments

The authors would like to express their gratitude to Gabriella Balestra for her helpful suggestions and comments about the applications of Simulated Annealing.

Problems

1. Evaluate the advantages and disadvantages of using the heuristic techniques described in Section 16.2 to solve the GPA problem.

2. Derive the set of equations necessary to solve the GPA problem.

3. Implement the SA algorithm for the GPA problem. Keeping constant all but one of the SA parameters to be tuned, analyze the behavior of the simulation results by varying one parameter at a time.

4. Implement a greedy algorithm (i.e., similar to the SA algorithm but accepting only moves leading to a better solution) for the GPA problem. Find the results by starting from different initial solutions and compare them with the results described in this chapter.

5. Apply the described SA solution to the GPA problem to networks of varying sizes and comment on the results.

6. Find a design problem related to communication networks and formulate it using a Linear Programming technique. Compare the solutions obtained using both linear programming solvers and SA algorithms.

7. Modify the previous problem by adding a nonlinear relation among the problem variables. Solve the problem using one of the Heuristic Techniques described in the chapter.

8. Choose and apply one of the heuristic techniques described in Section 16.2 to solve the GPA problem. Compare the obtained results with those found by the SA algorithm.

9. Is it possible to combine two different heuristic techniques to solve a given NP-hard problem in network design? (Bibliographic search on this topic may help.)

References

1. P.S. Szczepaniak, editor, Computational Intelligence and Applications, chapter, *Computational Intelligence: An Introduction*, by Pedrycz, W., Physica-Verlag, A Springer-Verlag Company, Heidelberg, Germany, 1999.
2. D. Fogel, Review of Computational Intelligence: Imitating Life, *IEEE Transactions on Neural Networks*, 6(6): 1562–1565, Nov. 1995.
3. D. Marks II, Intelligence: Computational versus Artificial, *IEEE Transactions on Neural Networks*, 4(5): 737–739, Sept. 1993.
4. J.C. Bezdek, On the relationship between neural networks, pattern recognition and intelligence, *Int. J. Approximate Reasoning*, 6: 85–107, 1992.
5. T. Dean, J. Allen, and Y. Aloimonos, *Artificial Intelligence Theory and Practice*, The Benjamin/Cummings Publishing Company, Inc., Redwood City, California, 1995.
6. D. Sriram and R. Adey, editors, Applications of Artificial Intelligence in Engineering Problems, *1st International Conference, Southampton University, U. K. April 1986*, Volume II, Springer-Verlag, Berlin, 1986.
7. D. Poole, A. Mackworth, and R. Goebel, *Computational Intelligence: A Logical Approach*, Oxford University Press, Inc., New York, New York, 1998.
8. M.R. Garey and D.S. Johnson, *Computers and Intractability. A Guide to the Theory of NP-Completeness*, W. H. Freeman and Company, New York, 1979.
9. J. Pearl, *Heuristics, Intelligent Search Strategies for Computer Problem Solving*, Addison-Wesley, Reading, Massachusetts, 1985 (reprint).
10. C.R. Reeves, editor, *Modern Heuristic Techniques for Combinatorial Problems*, John Wiley & Sons Inc., New York, 1993.
11. J.H. Holland, *Adaptation in Natural and Artificial Systems*, University of Michigan Press, Ann Arbor, 1975.

12. D.E. Goldberg, *Genetic Algorithms in Search, Optimization, and Machine Learning*, Addison-Wesley, Reading, Massachusetts, 1989.
13. K.S. Tang, K.F. Man, S. Kwong, and Q. He, Genetic Algorithms and Their Applications, *IEEE Signal Processing Magazine*, pages 22–36, November 1996.
14. D.E. Goldberg, K. Zakrzewski, B. Sutton, R. Gadient, C. Chang, P. Gallego, B. Miller, and E. Cantú-Paz, Genetic Algorithms: A Bibliography, Illigal Report 97011, Illinois Genetic Algorithms Laboratory (IlliGAL), December 1997.
15. D. Kalyanmoy and D.E. Goldberg, mGA in C: A Messy Genetic Algorithm in C, Illigal Report 91008, Illinois Genetic Algorithms Laboratory (IlliGAL), September 1991.
16. F. Glover, Heuristics for Integer Programming Using Surrogate Constraints, *Decision Sciences*, 8(1): 156–166, 1997.
17. F. Glover, Tabu search, part I, *ORSA Journal on Computing*, 1(3): 190–206, 1989.
18. N. Metropolis, A.W. Rosenbluth, M.N. Rosenbluth, A.H. Teller, and E. Teller, Equation of State Calculation by Fast Computing Machines, *Journal of Chemistry Physics*, 21: 1087–1091, 1953.
19. S. Kirkpatrick, C.D. Gelatt Jr., and M.P. Vecchi, Optimization by Simulated Annealing, *Science*, 220(4598): 671–680, 13 May 1983.
20. R. Ramaswami, N. Ramaswami, and K. Sivarajan, *Optical Networks: A Practical Perspective*, Morgan Kaufmann Publishers, Inc., San Francisco, California, 1998.
21. L. Kazovsky, S. Benedetto, and A. Willner, *Optical Fiber Communication Systems*, Artech House, Boston, London, 1996.
22. Godfrey R. Hill, Wavelength Domain Optical Network Techniques, *Proceedings of the IEEE*, 78(1): 121–132, January 1990.
23. E. Desurvire, *Erbium-Doped Fiber Amplifier Principles and Applications*, John Wiley & Sons Inc., New York, 1991.
24. P.E. Green Jr., *Fiber Optic Networks*, Prentice Hall, Englewood Cliffs, New Jersey, 1993.
25. C.S. Li, F.F.K. Tong, C.J. Georgiou, and M. Chen, Gain Equalization in Metropolitan and Wide Area Optical Netowrks Using Optical Amplifiers, In *Proceedings of IEEE INFOCOM '94*, pages 130–137, Toronto, Canada, June 1994.
26. B. Ramamurthy, J. Iness, and B. Mukherjee, Minimizing the Number of Optical Amplifiers Needed to Support a Multi-wavelength Optical LAN/MAN, In *Proceedings of the IEEE Infocom '97*, pages 990–997, Kobe, Japan, April 1997.
27. B. Ramamurthy, J. Iness, and B. Murkherjee, Optimizing Amplifier Placements in a Multi-wavelength Optical LAN/MAN: The Unequally Powered Wavelengths Case, *IEEE/ACM Transactions on Networking*, 6(6): 755–767, Dec. 1998.
28. A. Fumagalli, G. Balestra, and L. Valcarenghi, Optimal Amplifier Placement in Multi-wavelength Optical Networks Based on Simulated Annealing, In *All-Optical Networking: Architecture, Control, and Management Issues*, volume 3531 of *Proceedings of SPIE*, pages 268–279, Boston, Massachusetts, 3-5 Nov. 1998.
29. D.G. Brooks and W.A. Verdini, Computational Experience with Generalized Simulated Annealing over Continuous Variables, *American Journal of Mathematical and Management Sciences*, 8(3–4): 425-449, 1998.
30. P.J.M. van Laarhoven and E.H.L. Aarts, *Simulated Annealing: Theory and Applications*, D. Reidel Publishing Company, Holland, 1987.
31. L. Valcarenghi, Optical network design optimization based on simulated annealing, Master's thesis, The University of Texas at Dallas, May 1999.

17

Dynamic Channel Assignment Schemes Using Neural Networks

D. Tissainayagam, M. Palaniswami, and D. Everitt

CONTENTS

ABSTRACT The introduction of dynamic channel assignment in cellular systems has become imminent. Unfortunately, the best dynamic channel assignment schemes involve intensive computations. If they can be implemented on neural networks, then the computational complexity can be reduced. In this chapter, we compare the performance of four different neural networks that have been proposed to implement three different dynamic channel assignment schemes for a cellular system. Each approach is carefully explained and its strengths and drawbacks noted.

KEY WORDS: *Cellular system, Hopfield neural network, Feed-forward neural network, Winner-Take-all neural network, Fixed channel assignment, Dynamic channel assignment, Interference constraints.*

0-8493-1075-X/01/$0.00+$.50
© 2001 by CRC Press LLC

17.1 Introduction

Maximum performance of network resources in any telecommunication system is very important. Even more so in wireless telephony, because the allotted radio spectrum is limited and has to cope with an ever-increasing number of subscribers. Therefore, it is essential that the method of assigning a channel to an incoming call is as efficient as possible. A channel can be a single frequency band in a FDMA system (e.g., the AMPS standard) or a single time slot in a TDMA system (e.g., the GSM standard).

The more robust and adaptive channel assignment algorithms are computationally very expensive.[26] Thus, from a practical point of view, they are cumbersome for real-time applications where call-connection delays should be kept insignificant. One plausible way of rectifying this drawback in complex algorithms is to employ artificial neural networks to carry out channel assignments. The motivation for using neural networks stems from their proven inherent properties: learning and parallel processing capabilities. The expectation is that neural network-based assignment schemes will be as effective as their conventional counterparts, but faster.

Ideally, two calls in progress in the cellular system should be sufficiently separated in the frequency domain. Otherwise, electromagnetic interference would result in poor reception of the wanted signal. Practically, channel allocation methods take into account three different constraints that ought to be satisfied for "interference-free" operation of the system:[7]

- *Cochannel interference constraint*: The same channel cannot be assigned to certain pairs of cells simultaneously.
- *Adjacent-channel interference constraint*: Channels adjacent in the frequency domain cannot be assigned to adjacent cells simultaneously.
- *Cosite interference constraint*: Any pair of frequencies assigned to a cell must have a minimum separation in the frequency domain.

For negligible cochannel interference, cells that are physically far apart from each other are only allowed to support calls on the same channel simultaneously. This is the concept of *channel re-use* in cellular systems. The advantage in this concept is that a limited number of channels is enough to cover a large service area as the same frequencies are usable in spatially distant cells. On the other hand, channels that are currently in use in one cell are made unavailable to its adjacent cells so that the cochannel interference constraint is kept. These adjacent cells are known to fall into the *re-use group* of the call-cell. In the hexagonal representation of cells, any cell and the ring of six cells surrounding it comprise its re-use group. For this case, the *re-use group size* is defined to be 3. A re-use group size of 7 means that a two-cell buffer is wedged between cells that support calls on the same channel at the same time.

A drawback in dividing the coverage area into cells emerges when a mobile user crosses cell boundaries. Once a mobile user is beyond the range of the serving cell, a new connection with the present cell ought to be made and the previous connection severed. This linking procedure from one cell to another, apparently seamless to the oblivious subscriber, is termed a *handoff*. If a channel in the new cell cannot be found for some reason, the handoff is deemed a failure and the call may be cut off in mid-conversation. Hence, it is crucial that channel assignments are done by a very "smart" algorithm so that handoffs, when they are necessary, are mostly successful.

One solution on how to efficiently assign channels to calls is to implement Fixed Channel Assignment (FCA) schemes.[6, 7, 12, 20] Here, each cell is allotted a certain number of channels. This number is empirically determined by studying the long-term teletraffic in that cell. The allotments are carefully made so that there will be "interference-free" telecommunication in the cellular system. This FCA problem can be cast as a combinatorial optimization problem where the number of channels available to the system is minimized subject to cells' channel demands and interference criteria.

It has been shown that the FCA problem is a generalized graph coloring problem[22] and is therefore NP-complete. Many heuristic techniques have been devised for solving it.[27, 9, 8] The main disadvantage in FCA is its inflexibility. Most incoming calls will be blocked in cells experiencing congestion, even if system resources remain underutilized elsewhere. From a subscriber's perspective, this implies unfairness in accessibility to the cellular system and interrupted conversations due to unsuccessful hand-offs. Though FCA schemes are very popular, they will prove inadequate in the future due to their lack of adaptability to changes in the cellular system.

17.2 Dynamic Channel Assignment

In Dynamic Channel Assignment (DCA) methods, channels are not distributed to cells in advance. On the contrary, all channels that are in the system collectively form a "pool". From this set, channels are assigned, one at a time, on a call-by-call basis in real-time. Any channel can be assigned to a call in any cell as long as the interference conditions are not violated. DCA algorithms have been shown, at low to moderate traffic, to be superior to FCA strategies that are in many current FDMA-based systems. However, system performance of a DCA scheme at heavy traffic can be worse than that of some FCA schemes.[13, 14]

In DCA algorithms, any channel (in any cell) that is carrying a call is said to be *busy*. All other channels are *free*. However, all free channels cannot be assigned to incoming calls. For instance, two new callers from two cells, each falling into the other's re-use group, cannot be given two free channels of the same frequency band as that would breach the (cochannel) interference constraint. Any free channel that can be assigned without violating the interference constraints is an *available* channel. All busy channels, of course, are unavailable. Not all free channels are available, though all available channels are necessarily free.

DCA schemes assign channels to new calls and handoffs in such a way that the maximum number of available channels is left in the system at all times regardless of traffic densities. This issue is crucial because the grade of service provided to customers is dependent on the probability of their access to the system, which in turn depends on how efficiently channels are assigned by the system operator. Good channel assignment schemes that use the given spectrum to the maximum, should not depend on time-varying parameters of the system, such as the offered traffic across the cellular system.

Three DCA strategies are presented here:[8]

- Ordered Channel Search (OCS):
  1. Form a list of all the channels available to the call-cell.
  2. Select the first channel in the list.
  3. If there are no available channels, drop the call attempt and report "call blocked."

- MaxAvail:
  1. For each available channel in the call-cell, compute the *System-wide Channel Availability* (SWCA), defined as:

  $$SWCA = \sum_{allcells} \text{No. of available channels in a cell}$$

  by assuming that that channel is assigned to the call.
  2. Select the channel that maximizes this sum.
  3. In the case of a tie, select either one.
  4. If no channel is available in the call-cell, drop the call attempt and report "call blocked."

- Remax1:
  1. Call MaxAvail.
  2. If MaxAvail reports "call-blocked", go to step 3; otherwise, return selected channel by MaxAvail.
  3. Make a list of all the channels in the call-cell that are unavailable because of exactly one other call in progress.
  4. Call MaxAvail for each of these calls (interferers).
  5. For each of these interferers that are not reported "blocked" by MaxAvail, by assuming that the channel returned by MaxAvail

is assigned to it, and the corresponding freed channel is assigned to the new call, compute System-Wide Channel Availability.

6. Return that interferer which maximizes this sum, the channel to which it should be reassigned, and the channel that is to be assigned to the new call.

7. If no channel can be freed by assigning a single call, report "call blocked."

17.2.1 MaxAvail

In MaxAvail, there may be some available channels in the call-cell that are unavailable in one or more of its adjacent cells because the same channels are busy in some cells belonging to the re-use groups of those adjacent cells. Assigning such a channel to the incoming call in the call-cell decreases the SWCA by a lesser amount than assigning some other available channel. This is because the former choice was unavailable to some cells in the call-cell's re-use group even before. So, MaxAvail's selection is influenced by cells *not* belonging to the call-cell's re-use group, too. All the cells that belong to the re-use group of any one of the cells in the call-cell's re-use group, but are not in the call-cell's re-use group itself are termed *outer cells*. If a cell and its six neighbors form a re-use group, then the 12 cells surrounding that group are the outer cells. MaxAvail now comes down to selecting the available channel (in the call-cell) that is busy in the most number of outer cells.

17.2.2 Channel Rearrangements

The optimal spectral efficiency using any channel assignment scheme is given by the maximum packing strategy.[15] Compared to this upper limit, DCA algorithms are suboptimal. To improve on this, channel rearrangement methods can be incorporated into standard DCA techniques. Channel rearrangement methods attempt to free up a channel in the call-cell (when all channels are unavailable in it) by rearranging neighboring calls in progress. Firstly, channel selection without rearrangement is sought within the call-cell using any one of the standard DCA strategies. If this pursuit ends in failure, the call attempt is not dropped. Instead, an adjacent cell is chosen from which a busy channel is transferred to an available channel within that cell itself. If this procedure of call transfer (or intra-cell handoff) is done properly, it will result in an available channel to the call-cell which is assigned to the new caller. These rearrangement methods are limited in that they will only proceed to rearrange a fixed number of calls in progress to make way for new callers before reporting "call blocked." Hence, the augmented DCA schemes are also called *Limited Channel Rearrangement* (LCR) schemes. It is shown[8] that with at most 2 transfers, performance that is close to the theoretical maximum can be attained.

17.2.3 Remax1

As explained earlier, a channel in the call-cell is forced to remain unavailable if another channel on the same frequency is busy in one adjacent cell. If it can be transferred to some other channel within that cell itself, then it can be made available for assignment. This is the principle behind Remax1. In the absence of any available channels in the call-cell, the call is not blocked: instead a busy channel in a neighboring cell is sought under one constraint. That is, that this channel has to be busy in exactly one neighboring cell. There may be more than one channel satisfying the above criterion and they are all called *single interferers*. Transferring a single interferer to another channel in the same cell (intra-cell handoff) effectively frees a previously unavailable channel in the call-cell. Transferring busy channels that are not single interferers achieves nothing, as those channels will still remain unavailable to the call-cell. Once a list of all the single interferers is made, each one in it is tested for rearrangement within its own cell. The destination channel to which the single interferer is rearranged is chosen via Max-Avail. If MaxAvail reports "call blocked", that single interferer is discarded and the next one in the list is tested. Of all the single interferers that do not get blocked, the one with the maximum SWCA is selected as the channel to be assigned.

17.3 The Neural Network Approach

We will now consider several neural network architectures for performing dynamic channel assignment. Architectures based on feed-forward neural networks, winner-take-all neural networks, and modified Hopfield neural networks are now presented and their advantages and disadvantages discussed.

17.3.1 The Feed-Forward Neural Network Model 1

The model considered in References 2 and 3, has an input, an output, and one hidden layer. There are n input neurons and $\lceil \log_2 n \rceil$ output neurons if n channels are in the cellular system. The number of hidden layer neurons, an arbitrary parameter, is found by trial and error in each case. The DCA algorithm employed is the OCS without channel rearrangement.

The training procedure involves the Error Back Propagation Algorithm (EBPA). A momentum term is included for quicker learning. The desired output vector for an input vector is obtained by simulating the conventional OCS algorithm on a cellular system at a particular offered traffic load. This is done repeatedly until a specified number of input-output pairs have been generated. Each training pair is fed into the network and its synaptic weight matrix is updated according to the EBPA. Training is completed when the cumulative mean-squared-error of all pairs falls below a predetermined threshold value or when a certain number of updates has been reached.

The OCS algorithm is a far from optimum algorithm. Spectrally more efficient, albeit computationally costlier, DCA algorithms exist. The option of rearranging calls in a neighboring cell to make way for incoming calls is not considered. Channel rearrangement not only increases spectral efficiency but is also vital in a real cellular system as it reduces any unfairness in subscriber accessibility to the system by relieving congestion, to a limited extent, in high-traffic locales.

The channel chosen by the network is taken and verified to see whether it complies with the interference criteria. Verification of such nature is not strictly in tune with the principles of introducing artificial neural networks to cellular systems. Thus the demonstration of the reliability of trained feed-forward networks to assign channels is compromised.

There are values of parameters such as the learning rate, momentum coefficient, the number of training pairs, the number of hidden layer neurons and the number of stages used in the model that can only be found by trial and error. Also, the serial architecture prevents a parallel implementation of the neural model.

17.3.2   The Feed-Forward Neural Network Model 2

There are four separate networks in Reference 4, each having an input, an output, and one hidden layer. The first, *Normal Channel Selection* NN, tries to assign a channel to the incoming call in the call-cell. Failing that, the *Rearrange Cell* NN works out the cell in which the channel swapping is to occur. Then the *Call in Progress* NN and the *Selected Channel* NN respectively choose a busy channel and a free channel within the rearrange cell.

For large cellular systems, the training of the four NNs is reported to be difficult. The proposed solution is the formation of *channel groups*. The available "pool" of channels in the system is divided into groups and each of these channel groups has the required four NNs to assign any channel—possibly with rearrangement—within that group. The channel groups need not be mutually exclusive sets. A channel can be a member of more than one group. Besides, the number of channels in each group (the channel group size) need not be the same either.

Training of the four NNs is done separately. First, the conventional LCR algorithm, Remax1, is simulated using a set of sample input vectors and the corresponding output vectors for each NN are found. These input-output vector pairs are later used for training the appropriate NN. The favored method of learning was the classical EBPA with an additional momentum term for faster convergence.

The reluctant introduction of channel groups was necessary to keep the NNs to "trainable" sizes. As a consequence, when a rearrangement is warranted, it can only be effected between two channels belonging to the same channel group. Therefore, wise selection of channel groups is absolutely crucial to the on-line performance of feed-forward networks. Have the channel group sizes large (i.e., a single channel belonging to many groups), the probability of a successful rearrangement within any one group increases. But poor training, due to larger input, output, and hidden layers yields infeasible solutions. Have the channel group sizes small, and rearrangements end mostly in failure because of insufficient available channels within a group.

The quality of the training data obtained is very important, too. The performance of any neural model that undergoes supervised learning will only be as good as the training it receives. This leads to the question of how representative the collection of input data is of the entire input space. All possible input vectors cannot be presented to the MFNNs while training simply because of the sheer numbers involved.

The offered traffic load at which the training data are generated can be a matter of concern. If it is too high, then a significant number of channel assignments done in a low traffic environment will be wrong and vice versa.

17.3.3　The Winner-Take-All Neural Network Model

In this model, each channel in every cell is represented by a discrete, bipolar neuron with adjustable threshold values.[5, 16] Neuron (i, j), representing channel j in cell i, receives inhibitory inputs from neurons (r_i, j) where cell r_i falls into cell i's re-use group. Neuron (i, j) also receives excitory inputs from neurons (s_i, j) where cell s_i is an outer-cell to cell i. Besides, there are mutual inhibition between any pair of neurons within the same cell. This is done so to ensure that only one channel is selected as the output of this neural network.

Though the presented model resembles a continuous Hopfield Neural Network (HNN), it is not. First, there is no energy function that encapsulates the DCA problem. Therefore, synaptic weights cannot be read from the "weight matrix"; they are simply chosen—different in every system that the model is tested on—so that a "high" inhibiting input is larger in magnitude than all the "high" exciting inputs combined. Also, a neuron's output is forcibly reset to "low" if any one of the inhibiting inputs goes "high". All of the above and more in Reference 16 make us conclude that the actual architecture is more in line with a winner-take-all model which has an updating procedure similar to the continuous HNN.

No justification is given for any of the parameters that are relied on to produce desired results. Methods for assigning synaptic weight values are *ad hoc* rather than systematic. This makes the mathematical analysis of the networks dynamics almost impossible. Besides, the above architecture is neither parallelizable nor extendable to include channel rearrangements.

17.3.4 The Modified Hopfield Neural Network Model

Hopfield neural networks have been shown to be capable of solving combinatorial optimization problems.[1, 10, 21] For each problem an energy function that is dependent on the outputs of the neural network is constructed. The network is made to evolve in such a way that the energy function is always decreasing. In light of the knowledge, one DCA algorithm—MaxAvail—has been transformed into a quadratic minimization problem. The formulated energy function has its global minimum to be the channel that is selected by the MaxAvail algorithm. A proof of this and related issues such as precise selection of the arbitrary parameters that appear in the energy function are given in Reference 24.

The (modified) Hopfield network in Reference 23 has one neuron dedicated to each channel in the cellular system. As all channels are deemed common to all cells, the number of neurons in the neural network does not vary from cell to cell. The privilege had of being able to pool all the channels, thus avoiding the introduction of channel groups, is only because this architecture shows good scalable properties.

The energy function in Reference 23 dictates that all neurons be unipolar. In the interests of smoother convergence,[21] all neurons are made continuous, too. Convergence is carried out through a fixed number of random, asynchronous updates. To prevent neurons representing channels that are busy from going "low" any time during convergence, all neurons that are "high" before the first update are "locked" till the last update. This procedure admits "incremental optimization" and a great degree of scalability, two paramount qualities that are absent in the travelling salesman type problems. (The travelling salesman problem, which is NP-complete, had been attacked by Hopfield neural networks and met with mixed success.) At the end of the converging process, which is actually sped up by this locking mechanism, the Hamming distance between the input and output vectors are compared. If it is exactly 1, then a successful convergence is declared and the neuron that gives the extra "high" output represents the channel to be assigned. Otherwise, the call is blocked and cleared.

The notoriety of Hopfield networks to converge to a *local* minima is well known.[10, 21] Various heuristics have been successfully suggested in the literature to overcome this contingency. But, they are all problem specific. A somewhat universal and superior heuristic, known as *Simulated Annealing* is not considered here. The reason given is that such a refinement involves the direct calculation of the energy function periodically in real-time and not incorporable into a hardware design. Instead, a simple rule-of-thumb,

After every iteration, any one neuron that is not locked and "high" can spontaneously change its state to "low" with probability p, $(0 < p < 1)$,

is found to be adequate.

Once the neural model that is functionally equivalent to MaxAvail is perfected, that model is used as a building block to assemble an architecture that implements the Remax1 algorithm. The Remax1 algorithm involves MaxAvail in at most m cells, where m is the re-use group size. If the HNN in the call-cell reports "call blocked", the control passes simultaneously to at most $m - 1$ adjacent cells that are in the call-cell's list of single interferers. This enables parallelization of processing to a great degree as[23] shows. The cell in which the HNN first successfully converges to a destination channel is the rearrange-cell and the rearrangement of channels and assignment of the freed channel to the new call take place thereafter. The call is not blocked until all the adjacent cells' HNNs report failure.

17.4 Simulation

The following assumptions[4] are made throughout this study:

1. Offered traffic is Poisson and spatially homogeneous.
2. Only cochannel interference is modeled.
3. Adjacent-channel and cosite interference constraints are ignored.
4. Each call requires just one radio link. Mobile-to-mobile calls requiring two channels are not considered.
5. Only intercell handoffs with no retry are considered.
6. Intracell handoffs are effected only when rearrangement is sought.
7. Handoffs have no priority over new call attempts.
8. No channel packing is effected after call departures.
9. Blocked new call attempts and handoffs are dropped and cleared. There are no queues.

The performance of neural network based DCA or LCR algorithms at a particular traffic load is measured by the *New Call Blocking Probability* (B_N) and the *Handoff Blocking Probability* (B_H) which are defined, respectively, as:

$$B_N = \frac{\text{number of blocked calls in a cell}}{\text{number of call attempts to that cell}}$$

$$B_H = \frac{\text{number of blocked handoffs in a cell}}{\text{number of handoff attempts to that cell}}$$

One simulation consists of a fixed number of "events" (say, 1,000,000) at a certain offered traffic load. An event can be an arrival, departure, or a handoff. If an event is an arrival, then a cell is randomly chosen as the call-cell, the input vector to the HNN is prepared from its databank, and a channel assignment is thus sought. If the event is a departure, then a call in progress is randomly chosen from a randomly chosen cell and cleared. While call arrival

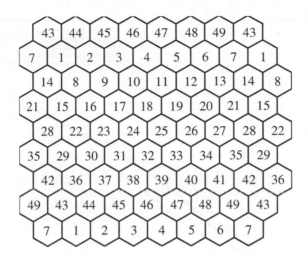

FIGURE 17.1
The 49 cell system with a re-use group of 7.

rate is a constant, call departure rate is directly proportional to the total number of calls in progress in the cellular system. When the event is a handoff, a randomly chosen call in progress in a randomly chosen cell is transferred to another randomly chosen cell in the same re-use group. For the duration of one simulation, the call-handoff rate is fixed at a fraction, α, of the call-departure rate. Here $0 \leq \alpha \leq 1$. For brevity, only results for the $\alpha = 0$ case are included here.

The performances of these neural networks are now evaluated in a 49 cell, 105 channel system (see Figure 17.1). The re-use group size is 7, meaning that a buffer zone of at least two cells must separate calls that are in progress on the same channel.

17.5 Application to a 49 Cell System

The new call-blocking probabilities are graphed against offered traffic for different neural networks in Figure 17.2. It is seen that the parallel HNN implementation of the Remax1 algorithm leads to the best relative performance. This is not just because every channel available in the cellular system is potentially at the disposal of every cell, but also because the HNN can be reliable enough to function like the conventional channel assignment algorithms. Besides, the HNN is "bias-free" in that it is independent of all training data as it only operates on unsupervised learning mode. To be sure, the Winner-Take-All model needs no training either, but it offers no guarantees on how well it approximates the conventional MaxAvail algorithm. The

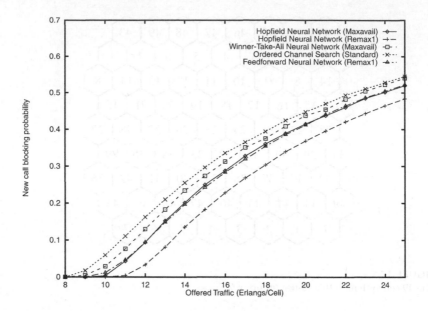

FIGURE 17.2
B_N vs. offered traffic.

competetive nature of its convergence allows it only to rate just better than the Ordered Channel Search scheme. On the other hand, the Feed-forward model, readily extendable to encompass Remax1, performs much better. Still it falls short of the HNN approach as its architecture and training protocol do not allow much scalability, necessitating channel groupings which decreases performance. Employing a feed-forward network to handle Remax1 and a Hopfield network to handle MaxAvail are roughly equivalent.

We may also compare the "power" of Remax1 as opposed to MaxAvail. Figure 17.3 suggests that at higher traffic loads, Remax1 is unable to increase the call-carrying capacity of the system, whereas MaxAvail does so. In fact, for the Remax1 algorithm, the percentage of channel assignments after rearrangement that are successful decreases with offered traffic even though the demand to rearrange steadily increases (Figure 17.4). The reason is that the call-cell's "competitors"—the other cells in its re-use group—are also assigning their calls by Remax1 and maximizing their capacity. Naturally this leads to a reduction in the number of available channels in the call-cell resulting in channel assignment failures. Moreover, there is a substantial deficit in the percentage of call connections successfully handled before rearrangement compared to the case when there is no rearrangement at all. This deficit is made up—but only just—by the initial 'failures' in the former case being subsequently assigned channels after rearrangement (Figure 17.4). The salient point is that having the option of rearranging channels means it will be exercised very frequently. A similar conclusion for smaller cellular systems is reached in Reference 25.

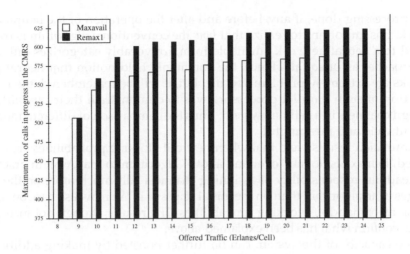

FIGURE 17.3
The maximum number of calls in progress.

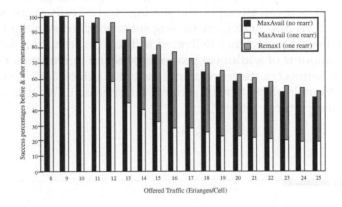

FIGURE 17.4
Percentage of successes of Remax1 before and after rearrangement compared to successes of MaxAvail.

17.6 Conclusion

The performance of neural networks to dynamically assign channels in real-time in a cellular mobile communication system was studied. We have seen that this varies from model to model even when they implement the identical channel assignment algorithm. The differences are caused not just by the neural architecture, but also by the learning strategy adopted, the mode of learning (supervised or unsupervised), the choice of parameters used, and the processing done, if any, before and after the operation of the neural network. It is important to realize that *how* the conventional algorithm is translated to a neural network amenable form, irrevocably categorizes different approaches as unequal. Hence, the commonly held notion that neural networks are extremely problem specific. But, it will be remembered, the introduction of such parallel processors was to demonstrate their *capability* to assign channels in a cellular system. Thus the comparison of different models is still valid and meaningful.

Simulation results yield that when new call blocking probabilities are compared, Remax1 is not an "over and above" algorithm to MaxAvail in practice. Instead, the responsibility of assigning channels falls much more on the last stages of an algorithm than on the initial stages of it. We may also base our conclusions on handoff-blocking probabilities. The results, though not included here, confirm what has been explained above.

The capacity of the system can be further boosted by making additional refinements to the employed channel-assignment scheme. For instance, at most two calls can be rearranged (Remax2) to make way for an incoming call. Directional antennas, frequency packing, cell splitting, giving priority to handoffs over new call attempts, and placing blocked calls in queues instead of clearing them also help in increasing subscriber accessibility to the network. But, the extra efficiency delivered will have to be substantial to justify the larger amounts of additional processing needed to implement them. For instance in Remax2, a disproportionately larger percentage of calls will be assigned channels only after the second rearrangement. We discern the law of diminishing returns setting in.

References

1. J. Bruck, On the convergence properties of the Hopfield model, in *Proc. of the IEEE*, Vol. 78, pp. 1579–1585, Oct 1990.
2. P.T.H. Chan, M. Palaniswami, and D. Everitt, Neural network based dynamic channel assignment for cellular communication systems, *IEEE Transactions on Vehicular Technology*, Vol. 43, pp. 279–288, May 1994.
3. P.T.H. Chan, M. Palaniswami, and D. Everitt, Feedforward neural network applications to dynamic channel assignment for cellular mobile radio systems *Proc. of International Joint Conference on Neural Network*, Nov 1992.
4. S. Terrill, M. Palaniswami, and D. Everitt, Limited channel rearrangement in cellular mobile communication systems using neural networks, in *Proc. of the Eighth Australian Teletraffic Research Seminar*, pp. 80–89, Dec 1993.
5. S. Terrill, D. Everitt, and M. Palaniswami, An overview of channel assignment techniques in mobile communication systems using artificial neural networks, in *Proc. of Australian Telecommunication, Networks and Applications Conference*, Vol. 2, pp. 679–684, Dec 1994.

6. D. Kunz, Channel assignment for cellular radio using neural networks, *IEEE Transactions on Vehicular Technology*, Vol. 40, pp. 188–193, Feb 1991.

7. N. Funabaki and Y. Takefuji, A neural network parallel algorithm for channel assignment problems in cellular radio networks, *IEEE Transactions on Vehicular Technology*, Vol. 41, pp. 430–436, Nov 1992.

8. K.N. Sivarajan, R.J. McEliece, and J.W. Ketchum, Channel assignment in cellular radio, in *Proc. of the 39th Vehicular Technology Conference*, pp. 846–850, May 1989.

9. M. Duque-Anton, D. Kunz, and B. Ruber, Channel assignment for cellular radio using simulated annealing, *IEEE Transactions on Vehicular Technology*, Vol. 42, pp. 14–21, Feb 1993.

10. J.J. Hopfield and D.W. Tank, Neural computation of decisions in optimization problems, *Biological Cybernatics*, Vol. 52, pp. 141–152, 1985.

11. Chee-Kit Looi, Neural network methods in combinatorial optimization, *Computers Ops. Res.*, Vol. 19, No. 3, pp. 191–208, 1992.

12. Y. Takefuji and K.C. Lee, An artificial hysteresis binary neuron: a model suppressing the oscillatory behaviours of neural dynamics, *Biological Cybern.*, Vol. 64, pp. 353–356, 1991.

13. D. Everitt and D. Manfield, Performance analysis of cellular mobile communication systems with dynamic channel assignment, in *IEEE Journal on Selected Areas in Communications*, Vol. 7, no. 8, pp. 1172–1180, Oct 1989.

14. D. Everitt, Traffic capacity of cellular mobile communication systems, *Computer Networks and ISDN Systems*, Vol. 20, pp. 447–454, 1990.

15. D.E. Everitt and N.W. Macfadyen, Analysis of multicellular mobile radio telephone systems with loss, *British Telecom Technology Journal*, Vol. 1, no. 2, pp. 37–45, Oct 1983.

16. S. Terrill, Channel assignment in cellular mobile communication systems using neural networks, MEngSc thesis, The University of Melbourne, Victoria, Australia, June 1995.

17. A. Gamst and W. Rave, On frequency assignment in mobile automatic telephone systems, in *Proc. GLOBECOM '82*, pp. 309–315.

18. A. Gamst, Homogeneous distribution of frequencies in a regular hexagonal cell system, *IEEE Transactions on Vehicular Technology*, Vol. 31, pp. 132–144, Aug 1982.

19. A. Gamst, Some lower bounds for a class of frequency assignment problems, *IEEE Transactions on Vehicular Technology*, Vol. 35, no. 1, pp. 8–14, Feb 1986.

20. J.-S. Kim, S.H. Park, P.W. Dowd, and N.M. Nasrabadi, A modified Hopfield network approach for cellular radio channel assignments, *Proc. of the 45th IEEE Vehicular Technology Conference*, pp. 589–593, 1995.

21. K.A. Smith, Solving combinatorial optimization problems using neural networks, PhD dissertation, The University of Melbourne, Victoria, Australia, March 1996.

22. H. Tamura, M. Sengoku, S. Shinoda, and T. Abe, Channel assignment problem in a cellular mobile system and a new coloring problem of networks, *IEICE Transactions*, Vol. 74, no. 10, pp 2983–2989, 1991.

23. D. Tissainayagam, D. Everitt, and M. Palaniswami, Hopfield neural network approach to limited channel rearrangement in cellular mobile radio systems, *Proc. of Australian Telecommunication Networks and Applications Conference*, Vol. 1, pp. 321–326, Dec 1996.

24. D. Tissainayagam, D. Everitt, and M. Palaniswami, A neural network driven solution to a channel assignment problem in wireless telephony, *Proc. of IEEE International Conference on Neural Networks*, Vol. 1, pp. 133–137, June 1997.

25. D. Tissainayagam, D. Everitt, and M. Palaniswami, A performance comparison of neural network based dynamic channel assignment in cellular mobile radio systems, *Proc. of IEEE Singapore International Conference on Networks,* pp. 133–147, Apr 1997.
26. D. Tissainayagam, M. Palaniswami, and D. Everitt, Computational complexities of neural networks based dynamic channel assignment algorithms, *Proc. of IEEE International Conference on Intelligent Processing Systems,* Vol. 1, pp. 10–14, Aug 1998.
27. F. Box, A heuristic technique for assigning frequencies to mobile radio nets, *IEEE Transactions on Vehicular Technology,* Vol. 27, no. 2, pp. 57–64, 1978.

18

Mobility Profile Prediction Using Fuzzy Inference in Cellular Networks

Xuemin Shen and Jon W. Mark

CONTENTS

ABSTRACT Predicting the probabilities that a mobile user will be active in other cells at future moments in a cellular system is an important issue in cellular networks. With this information, mobile switching centers can predict future resource demands and assist base stations to maintain a balance between guaranteeing quality of service (QoS) to mobile users and maintaining maximum resource utilization. This chapter describes a novel adaptive fuzzy logic inference system to estimate and predict the probability information for direct sequence code division multiple access (DS/CDMA) wireless communications networks. The estimation is based on measured pilot signal strengths at the mobile users from a number of the nearby base stations, and

0-8493-1075-X/01/$0.00+$.50
© 2001 by CRC Press LLC

449

the prediction is obtained with recursive least square (RLS) algorithm. Numerical results are presented to demonstrate the performance of the proposed technique under various path losses and channel-shadowing conditions. The proposed technique can achieve simplicity, accuracy, and low cost.

KEY WORDS: *Base station, cellular networks, channel shadowing, code division multiple access (CDMA), fuzzy logic, fuzzy inference system, handoff, mobility management, propagation path loss, recursive least square (RLS), resource management, RLS prediction, user mobility profile, wireless communications.*

18.1 Introduction

A salient feature of a wireless communication network is the freedom it provides mobile users to roam without suffering service interruption. Service continuity can be maintained through efficient handover, a mechanism by which a mobile user's connection through its currently serving base station (BS) is transferred to a neighboring BS. *A priori* knowledge of the user's movement pattern will help the design of an effective and efficient handoff mechanism.

Handoff is often initiated either by cell boundary crossing or poor link quality in the current connection. One way to deal with user mobility is to treat each handoff call as a new call in call admission control. This approach cannot guarantee a low handoff call dropping rate. It also tends to increase the number of new calls and, hence, computational intensity in call admission. Another way is to reserve resource in all the cells for a mobile user once the user is admitted to the network. This approach achieves a low rate of handoff call dropping at the expense of standby capacity, which reduces system efficiency and hence increases the blocking rate for new calls. Because of the limited radio spectrum, it is critical that wireless networks make efficient use of the radio frequency bandwidth. Several techniques have been proposed to balance the tradeoff between the satisfaction of QoS and the maximization of resource utilization. The shadow cluster[1] and virtual (connection) tree[2-3] approaches make use of statistical multiplexing of data traffic to and from mobile users so that a higher resource utilization can be attained without increasing the call blocking and dropping probabilities. However, statistical multiplexing requires a knowledge of the user movement patterns and trends. In the previous work reported in the literature, the probability that a mobile user will reside in a particular cell is assumed known.

Mobility information plays an important role in the design of cellular systems. Previous research efforts on mobility information have focused on statistics such as mobility model,[4-5] user location tracking and trajectory prediction,[6] channel holding time,[7] cell boundary crossing rate,[8-10] mean handover rate,[4, 11] and cell residence time.[4, 12] In this chapter, we are concerned with the determination of mobility information that a mobile user is to hand-off to a particular BS at future moments. The mobility information can be used to assist user mobility management (traffic routing),[2] to manage network resources (resource allocation, call admission control, congestion, and flow control),[13] and to analyze handoff algorithms in integrated wired/wireless networks.[14] In general, if a mobile user is closer to a base station (BS), then the propagation path attenuation from the BS to the mobile user is smaller, and vice versa. Hence, if the BS transmits a pilot signal with constant transmitted power, then the received power of the signal at the mobile user carries the information of the distance between the mobile user and the BS. Since the probability that the mobile user will be active in a particular cell at a future moment is a function of the current distances between the mobile user and its nearby BSs, this probability can be estimated based on real-time measurements of the received pilot signal power at the mobile user from the BSs. Furthermore, the probability depends on the mobile user's movement pattern (such as movement trajectory). Although the movement patterns of mobile users are random in nature, the movement of each mobile user has a relatively smooth trajectory most of the time. That is, the location of a mobile user at a future moment depends on its locations at the current moment and previous moments. As a result, it is possible to predict the mobility information based on the current and previous measurement data. If the predictions of future mobility information can be obtained with reasonable accuracy, then the network resource management will become substantially efficient in terms of user QoS and resource utilization.[1, 6]

The challenges in estimating and predicting the mobility information based on the pilot signal power measurements come from the following facts: (a) Normally there is no one-to-one relation between the distance and the probability. Even when such a relation exists for some special environments, it is very difficult (if not impossible) to describe the relation accurately, e.g., using mathematical expressions; (b) There exists a relatively slow fluctuation of the received signal level due to scattering in the propagation medium between the BS and the mobile user. The shadowing process randomizes the relation between the received pilot signal power and the distance from the mobile user to the BS. On the average, the larger is the power, the smaller is the distance. However, this relation may not hold for every measurement; (c) The received signals are contaminated by the multiple access interference (MAI) due to other users in the system and unavoidable background noise. That is, the measured data are not accurate. As a result, it is impossible to accurately derive the probability information based on the measurements. To tackle this difficulty, an adaptive fuzzy inference prediction system is presented in this chapter. The system deals with (a) the uncertainty inherent in the relation between the distance and the probability and (b) the random shadowing effect by using training data from real measurement or from statistical models of practical propagation environments. To handle the measurement error, the system incorporates the degree of certainty (or accuracy) of the measurements by giving a larger degree of importance to the data with higher measurement accuracy.

A fuzzy inference system was proposed in Reference 15 to estimate the current mobility information based on the real-time measurements. The main concern of this chapter is to predict the mobility information of future moments with an adaptive fuzzy inference approach. The remainder of this chapter is organized as follows. Section 18.2 describes the system model which uses direct sequence code division multiple access (DS/CDMA). By using the pilot signals in the forward channel (down link), no extra signaling is needed for obtaining user mobility information. After giving the motivation of applying the fuzzy inference approach for mobility information acquisition, Section 18.3 is devoted to the design of an adaptive fuzzy inference prediction system which combines fuzzy inference logic with a recursive least square (RLS) predictor. Numerical results and discussions on the performance and applications of the proposed technique are presented in Section 18.4. Section 18.5 gives some concluding remarks of this work.

18.2   Mobility Information Model

We consider a wireless communication network operating in a frequency division duplex (FDD) mode. Mobile users in each cell share the radio frequency spectrum through the DS/CDMA protocol. The same total frequency bandwidth is reused in every cell to increase the radio spectral efficiency and to eliminate the need for frequency coordination. A mobile can transmit to and receive signals from more than one BS at any time. A CDMA system employs soft handoff, which makes before break. During transition from one cell to a neighboring cell, the mobile user establishes a communications link with the new BS while, at the same time, keeping its communications link with the original BS. The original communications link is terminated only after the mobile user has firmly established itself in the new cell. In the forward link, each base station transmits a distinct pilot signal for pseudorandom noise (PN) code and carrier synchronization at the receiver of the mobile user. Prior to any transmission, the mobile user monitors the received pilot signal power levels from nearby base stations. It chooses its home BS according to the maximum pilot signal power received. The network uses mobile user assisted soft handoff as in the IS-95 proposal.[16] While tracking the signal from the home BS, the user searches for all the possible pilots and maintains a list of all pilots whose signals are above a prescribed threshold. This list is transmitted to a mobile switching center (MSC) periodically through the home BS. The MSC uses the information to make decision on when the soft handoff should start. In addition, the MSC uses the information to estimate and predict the probabilities that the mobile will locate in a particular cell at the future moments.

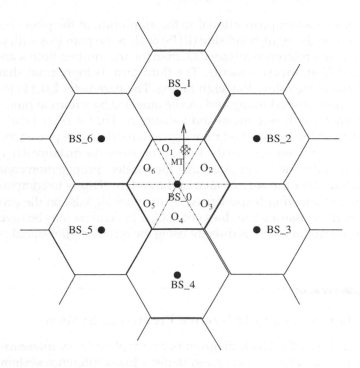

FIGURE 18.1
The hexagonal cell layout.

Consider a uniform grid of hexagonal cells, as shown in Figure 18.1, where the mobile user under consideration is located in the subregion O_1 of cell_0. For each mobile user, we will focus on the mobility information related only to its home BS (denoted by BS_0) and to the six first-tier neighboring BSs (denoted by BS_1, BS_2, . . . , BS_6). The index i will be used throughout this chapter to denote variables related to the home BS ($i = 0$) and to the neighboring BSs ($i = 1, 2, . . . , 6$). The time t will be discretized and represented as $t_n (= n\Delta t)$, $n = 1, 2, . . . $, where Δt is the time interval over which the received pilot signals are measured once and then the estimation and prediction of the probabilities are updated. Let $d_i(t_n)$ denote the distance at time t_n between the mobile user and BS_i. Given the coverage areas of the BSs, the probabilities that the mobile user will be in cell_i at the next moment t_{n+1} mainly depends on the distances $d_i(t_n)$, the velocity, and the direction of the user movement. The larger the distance $d_i(t_n)$, the smaller the probability that the user will be in cell_i. By measuring the power of the pilot signal from BS_i received at the mobile user, the distance $d_i(t_n)$ can be estimated. At t_n, the local mean of the pilot signal amplitudes received at each mobile user can be modeled by[17]

$$a_{n,i} = \gamma_i \cdot [d_i(t_n) / D_0]^{-k} \cdot 10^{\xi_i(t_n)/10}, \quad i = 0, 1, . . . , 6 \qquad (18.1)$$

where γ_i is a constant proportional to the amplitude of the pilot signal. The second term on the right-hand side (RHS) of (1) is the path loss with path loss exponent κ, and reference distance D_0 from the transmitter. Both κ and D_0 are determined from measurements. The third term is lognormal shadowing which characterizes slow Rayleigh fading. The parameter $\xi_i(t_n)$ is to characterize the effect of shadowing and can be modeled by a normal random variable (for any t_n) with zero mean and variance σ^2. For $i \neq j$, $\xi_i(t_n)$ and $\xi_j(t_n)$ are independent. If the transmitted pilot signals have the same power, then $\gamma_i = \gamma$ for $i = 0, 1, \dots, 6$. Here we use the relation between the distance $d_i(t_n)$ and the probability, under the assumption that the wireless propagation condition is homogeneous over the service area of the system. This is to compensate for the effect of shadowing (experienced by the pilot signals) on the probability estimation. If the assumption does not hold, other criteria may be used to estimate the probability, such as directly using the received pilot signal powers.

18.3   Adaptive Fuzzy Inference Prediction System

Figure 18.2 shows the block diagram of an adaptive fuzzy inference prediction system. It consists of two subsystems: a fuzzy inference system and an RLS predictor. The fuzzy inference system estimates the probability that a mobile user will be active in cell i at time t_n based on the measured pilot signal strengths at time t_n. Before measurements at t_{n+1} are available, the RLS predictor predicts the probability that the mobile user will be active in cell i at time t_{n+N} for $N = 1, 2, \dots$ based on the estimates from the fuzzy inference system up to time t_n. As depicted in Figure 18.3, only the one-step prediction is of interest here, i.e., $N = 1$. The predicted probability information can be used in resource management to handle user mobility in advance of the mobile user's next move. The design of the subsystems are given in Sections 18.3.1 and 18.3.2.

18.3.1   The Fuzzy Inference System

The fuzzy inference system is a special expert system. It employs a knowledge base, expressed in terms of fuzzy inference rules, and an appropriate inference engine to estimate the probability of a mobile user being active in cell_i at t_n based on the measurement data $a_{n,i}$. The knowledge base can be designed to take into account (a) the wireless propagation environment such as the one described by equation (18.1), (b) intuitive understanding of the general relation between the distance and the probability, and (c) measurement errors. The system is capable of utilizing knowledge elicited from human operators. The knowledge is expressed by using natural language, a cardinal element of which is linguistic variables.[18, 19] Let the linguistic variable $a_{n,i}$ be

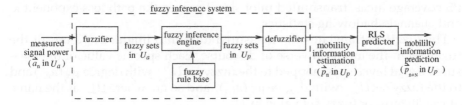

FIGURE 18.2
The adaptive fuzzy inference prediction system.

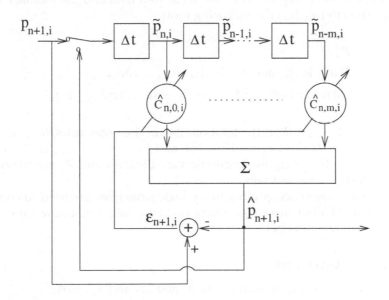

FIGURE 18.3
The structure of the RLS predictor.

the received signal level from BS_*i* at time t_n, then the corresponding universe of discourse is the set of all possible received signal levels. We choose the term set of $a_{n,i}$, denoted by $U_{a_{n,i}}$, to contain the following elements: extremely small (ES), very small (VS), small (S), small to medium (SM), medium to large (ML), large (L), very large (VL), and extremely large (EL). Let the linguistic variable $p_{n,i}$ be the probability that the mobile user will be active in cell_*i* at epoch t_n, with the universe of discourse being the interval [0, 1]. We choose the term set of $p_{n,i}$, denoted by $U_{p_{n,i}}$, to be the set containing the following elements: zero (ZE), extremely small (ES), very small (VS), small (S), small to medium (SM), medium to large (ML), large (L), very large (VL), extremely large (EL), and one (OE). The number of terms in $U_{a_{n,i}}$ and $U_{p_{n,i}}$, respectively, is selected so as to achieve a compromise between the complexity and the fuzzy inference system performance. The membership functions of the input (the received signal levels) and the output (the probabilities) depend on the

BS coverage areas, transmitted pilot signal power, the path loss exponent κ, and channel-shadowing statistics.

The fuzzifier translates the measured data into linguistic values of the fuzzy set in the input universe of discourse. Each specific value of the measured signal level $a_{n,i}$ is mapped to the fuzzy set $U_{a_{n,i}}^1$ with degree $\mu_{x_i}^1(a_{n,i})$ and to the fuzzy set $U_{a_{n,i}}^2$ with degree $\mu_{x_i}^2(a_{n,i})$, and so on, where $U_{a_{n,i}}^J$ is the name of the Jth term or fuzzy set value in $U_{a_{n,i}}$.

The fuzzy rule base is the control policy knowledge base, characterized by a set of linguistic statements in the form of IF-THEN rules that describe the fuzzy logic relationship between the measured data and the mobility information. The kth rule has the following form.

R_k:

If $a_{n,0}$ is A_{0k} and $a_{n,1}$ is A_{1k} and $\ldots$ and $a_{n,6}$ is A_{6k}

then $p_{n,0}$ is P_{0k} and $p_{n,1}$ is P_{1k} and $\ldots$ and $p_{n,6}$ is p_{6k}

where $k = 1, 2, \ldots, K$, K, is the total number of the fuzzy rules, $(a_{n,0}, a_{n,1}, \ldots, a_{n,6}) \in U_{a_{n,0}} \times U_{a_{n,1}} \times \ldots \times U_{a_{n,6}} \triangleq U_a$ and $(p_{n,0}, p_{n,1}, \ldots, p_{n,6}) \in U_{p_{n,0}} \times U_{p_{n,1}} \times \ldots \times U_{p_{n,6}} \triangleq U_p$ are linguistic variables, A_{ik} and P_{ik} are fuzzy sets in $U_{a_{n,i}}$ and $U_{p_{n,i}}$, respectively.

In the fuzzy inference engine, fuzzy logic principles are used to combine the fuzzy IF-THEN rules in the fuzzy rule base into a mapping from fuzzy sets in U_a to fuzzy sets in U_p.

Given Fact:

$a_{n,0}$ is $\tilde{A}_0$ and $a_{n,1}$ is $\tilde{A}_1$ and $\ldots$ and $a_{n,6}$ is $\tilde{A}_6$

Consequence:

$p_{n,0}$ is $\tilde{P}_0$ and $p_{n,1}$ is $\tilde{P}_1$ and $\ldots$ and $p_{n,6}$ is $\tilde{P}_6$

where $\tilde{A}_i$ and $\tilde{P}_i (i = 0, 1, \ldots, 6)$ are linguistic terms for $a_{n,i}$ and $p_{n,i}$, respectively. The fuzzy rule base can be created from training data sequence (e.g., measured input-output pairs). To avoid tedious field trials, the training data can be generated in computer simulation based on propagation model and cell structure. Given a set of desired input-output data pairs, a set of fuzzy IF-THEN rules can be generated. In addition, a degree which reflects the expert's belief of the importance of the rule can be assigned to each rule. For example, the importance of a rule increases if the corresponding input data has a higher measurement accuracy. The measurement accuracy increases as the received signal-to-interference-and-noise ratio (SINR) increases. With the same interference-and-noise component for all received pilot signals, the differences among the SINR values are proportional to the differences among the received power values of the pilot signals. If the mobile is closer to BS_i than to BS_j, the average received signal power from BS_i is larger than that from BS_*j*. Hence, the measured data for BS_*i* should be weighted more (i.e., have a larger degree) than that for BS_*j*. The degree assigned to rule *k* is calculated by using product operations

$$Q_k = \mu_k \prod_{i=0}^{6} \mu_{I_{ik}}(a_{n,i}) \prod_{i=0}^{6} \mu_{O_{ik}}(p_{n,i}) \tag{18.2}$$

where I_{ik} denotes the input region of rule k for $a_{n,i}$, O_{ik} the output region for $p_{n,i}$, $\mu_{I_{ik}}(a_{n,i})$ is the degree of $a_{n,i}$ in I_{ik} obtained from the membership functions, $\mu_{O_{ik}}(p_{n,i})$ is the degree of $p_{n,i}$ in O_{ik}, and μ_k is the degree of the data vector $(a_{n,0}, a_{n,1} \ldots, a_{n,6})$ assigned by human operators. When there is more than one rule in one box of the fuzzy rule base, the rule that has the largest degree is chosen.

The defuzzifier performs a mapping from fuzzy sets $(p_{n,0}, p_{n,1}, \ldots, p_{n,6}) \in U_p$ (the output of the inference engine) to a crisp point $(\tilde{p}_{n,0}, \tilde{p}_{n,1}, \ldots, \tilde{p}_{n,6}) \in U_p$. Among the commonly used defuzzification strategies, the center average defuzzification method yields a superior result.[20] Let $\tilde{p}_{n,i}$ denote the estimate (generated by the fuzzy inference system at time t_n) of the true probability $p_{n,i}$. The formula for the estimate at the defuzzifier output is

$$\tilde{p}_{n,i} = \frac{\sum_{k=1}^{K} \overline{Q}_k \prod_{j=0}^{6} \mu_{I_{jk}}(a_j) \overline{p}_{ik}}{\sum_{k=1}^{K} \overline{Q}_k \prod_{j=0}^{6} \mu_{I_{jk}}(a_j)} \tag{18.3}$$

where $\overline{p}_{ik}$ is the center value of the output region of rule k, and $\overline{Q}_k$ is the degree (normalized to 1) of rule k.

18.3.2   The RLS Predictor

For each mobile user, there is a strong correlation among its locations at adjacent time moments if the product of the mobile user velocity and the time interval Δt is small. As a result, there may exist a strong correlation among the $(p_{n,0}, p_{n,1}, \ldots, p_{n,6})$ values for some consecutive discrete time moments. This makes it possible to predict the future probability values based on its current and previous values. The RLS algorithm is used for the prediction here because (a) it is easily implemented using a tapped-delay line and (b) a forgetting factor can be introduced to take into account the fact that the correlation between two locations fades as the time interval separating the corresponding time moments increases. The RLS predictor takes the probability estimates up to time t_n from the fuzzy inference system and processes the data to predict the probability that a mobile user will be active in neighboring cells at a future moment t_{n+N}, where $N = 1, 2, 3 \ldots$. Figure 18.3 shows the structure of the RLS predictor for $N = 1$, which is basically a tapped-delay-line filter with $(m + 1)$ taps. The tap coefficients are obtained using the RLS algorithm.[22, 21] In Figure 18.3, the values $(\tilde{p}_{n,i}, \tilde{p}_{n-1,i}, \ldots, \tilde{p}_{n-m,i})$ are available from the fuzzy inference system. We use these and future values, shown by $\tilde{p}_{n+1,i}$, to train the linear least square filter. After the tap coefficients have been trained, the system can be used as an RLS to perform the prediction. When $\hat{p}_{n+1,i}$ becomes available, $\hat{p}_{n+1,i}, \tilde{p}_{n,1}, \ldots, \tilde{p}_{n-m+1,i}$ are used to refresh the contents of the tapped-delay line and perform the next set of prediction steps.

At time t_n, the input vector of the predictor is

$$V_n = (\vec{p}_n, \vec{p}_{n-1}, \ldots, \vec{p}_{n-m})^T$$

where the superscript "T" denotes transposition, and $\vec{p}_{n-l} = (\tilde{p}_{n-l,0}, \tilde{p}_{n-l,1}, \ldots, \tilde{p}_{-l,6})$ $l = 0, 1, \ldots m$, is the fuzzy inference system output at t_{n-l}. All the elements are set to zero for the initial moments $n \leq l$. The corresponding tap coefficient vector is

$$C_n = (\vec{c}_{n,0}, \vec{c}_{n,1}, \ldots, \vec{c}_{n,m})^T$$

where $\vec{c}_{n,l} = (c_{n,l,0}, c_{n,l,1}, \ldots, c_{n,l,6})$, corresponding to $\vec{p}_{n-l}$. C_n should be chosen (optimized) to minimize the mobility information estimation error. Over a short time duration, the RLS predictor can be described by the following system model equations

$$c_{n+1} = c_n \tag{18.4}$$

$$\vec{p}_{n+1} = V_n^T C_n + w_n \tag{18.5}$$

where C_n is the optimal tap-coefficient vector under the constraint of a finite tap number, $\vec{p}_{n+1}$ is the output of the fuzzy inference system at time t_{n+1} and is used as the desired output of the predictor, w_n is the measurement error with zero mean and finite variance to capture the effects due to the stochastic nature of wireless propagation environments, random movement of the mobile user, etc. Equation (18.4) is adequate over a short duration of time (a small number of Δt intervals) if Δt is relatively small compared with the average channel-shadowing duration. However, the model is inadequate over a long time interval, which is to be compensated for by introducing an exponential forgetting factor to the filtering algorithm. In Figure 18.3, the ith component ($i = 0, 1, \ldots, 6$) of the predictor output is given by

$$\hat{p}_{n+1,i} = \sum_{l=0}^{m} \hat{c}_{n,l,i} \tilde{p}_{n-l,i} = V_{n,i}^T \hat{C}_{n,i} \tag{18.6}$$

where

$$V_{n,i} = (\tilde{p}_{n,i}, \tilde{p}_{n-1,i}, \ldots, \tilde{p}_{n-m,i})^T$$

is the ith element of V_n and

$$\hat{C}_{n,i} = (\hat{c}_{n,0,i}, \hat{c}_{n,1,i}, \ldots, \hat{c}_{n,m,i})^T$$

is an estimate of the ith tap-coefficient vector $C_{n,i}$ at t_n, under the assumption that $c_{n,i}$ is time-invariant over a small number of Δt intervals. The estimate $\hat{C}_{n,i}$ is computed based on the fuzzy inference system output up to t_n. The ith estimation error component at t_{n+1} is defined as

$$\epsilon_{n+1,i} \triangleq \tilde{p}_{n+1,i} - \hat{p}_{n+1,i} = \tilde{p}_{n+1,i} - V_{n,i}^T \hat{C}_{n,i} \tag{18.7}$$

where $\tilde{p}_{n+1,i}$ is the estimated probability from the fuzzy inference system at time t_{n+1}, and $\hat{p}_{n+1,i}$ is a least square estimate of $\tilde{p}_{n+1,i}$ obtained based on $\tilde{p}_{n+1,i}, l = 0, 1, \ldots, \dot{m}$.

In the RLS algorithm, the estimation error vector sequence $\{\vec{\epsilon}_n\} = \{(\epsilon_{n,0}, \epsilon_{n,1}, \ldots, \epsilon_{n,6})\}$ is considered to be a deterministic process. The algorithm starts with an initial estimate $\hat{C}_{0,i}$ and uses the information contained in new data samples to update the old estimates. Therefore, the length of observable data is variable. The design criterion is to adaptively estimate the tap-coefficient vector $\hat{C}_{n,i}$ such that the weighted squared error (cost function) at t_{n+1}, defined as

$$j_{n+1,i} = \sum_{j=0}^{n+1} \lambda^{n+1-j} |\epsilon_{n+1-j,i}|^2 \tag{18.8}$$

$$= \sum_{j=0}^{n+1} \lambda^{n+1-j} |\tilde{p}_{n+1-j,i} - V_{n-j,i}^T \hat{C}_{n-j,i}|^2$$

is minimized. In Equation 18.8, λ^{n+1-j} is an exponential forgetting factor taking into account that the correlation between V_n and V_{n+N} decreases as N increases. If $\lambda = 1$, then all the estimates $V_{n-j}, j = 0, 1, \ldots, n$, are to be treated equally; if $\lambda < 1$, then the estimate obtained at earlier times (with larger j values) are to have a smaller influence than more recent estimates (with smaller j values). The RLS algorithm with a constant λ for updating the estimate of the tap-coefficient, $\hat{C}_{n+1,i}$, can be summarized as

$$\hat{C}_{n+1,i} = \hat{C}_{n,i} + G_{n+1,i}(H_{n+1,i} - V_{n,i}^T \hat{C}_{n,i}) \tag{18.9}$$

$$G_{n+1,i} = H'_{n,i} V_{n,i} (1 + V^T_{n,i} H'_{n,i} V_{n,i})^{-1} \tag{18.10}$$

$$H_{n+1,i} = (H_{n,i} - G_{n+1,i} V^T_{n,i} H_{n,i}) / \lambda \tag{18.11}$$

where the $(m + 1)$-by-$(m + 1)$ matrix $H_{n+1,i}$ is defined as $H_{n+1,i} \triangleq [\Sigma^n_{l=1} \lambda^{n-l} V_{n,i} V^T_{n,i}]^{-1}$ and $H'_{n,i} = H_{n,i} / \lambda$. The initial values of $\hat{C}_{n,i}$ and $H_{n,i}$ can be chosen as

$$\hat{C}_{0,i} = \vec{0}, H_{0,i} = \delta I$$

for a soft-constrained initialization, where $\delta \gg 1$ is a large positive constant, and I is the identity matrix of dimension $(m + 1)$.

The above discussion shows how to predict the probability information for the time moment t_{n+1} based on the measurement data up to time t_n. If it is desirable to further predict the probability information for t_{n+2} based on the measurement data up to time t_n, one way is to update the input vector of the RLS predictor to

$$\hat{V}_{n+1} = (\vec{\hat{p}}_{n+1}, \vec{p}_n, \dots, \vec{p}_{n+1-m})^T$$

where the first component is the previous output of the RLS predictor and all other components are the previous outputs of the fuzzy inference system. The ith element of the probability vector $\vec{p}_{n+2}$, $\hat{p}_{n+2,i}$, can then be obtained from Equation (18.6) using the same tap coefficient $\vec{\hat{C}}_{n,i}$.

18.3.3   Discussions

Several issues regarding the adaptive fuzzy inference system and its applications need to be discussed:

i) The complexity of the multiple-input $(a_{n,i})$ multiple-output $(p_{n,i})$ fuzzy inference system (for $i = 0, 1, \dots, 6$) may be a concern. However, in practice, the complexity can be significantly reduced if (a) we make use of the relation $\Sigma_i p_{n,i} = 1$ and (b) the number of BSs to which the mobile user has a potential to handoff at t_{n+1} is limited to less than 6 by neglecting the BSs which have weak pilot signal power at the mobile user. For example, for the mobile user shown in Figure 18.1, it is reasonable to limit the future BSs that the mobile user will communicate with (at t_{n+1}) to BS_1, BS_2, and BS_6, since the mobile user is located in subregion R_1 of cell_0 at time t_n.

ii) The implementation cost for the fuzzy inference system is low in the sense that: (a) It is a one-pass build-up procedure that does not require time-consuming on-line training; (b) it makes use of the available pilot signal power measurement and transmission of the measured data to the MSC in the wireless system (no extra signaling and measurement are necessary); (c) the required real-time measurement and computation are a linear function of the number of mobile users. As a result, the proposed system is practical even when the number of mobile users is large.

iii) Based on the propagation model, Equation (18.1), the probabilities are estimated and predicted using the received pilot signal powers. There are other (handoff initiating) criteria, such as carrier-to-interference ratio, which can affect the estimation and prediction. The fuzzy inference system proposed here can be directly extended to situations using other handoff initiating criteria. By defining the relation between the mobility information and the criterion under consideration, the same training procedure can be used to establish the fuzzy rule base according to the criterion employed. If the relation can be characterized by a statistical model, then the training data can be generated from computer simulation; otherwise, if such a model is not available, field measurements should be carried out to obtain the data.

iv) The fuzzy IF-THEN rules in the fuzzy rule base have probabilities, $\tilde{P}_i$ ($i = 0, 1, \ldots, 6$), in the conclusion part. As the probabilities are expressed in the linguistic terms, it is difficult to measure whether they obey the rules of probability. However, proper steps should be taken to ensure that $\tilde{P}_i$ has the basic properties of probability to some degree. Such steps include: (a) the fuzzy set is defined within the universal set [0, 1]; and (b) for each rule, the sum of the center value of $\tilde{P}_i$ for i from 0 to 6 should be equal to 1.

v) In the proposed adaptive fuzzy inference system as shown in Figure 18.2, the fuzzy inference system and the RLS predictor are connected in tandem. The proposed structure offers the advantage of implementation simplicity. Another possible approach is to integrate the fuzzy inference system with the RLS predictor, i.e., to put the RLS predictor inside the fuzzy inference system, as suggested in Reference 18. By combining the RLS algorithm with each fuzzy inference rule, the prediction accuracy may be increased. However, this is achieved at an increased implementation complexity, which may not be practical for real-time prediction especially when the number of the neighboring BSs taken into consideration is large.

vi) With the RLS predictor, the adaptive fuzzy inference system can predict the probability vectors a few steps into the near future. The mobility information is particularly useful in prediction-based wireless network resource management such as call admission control and rate-based flow control. Due to user mobility, QoS provisioning in wireless/wired network environments is technically very challenging. The shadow cluster and VCT approaches[1-3] have been proposed as an effective way to manage network resources. In the approach, base stations reserve resources in advance for handoff calls according to predicted mobility information, which reduces the chance of handoff dropped calls and ensures QoS provisioning for mobile users. In other words, if the resource reserved for a mobile user in a neighboring cell is weighted by the probability of the user handing off to the cell, a large statistical multiplexing gain can then be achieved in the resource management when the network operates in the neighborhood of its full capacity, taking into account a large number of mobile users. This, on the other hand, will allow the network to accept more new call requests without breaking the QoS commitments made to the mobile users already admitted to the network. The prediction of the probability will also allow the network to allocate its resources dynamically to mobile users with different QoS requirements. For instance, if it is predicted that a mobile user with real-time traffic has a high chance to handoff to a neighboring cell, then the future home BS can reserve enough resources for the mobile user by allocating less resources to non real-time traffic sources in the cell through rate-based flow control.

18.4 Numerical Results

This section first gives the details of how the simulation environment is set up, then presents the performance of the fuzzy inference subsystem, and finally evaluates the adaptive fuzzy inference system for predicting the mobility information.

18.4.1   The Simulation System

The microcellular network under consideration has a hexagonal cell structure as shown in Figure 18.4. The BS is located at the center of each cell. The probability $p_{n,i}$ that a mobile user will be active in cell_i at t_n depends on its location (x_{MT}, y_{MT}) at t_n. In order to reduce the complexity of the estimation, the following assumptions are made:

 i) Limit the number of BSs for handoff to 3. For the mobile user shown in Figure 18.4, since $y_{MT} > 0$, let $p_{n,4} = p_{n,5} = p_{n,6} = 0$;

 ii) Further reduce the number of BSs for handoff to 2. For the mobile user shown in figure 4, since $x_{MT} > 0$ (i.e., the mobile user is located on the right side of cell_0), let $p_{n,3} = 0$;

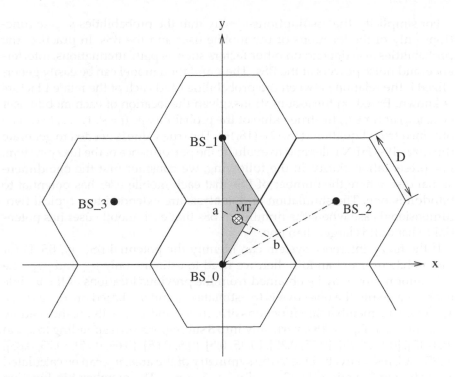

FIGURE 18.4
A mobile user located at (x_{MT}, y_{MT}) in the cellular system.

iii) The probability that the mobile user will remain in cell_0 depends on y_{MT}

$$p_{n,0} = 1 - y_{mT} / \sqrt{3D};\qquad(18.12)$$

iv) $p_{n,1}$ and $p_{n,2}$ can be solved by

$$p_{n,1} + p_{n,2} = 1 - p_{n,0}\qquad(18.13)$$

$$p_{n,1} / p_{n,2} = d_2 / d_1\qquad(18.14)$$

where the distances d_1 (from point a to the mobile user) and d_2 (from the mobile user to point b) depend on x_{MT} and y_{MT}. From geometry we have

$$d_1 = \frac{x_{MT}(y_{MT} / \sqrt{3} - x_{MT})}{\sqrt{x_{MT}^2 + y_{MT}^2}\sin(\alpha)}\qquad(18.15)$$

$$d_2 = d_1 + 2\sqrt{x_{MT}^2 + y_{MT}^2}\sin(\alpha)\qquad(18.16)$$

where $\alpha = \pi/6 - \arctan(x_{MT}/y_{MT})$.

For simplicity, the assumptions specify that the probabilities $p_{n,i}$ are functions only of the locations of the mobile user and the BSs. In practice, the probabilities also depend on other factors such as path attenuations, interference, and noise powers at the BSs. The simulation model can be easily generalized if the relation between the probabilities and each of the related factors is known. Based on the assumptions, given the location of each mobile user (x_{MT}, y_{MT}) at time t_n, the true value of the probability $p_{n,i}(i = 0, 1, \ldots, 6)$ can be obtained from Equations (18.12)–(18.16). The true value is needed to generate the fuzzy IF-THEN rules and to evaluate the performance of the fuzzy system. For presentation clarity, in the following, we consider first the one-dimensional case where the number of BSs that each mobile user has potential to handoff is one. The simulation and analysis are extended to a typical two-dimensional case where the number of BSs that each mobile user has potential to handoff is larger than one.

If the fuzzy inference system can identify the potential BS(e.g., BS_1) for the mobile user to handoff, then we need to estimate only $p_{n,0}$ and $p_{n,1}$. The side information may be obtained from the previous locations of the mobile user. As a result, the objective is to estimate $p_{n,0}$ and $p_{n,1}$ based on $a_{n,0}$ and $a_{n,1}$. To obtain the membership functions of $a_{n,i}$ ($i = 0$ and 1), we divide the shadow area shown in Figure 18.4 vertically into 8 subregions corresponding to $p_{n,0}$ in [0, 0.15], [0.15, 0.25], [0.25, 0.35], [0.35, 0.5], [0.5, 0.65], [0.65, 0.75], [0.75, 0.85], [0.85, 1.0], respectively. Due to the symmetry of the area, $p_{n,0}$ can be calculated according to Equation (18.12), and $p_{n,1} = 1 - p_{n,0}$. The membership function of $a_{n,i}$ is determined based on the mean and variance of $a_{n,i}$ for each subregion. The membership function of $p_{n,0}$ is determined based on the probability values for each subregion. In simulations, we consider 50,000 mobile users uniformly distributed in the shadow area shown in Figure 18.4. The simulation parameters are: $D_0 = 100$ meters, $D = 1,500$ meters, $\kappa = 2, 4, 6, \gamma_{n,i} = 1$(normalized), and $\sigma = 1, 2, \ldots, 6$ dB, respectively. Figures 18.5 and 18.6 show the membership functions of $a_{n,0}$ (for $\sigma = 2$ dB) and $p_{n,0}$, respectively. Graphs of these functions have triangular shapes. The overlapping of the triangular

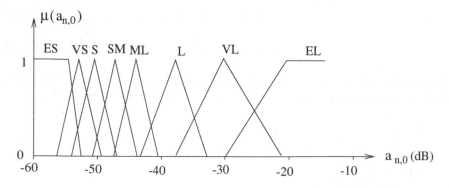

FIGURE 18.5
Membership function of $a_{n,i}$ ($i = 0$ and 1) for $\sigma = 2$ dB.

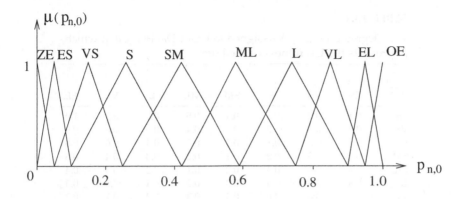

FIGURE 18.6
Membership function of $p_{n,i}$ ($i = 0$ and 1).

shapes possess a natural capability to express and deal with observation and measurement uncertainties (crisp points do not have this capability). The triangular shapes are used because it is relatively easy to determine their parameters. Each triangle is defined by 3 parameters, which are further reduced to 2 parameters taking into account the symmetry. Each triangle in the membership functions corresponds to a subregion of the shadow area in Figure 18.4. The membership function for $p_{n,0}$, $\mu(p_{n,0})$, is defined by intuition from the intervals of $p_{n,0}$. The membership function of $a_{n,0}$, $\mu(a_{n,0})$, is determined from the 50,000 pairs of training data. The center of each triangle is the average value of the received signal amplitude from all the mobile users in the corresponding subregion, and the width of each triangle is proportional to the standard deviation of the received signal amplitude from the mobile users in the subregion. As a result, the service area should be divided carefully into subregions, so that (a) the probability associated with each subregion has a unique range and the range should not be too large, and (b) the triangles in membership function $\mu(a_{n,0})$ do not overlap too much, otherwise they become indistinguishable and therefore redundant. From the training data, the membership function, $\mu(a_{n,0})$, is first determined. With the membership functions $\mu(a_{n,0})$ and $\mu(p_{n,0})$, each input-output training data pair contributes to the kth fuzzy IF-THEN rule, as long as the training data has a non-zero degree in the input region for the signal amplitude of the rule. The fuzzy rule base can then be generated using all the information from the 50,000 training data pairs. Table 18.1 gives the degree $\mu(a_{n,0}, a_{n,1})$ of expert's belief on each input data pair $(a_{n,0}, a_{n,1})$. If a mobile user is closer to BS_i ($i = 0$ or 1), then $a_{n,1}$ is large and the effect of shadowing, MAI, and background noise is relatively small. That is, we have a high confidence level about the measurement accuracy of $a_{n,i}$. Therefore, we assign a large value to $\mu(a_{n,0}, a_{n,1})$ corresponding to a data pair $(a_{n,0}, a_{n,1})$ which has one large component. When the mobile user is close to the cell boundary, the shadowing, MAI, and noise have a relatively large effect on both $a_{n,0}$ and $a_{n,1}$; therefore, we assign a small value to $\mu(a_{n,0},$

TABLE 18.1

The Degree $\mu\,(a_{n,\,0},\,a_{n,1})$ Assigned to Input Data $(a_{n,0},\,a_{n,1})$ which Represents the Usefulness of the Data

$\dfrac{a_{n,0}}{a_{n,1}}$	ES	VS	S	SM	ML	L	VL	EL
ES	0.1	0.2	0.3	0.4	0.5	0.8	0.9	1.0
VS	0.2	0.1	0.2	0.3	0.4	0.6	0.8	0.9
S	0.3	0.2	0.1	0.2	0.3	0.4	0.6	0.8
SM	0.4	0.3	0.2	0.1	0.2	0.3	0.4	0.6
ML	0.5	0.4	0.3	0.2	0.1	0.2	0.3	0.4
L	0.8	0.6	0.4	0.3	0.2	0.1	0.2	0.3
VL	0.9	0.8	0.6	0.4	0.3	0.2	0.1	0.2
EL	1.0	0.9	0.8	0.5	0.4	0.3	0.2	0.1

$a_{n,1})$ corresponding to a data pair which has a small value for $|a_{n,0}-a_{n,1}|$. The degree $\mu(a_{n,0},\,a_{n,1})$ increases linearly as the difference between $a_{n,0}$ and $a_{n,1}$ increases. The overall degree of expert's belief on each training data set $\{a_{n,0},\,a_{n,1},\,p_{n,0}\}$ to rule k is determined by

$$\mu_k \,=\, (\mu\,(a_{n,\,0},\,a_{n,\,1})\,/\,(\sigma_{A_{0k}}\sigma_{A_{1k}})) \tag{18.17}$$

where $\sigma_{A_{0k}}$ and $\sigma_{A_{1k}}$ are the standard deviations of $a_{n,0}$ and $a_{n,1}$, respectively, for the input region of rule k. The standard deviation characterizes the degree of uncertainty in each measured $a_{n,i}$ value and depends on the value of σ and the cell structure. From Equation (18.2), the degree Q_k assigned to rule k is then

$$Q_k \,=\, \mu(a_{n,\,0},\,a_{n,\,1})[\mu_{I_k}(a_{n,\,0})\,/\,\sigma_{A_{0k}}][\mu_{I_k}(a_{n,\,1})\,/\,\sigma_{A_{1k}}]$$
$$\cdot\,\mu_{O_k}(p_{n,\,0})\mu_{O_k}(p_{n,\,1}). \tag{18.18}$$

18.4.2   Performance of the Fuzzy Inference Subsystem

In the following, two cases are considered: (a) one-dimensional space where the number of BSs that each mobile user has potential to handoff is one; and (b) two-dimensional space where the number of BSs is two.

18.4.2.1   One-Dimensional Space

The fuzzy rule base generated based on the 50,000 pairs of training data for $\kappa = 4$ and $\sigma = 2$ dB is shown in Table 18.2 and the degree $\overline{Q}_k$ (normalized to 1) associated with rule R_k is shown in Table 18.3. In Table 18.2, there is no rule for input data pair $(a_{n,0},\,a_{n,1})$ where both $a_{n,0}$ and $a_{n,1}$ are small or large, due to the fact that no training data pair falls in the domain. The corresponding

TABLE 18.2

The Fuzzy Rule Base for $p_{n,0}$ with $\kappa = 4$ and $\sigma = 2$ dB (One Dimension)

$\dfrac{a_{n,0}}{a_{n,1}}$	ES	VS	S	SM	ML	L	VL	EL
ES					SM	S	VS	ZE
VS				SM	SM	S	VS	ZE
S			SM	SM	SM	S	VS	ES
SM		ML	SM	ML	SM	S	S	VS
ML	ML	ML	ML	SM	SM			
L	L	L	L	ML				
VL	VL	VL	VL					
EL	OE	OE	OE	OE				

TABLE 18.3

The Degree Associated with Each Rule for $p_{n,0}$ with $\kappa = 4$ and $\sigma = 2$ dB (One Dimension)

$\dfrac{a_{n,0}}{a_{n,1}}$	ES	VS	S	SM	ML	L	VL	EL
ES	0.00	0.00	0.00	0.00	0.58	0.91	1.00	0.68
VS	0.00	0.00	0.00	0.25	0.57	0.75	0.88	0.63
S	0.00	0.00	0.10	0.29	0.34	0.47	0.64	0.38
SM	0.00	0.44	0.28	0.11	0.21	0.30	0.38	0.19
ML	0.71	0.46	0.34	0.22	0.08	0.20	0.00	0.00
L	1.00	0.74	0.48	0.33	0.17	0.00	0.00	0.00
VL	0.92	0.85	0.63	0.36	0.00	0.00	0.00	0.00
EL	0.70	0.60	0.52	0.27	0.00	0.00	0.00	0.00

degree in Table 18.3 has value equal to zero. In Table 18.3, it can be seen that, in general, a large difference between the $a_{n,0}$ and $a_{n,1}$ values results in a large value of the degree. However, the relation between the degree and $|a_{n,0} - a_{n,1}|$ is nonlinear and is different from that shown in Table 18.1 because the degree of each rule depends on the membership functions of $a_{n,i}$ and $p_{n,0}$, the standard deviations of $a_{n,0}$ and $a_{n,1}$, etc., in addition to the degree $\mu(a_{n,0}, a_{n,1})$.

Figure 18.7 shows the comparison between the true probability $p_{n,0}$ and the fuzzy inference system output $\tilde{p}_{n,0}$ for $\kappa = 4$, $\sigma = 2$ dB, whereas Figure 18.8 shows the corresponding estimation error $p_{n,0} - \tilde{p}_{n,0}$ versus $p_{n,0}$. Table 18.4 gives the mean and standard deviation of the estimation error $(p_{n,0} - \tilde{p}_{n,0})$ for various σ values, where the true probability $p_{n,0}$ is obtained based on the mobile user location (x_{MT}, y_{MT}) and the estimated probability $\hat{p}_{n,0}$ is obtained by the fuzzy inference system according to the measurement data $a_{n,0}$ and $a_{n,1}$. Due to the geometrical symmetry, estimation of $p_{n,1}$ and the estimation accuracy are the same as those of $p_{n,0}$. From the simulation results given in Figures 18.7 and 18.8 and in Table 18.4, it is observed that: (a) The estimator is unbiased since the mean of the estimation error is very small and can take on positive or negative values; (b) As the value of σ increases, there is an increase in

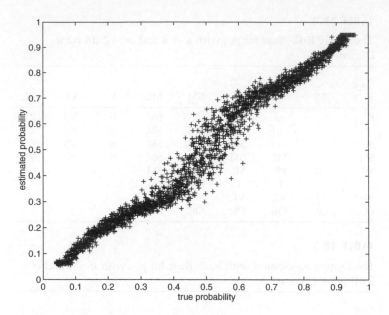

FIGURE 18.7
The estimated probability $\tilde{p}_{n,0}$ versus true probability $p_{n,0}$ with $\kappa = 4$ and $\sigma = 2$ dB (one dimension).

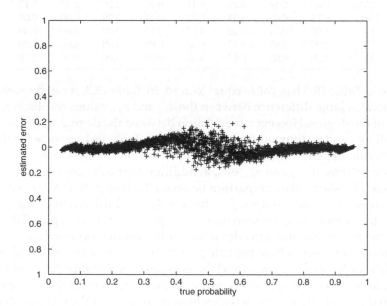

FIGURE 18.8
The estimation error $(p_{n,0} - \tilde{p}_{n,0})$ versus true probability $(p_{n,0})$ with $\kappa = 4$ and $\sigma = 2$ dB (one dimension).

TABLE 18.4

The Mean and Standard Deviation of the
Estimation Error $(p_{n,0} - \tilde{p}_{n,0})$ of the Fuzzy
Inference System Given $\kappa = 4$
(One Dimension)

σ (dB)	Mean	Standard deviation
1	$-3.55e-4$	3.14e-2
2	4.10e-4	4.08e-2
3	$-3.39e-3$	4.80e-2
4	1.13e-3	6.25e-2
5	$-1.75e-3$	7.32e-2
6	$-5.29e-3$	8.32e-2

TABLE 18.5

The Mean and Standard Deviation of the
Estimation Error $(p_{n,0} - \tilde{p}_{n,0})$ of the Fuzzy
Inference System Given $\sigma = 2$ dB (One
Dimension)

κ	Mean	Standard deviation
2	8.20e-4	6.94e-2
4	4.10e-4	4.08e-2
6	3.00e-4	3.92e-2

the degree of shadowing effect of the propagation channel, resulting in an increased estimation error; (c) Given a certain σ value (such as $\sigma = 2$ dB in Figures 18.7 and 18.8), the estimation error is relatively small when the mobile user is close to one BS (i.e., $p_{n,0}$ is very small or very large), where the shadowing has less effect on degrading the performance of the fuzzy inference system. In other words, the effect of the shadowing on the estimation accuracy increases as the mobile user moves to the cell boundary, even though the area close to the cell boundary is very important for making handoff decisions. The reduced accuracy is due to the reduced confidence level on the measured data, which is a direct result of the near-far problem inherent in CDMA systems.

Table 18.5 illustrates how the first and second order statistics of the estimation change as the path loss exponent, κ, changes, where $\sigma = 2$ dB is used for the different κ values. It is observed that, as the value of κ increases, both the mean and the standard deviation decreases. This is because a larger κ value means a faster attenuation of the received signal level as the distance between the mobile user and the BS increases. Correspondingly, the degree of the randomness in the received signal level due to different x_{MT} values in each subregion is reduced, resulting in a better estimation. On the other hand, variations in the value of κ do not significantly change the accuracy of the estimation as long as σ is fixed. From Tables 18.4 and 18.5, the parameter σ plays a more important role in the estimation accuracy than the parameter κ, because the shadowing characterized by σ is the main source which introduces randomness to the received signal levels.

18.4.2.2 Two-Dimensional Space

In this case, the estimation of $p_{n,0}, p_{n,1}$ and $p_{n,2}$ is based on the three strongest pilot signals. Due to the symmetry of the area, $p_{n,0}, p_{n,1}$, and $p_{n,2}$ can be obtained by Equations (18.12)–(18.14). During the simulations, it is observed that the changes of $a_{n,2}$ are relatively small compared with those of $a_{n,0}$ and $a_{n,1}$ when the mobile user moves from one subregion to another, mainly due to small distance changes from BS_3 to each subregion defined according to $p_{n,0}$. As a result, the estimation of $p_{n,0}$ relies only on $a_{n,0}$ and $a_{n,1}$ for the simplicity of the fuzzy inference system. The estimation procedure is similar to that of estimating $p_{n,0}$ in the one-dimensional case except that $p_{n,2}$ can take on non-zero value. Based on training data from 50,000 mobile users uniformly distributed in the shadow area of Figure 18.4, Table 18.6 gives the decision rules for $\kappa = 4$ and $\sigma = 2$ dB, and Table 18.7 gives the corresponding degree assigned to each rule. Figures 18.11 and 18.12 show the estimation error $p_{n,0} - \tilde{p}_{n,0}$ versus $p_{n,0}$ for $\kappa = 4$ and σ equals to 2 dB and 4 dB, respectively. Table 18.8 gives the mean and standard deviation of the estimation error for various σ values. Due to the geometrical symmetry, estimation of $p_{n,1}$ and the estimation accuracy are the same as those of $p_{n,0}$. We have some observations similar to those in the one-dimensional case: (a) the estimator is unbiased; (b) given a certain σ value, the estimation error is relatively small when the mobile user is close to one BS; (c) the effect of the shadowing on the estimation accuracy increases as the mobile user moves to the cell boundary; (d) a larger value of σ results in a larger estimation error. Comparing Tables 18.2 and 18.6, we see that $p_{n,0}$ in the two-dimensional case is the same as, or very close to, the corresponding value in the one-dimensional case when $p_{n,0}$ is relatively large (say, larger than 0.5), where $p_{n,2}$ is equal to or close to zero; on the other hand, when $p_{n,0}$ is relatively small (say, less than 0.5), the value of $p_{n,0}$ in the two-dimensional case is slightly smaller than that in the one-dimensional case due to the

TABLE 18.6

The Fuzzy Rule Base for $p_{n,0}$ with $\kappa = 4$ and $\sigma = 2$ dB
(Two Dimension)

$\dfrac{a_{n,0}}{a_{n,1}}$	ES	VS	S	SM	ML	L	VL	EL
ES				VS	S	VS	VS	ZE
VS			S	S	VS	VS	S	ZE
S			ML	SM	SM	VS	VS	ES
ML	ML	ML	S	S	VS	VS	S	
ML	ML	ML	S	ML	SM	VS	VS	
L	L	L	L	L	ML	ML		
VL	VL	VL	VL	VL				
EL	OE	OE	OE	EL				

TABLE 18.7

The Degree Associated with Each Rule for $p_{n,0}$ with $\kappa = 4$ and $\sigma = 2$ dB (Two Dimension)

$\dfrac{a_{n,0}}{a_{n,1}}$	ES	VS	S	SM	ML	L	VL	EL
ES	0.00	0.00	0.00	0.59	0.84	0.73	0.58	0.52
VS	0.00	0.00	0.39	0.36	0.55	0.50	0.47	0.39
S	0.00	0.00	0.17	0.30	0.44	0.36	0.39	0.31
SM	0.86	0.57	0.33	0.14	0.30	0.29	0.29	0.00
ML	0.75	0.45	0.41	0.22	0.11	0.15	0.15	0.00
L	1.00	0.79	0.54	0.33	0.23	0.05	0.00	0.00
VL	1.00	0.87	0.64	0.38	0.00	0.00	0.00	0.00
EL	0.69	0.61	0.49	0.25	0.00	0.00	0.00	0.00

TABLE 18.8

The Mean and Standard Deviation of the Estimation Error $(p_{n,0} - \tilde{p}_{n,0})$ of the Fuzzy Interference System Given $\kappa = 4$ (Two Dimension)

σ (dB)	Mean	Standard deviation
1	3.35e-3	5.95e-2
2	1.39e-3	6.36e-2
3	-1.92e-3	7.66e-2
4	1.99e-3	9.04e-2
5	-4.76e-3	1.29e-1
6	2.51e-3	1.44e-1

fact that $p_{n,2}$ is larger in the former situation. Comparing the estimation accuracy in the two-dimensional case (Figures 18.11 and 18.12 and Table 18.8) with that in the one-dimensional case (Figures 18.8 and 18.10, and Table 18.4), we see that the side information that the mobile user will communicate with either BS_0 or BS_1 (but not BS_2) improves the performance of the fuzzy inference system.

18.4.3 Evaluation of the Adaptive Fuzzy Inference Prediction System

In order to evaluate the overall system performance, 500 mobile users are simulated with movement patterns characterized by the following:

i) The initial location of each mobile user is uniformly distributed in the sub-regions O_1 of cell_0 and O_4 of cell_1 as shown in Figure 18.1;

ii) Each mobile user has a constant velocity uniformly distributed in[10, 30] meters per second;

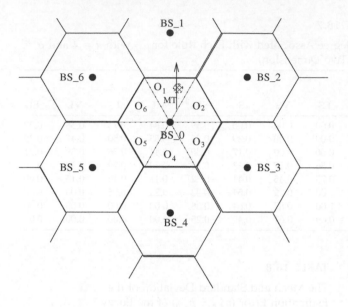

FIGURE 18.9
The estimated probability $\tilde{p}_{n,0}$ versus true probability $p_{n,0}$ with $\kappa = 4$ and $\sigma = 4$ dB (one dimension).

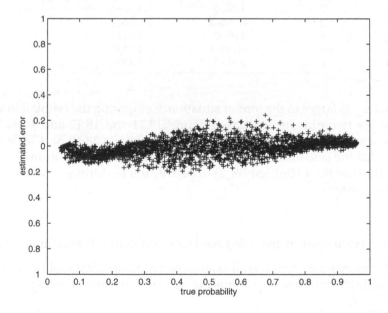

FIGURE 18.10
The estimation error $(p_{n,0} - \tilde{p}_{n,0})$ versus true probability $(p_{n,0})$ with $\kappa = 4$ and $\sigma = 4$ dB (one dimension).

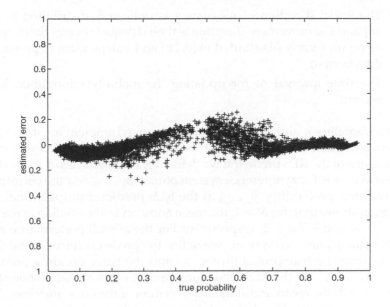

FIGURE 18.11
The estimation error $(p_{n,0} - \tilde{p}_{n,0})$ versus true probability $(p_{n,0})$ with $\kappa = 4$ and $\sigma = 2$ dB (two dimension).

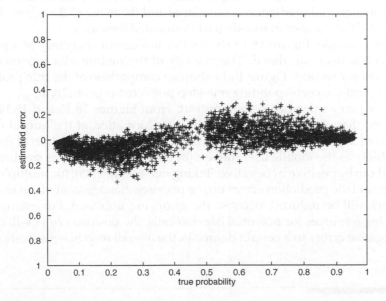

FIGURE 18.12
The estimation error $(p_{n,0} - \tilde{p}_{n,0})$ versus true probability $(p_{n,0})$ with $\kappa = 4$ and $\sigma = 4$ dB (two dimension).

iii) The initial direction of movement is uniformly distributed in [0, 2π] and the movement direction is then changed several times each being uniformly distributed in [0, 2π] and independent of previous direction(s);

iv) The time interval Δt for updating the mobility information is 1 second.

The parameters κ and σ of the propagation environment are 4 and 2 dB, respectively, and the parameter m of the RLS algorithm is 7. To evaluate the performance of the RLS predictor, we define the estimation error as the difference between the fuzzy inference system output $\tilde{p}_{n+N,0}$ and the corresponding predicted probability $\hat{p}_{n+N,0}$ at the RLS predictor output. Computer simulation shows that, for $N = 1$, the mean and standard deviation of the error are $2.10e - 4$ and $9.53e - 2$, respectively. For the overall performance of the adaptive fuzzy inference system, we define the prediction error as the difference between the true probability $p_{n+N,0}$ and the corresponding predicted probability $\hat{p}_{n+N,0}$ at the RLS predictor output. Computer simulation shows that, for $N = 1$, the mean and standard deviation of the error are $9.76e - 3$ and $9.90e - 2$, respectively. The standard deviation of the prediction error is close to the corresponding value of the estimation error, but is larger than the corresponding value in Table 18.4. This is because the prediction error includes the estimation error of the RLS predictor and the error of the fuzzy inference system. From a detail analysis of the simulation result, it is concluded that both the estimation and prediction are unbiased and the mean of the errors should decrease if the number of mobile users simulated increases.

As an example, Figure 18.13 shows the movement trajectory of a particular mobile user simulated. The velocity of the mobile user movement is 24 meters per second. Figure 18.14 shows a comparison of the true probability $p_{n+1,0}$ and the corresponding one-step predicted probability $\hat{p}_{n+1,0}$ at the adaptive fuzzy inference system output. From Figures 18.13 and 18.14, it is observed that: (a) the predicted mobility information at the output of the adaptive fuzzy inference system can track quite well the variation of the true probability as the mobile user moves; (b) the prediction error is smaller than 0.1 and can be positive or negative. Taking into account statistic multiplexing, the effect of the prediction errors on the resource management of the wireless network will be reduced, because the errors are unbiased. For example, in reserving resources for potential handoff calls, the positive errors will cancel the negative errors to a certain degree in the overall resource reservation.

18.5 Conclusions

An adaptive fuzzy inference prediction system is developed to predict the probabilities that a mobile user will be active in the nearby cells at future

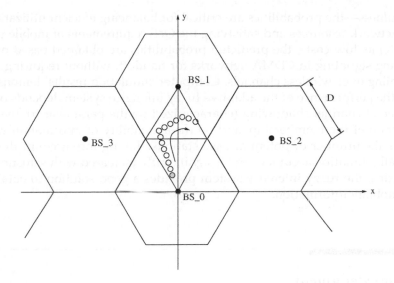

FIGURE 18.13
The movement trajectory of a mobile user simulated.

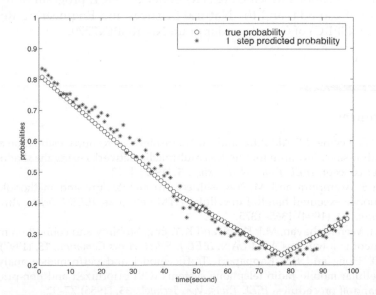

FIGURE 18.14
The comparison of the true probability and the corresponding one-step predicted probability for the mobile user.

moments using the real-time measurement data of the pilot signal powers received at the mobile user from the BSs. The advantages of the adaptive fuzzy inference system lie in (a) its simplicity—it is a one-pass build-up procedure that does not require time-consuming on-line training, (b) its usefulness—the probabilities are critical for balancing efficient utilization of the network resources and satisfying the QoS requirements of mobile users, and (c) its low cost—the predicted probabilities are obtained based on the existing signaling in CDMA networks for handoff, without requiring extra signaling over wireless channels. Computer simulation results demonstrate that the performance of the adaptive fuzzy inference system depends on the degree of channel shadowing (characterized by the parameter σ), the construction of the membership, and on the availability of information which limits the number of potential base stations. Taking into account that the overall estimation accuracy can be significantly increased with statistic multiplexing, the fuzzy inference system provides a good solution to obtaining the mobility information.

Acknowledgment

This work has been supported by a grant from the Canadian Institute for Telecommunications Research (CITR) under the NCE program of the Government of Canada, and the Natural Sciences and Engineering Research Council (NSERC) of Canada under grant No. RGPIN7779.

References

1. D.A. Levine, I.F. Akyildiz, and M. Naghshineh, A resource estimation and call admission algorithm for wireless multimedia networks using the shadow cluster concept, *IEEE Trans. Networking*, **5**, (1997) 1–12.
2. A.S. Acampora and M. Naghshineh, An architecture and methodology for mobile-executed handoff in cellular ATM networks, *IEEE J. Select. Areas Commun.*, **12**, (1994) 1365–1375.
3. M. Veeraraghavan, M.J. Karol, and K.Y. Eng, Mobility and connection management in a wireless ATM LAN, *IEEE J. Select. Areas Commun.*, **15**, (1997) 50–68.
4. D. Hong and S.S. Rappaport, Traffic model and performance analysis for cellular mobile radio telephone systems with prioritized and non-prioritized handoff procedures, *IEEE Trans. Veh. Technol.*, **35**, (1986) 77–92.
5. E. Del Re, R. Frantacci, and G. Giambene, Handover and dynamic channel allocation techniques in mobile cellular networks, *IEEE Trans. Veh. Technol.*, **44**, (1995) 229–237.
6. T. Liu, P. Bahl, and I. Chlamtac, Mobility modeling, location tracking, and trajectory prediction in wireless ATM networks, *IEEE J. Select. Areas Commun.*, **16**, (1998) 922–936.
7. R.A. Guerin, Channel occupancy time distribution in a cellular radio system, *IEEE Trans. Veh. Technol.*, **36**, (1987) 89–99.

8. G. Morales-Andreas and M. Villen-Altamirano, An approach to modeling subscriber mobility in cellular radio networks, in: *5th World Telecommun. Forum,* Geneva, Switzerland, 1987, pp. 185–189.

9. R. Thomas, H. Gilbert, and G. Mazziotto, Influence of the movement of the mobile station on the performance of a radio cellular network, in: *3rd Nordic Seminar,* Copenhagen, Denmark, 1988, paper 9.4.

10. I. Seskar, S. Maric, J. Holtzman, and J. Wasserman, Rate of location area updates in cellular systems, in: *Proc. 42nd IEEE Veh. Technol. Conf. (VTC '92),* Denver, USA, 1992, pp. 694–697.

11. S. Nanda, Teletraffic models for urban and suburban microcells: cell sizes and handoff rates, *IEEE Trans. Veh. Technol.,* **42**, (1993) 673–682.

12. M.M. Zonoozi and P. Dassanayake, User mobility modeling and characterization of mobility pattern, *IEEE J. Select. Areas Commun.,* **15** (1997) 1239–1252.

13. M. Naghshineh and M. Schwartz, Distributed call admission control in mobile/wireless networks, *IEEE J. Select. Areas Commun.,* **14**, (1996) 711–717.

14. R. Vijayan and J. M. Holtzman, A model for analyzing handoff algorithms, *IEEE Trans. Veh. Technol.,* **42**, (1993) 351–356.

15. X. Shen and J. W. Mark, Mobility information for resource management in wireless ATM networks, *Computer Networks,* **31**, (1999) 1049–1062.

16. *An Overview of the Application of Code Division Multiple Access (CDMA) to Digital Cellular Systems and Personal Cellular Networks,* QUALCOMM Inc., 1992.

17. T.S. Rappaport, *Wireless Communications: Principles and Practice,* Prentice Hall, 1996.

18. L.X. Wang, *Adaptive Fuzzy Systems and Control: Design and Stability Analysis,* Prentice-Hall, Englewood Cliffs, NJ, 1994.

19. G.J. Klir and B. Yuan, *Fuzzy Sets and Fuzzy Logic: Theory and Applications,* Prentice-Hall, Englewood Cliffs, NJ, 1995.

20. M. Braae and D.A. Rutherford, Fuzzy relations in a control setting, *Kybernetes,* 7, (1978) 185–188.

21. L. Ljung and T. Söderström, *Theory and Practice of Recursive Identification,* MIT Press, Cambridge, MA, 1983. Chapter 2.

22. S. Haykin, *Adaptive Filter Theory,* 2nd Edition, Prentice Hall, Englewood Cliffs, NJ, 1991. Chapter 13.

19

Intelligent Agents in Telecommunication Networks

Costas Tsatsoulis and Leen-Kiat Soh

CONTENTS

0-8493-1075-X/01/$0.00+$.50
© 2001 by CRC Press LLC

KEY WORDS: *Telecommunication networks, intelligent agents, multiagent systems, network management, mobile agents, intelligent interface agents, swarm intelligence, ants, network management architecture, network diagnosis, network traffic control and routing, network mobility platform, network configuration, network monitoring and accounting, service management and provisioning, negotiating agents.*

19.1 Introduction

Telecommunication networks today usually exist in a large, heterogeneous environment. The network components feature different operating systems, platforms, communication languages, and vendors. These components may be incompatible and require a channel or link to facilitate cooperation and coordination within the networks. In addition to such multiplicity in telecommunication networks, we are witnessing a growing enterprise of data, in terms of both demand and supply. The importance and the need for data collection and the subsequent data distribution are becoming, more than ever, paramount. Interactions among networks or network components are inevitable due to the proliferation of data. This phenomenon has brought on a flood of various on-line activities such as advertisement, sale, research, information gathering, information passing, etc., which results in network traffic congestions and strains network management. Subsequently, this has demanded a better handling of the collection, processing, distribution, and understanding of data within telecommunication networks.

When the networks were more constrained and localized, a centralized management approach was adequate to handle various system administration and traffic control tasks. However, today's telecommunication networks are constantly expanding with many distributed activity centers. This development naturally points to a distributed approach to address issues such as routing, switching, configuration, accounting or monitoring, performance, security, and reliability in telecommunication networks. Instead of one centralized and usually very large system that assumes the complete control and intelligence of the network, a number of smaller systems, or agents, can be used to help manage the network in a cooperative manner. This has motivated the multiagent systems (MAS) in telecommunication networks.

In this chapter, we will present an overview of the application of intelligent agents to telecommunication networks. In Section 19.2 we discuss what constitutes an agent and its intelligence. In Section 19.3 we talk about how intelligent agents can help manage telecommunication networks, ranging from increasing the reliability and security of the networks to providing better human-computer interface tools and fault diagnosis. In Section 19.4 we mention several research projects and approaches that are investigating and exploring new technologies in intelligent agents in telecommunication networks. In Section 19.5 we discuss negotiation-based communications among agents and how it applies to agents in telecommunication networks. In Section 19.6 we present a design example of an intelligent agent in telecommunication networks. Finally, we conclude the discussion in Section 19.7.

19.2 Agents

In this section, we report on some contemporary taxonomies of agents and some criteria required for agenthood.

Nwana (1996) provided a comprehensive discussion of agent characteristics and typology. The author identified a minimal list of three primary attributes: autonomy, learning, and cooperation. Autonomy refers to the principle that agents can operate on their own without the need for human guidance. For agent systems to be truly smart, agents would have to learn as they react or interact with their external environment. In order to cooperate, agents need to possess a social ability, i.e., the ability to interact with other agents and possibly humans via some communication language. From agents that possess these attributes, the author further distinguished collaborative agents, interface agents, collaborative learning agents, and smart agents. In addition, agents can sometimes be labeled by their applications such as information agents, reactive agents, mobile agents, and hybrid agents.

Woodridge and Jennings (1995) provided two notions of agenthood. A weak agent enjoys the following properties: autonomy, social ability, reactivity, and proactiveness. To be reactive or responsive, agents must perceive their environment and respond in a timely fashion to changes that occur in it. To be proactive, agents do not simply act in response to their environment, but they can also exhibit opportunistic, goal-directed behavior by taking the initiative. A reactive (or responsive), proactive and social agent is also a flexible agent. A stronger notion of agenthood involves ascribing to agents humanistic concepts such as the belief-desire-intention (BDI) model (Shoham 1993, Rao and Georgeff 1995) and human characteristics such as trust and competence (Maes 1994) or emotions (Bates 1994).

In Jennings et al. (1998), the authors defined an agent as a computer system, situated in some environment, that is capable of flexible autonomous action in order to meet its design objectives. The three key defining concepts are situatedness, autonomy, and flexibility. According to the authors, situatedness means that the agent receives sensory input from its environment and its actions affect the environment. By flexible, the authors indicated that an agent is responsive, proactive, and social.

In addition, others have identified collaborative interface agents (Lashkari et al. 1994), infrastructure agents (Eaton et al. 1998), task-specific and performative agents (King 1995) such as search agents, navigation agents, information agents

(Papazoglou et al. 1992), softbots (Etzioni and Weld 1994, Etzioni et al. 1995), knowbots (Hylton et al. 1996, Hylton and van Rossum 1997), infobots, autobots, webbots, and others. Magedanz et al. (1996) taxonomically divided agents into local agents and networked agents for single-agent systems, DAI-based agents and mobile agents for multiagent systems. There are also advisory agents that work as personal assistants to users in email management, news reading, Web browsing, and searching (Lieberman 1995, Maes 1994).

Many attributes sufficient or necessary for defining an agent have been proposed and identified. Here we provide a list of these attributes (which is by no means complete): (1) autonomy—an agent can operate without direct intervention of human users or without having to depend on human users after creation and deployment, (2) communication ability—an agent can interact with other agents or human users to collaborate, negotiate, or coordinate in order to define and perform its tasks, (3) reactivity—an agent can perceive its environment and responds to cues automatically without delay and having to consult its human user, (4) mobility—an agent can move across operating environments and perform tasks remotely, (5) adaptivity—an agent can learn from its experience in dealing with its environments, including modeling human users, and (6) proactivity—an agent can reason, plan, and execute tasks by taking the initiative without prompting. Other attributes include (1) veracity—an agent will not knowingly lie (Galliers 1988), (2) benevolence—an agent will carry out its tasks faithfully (Rosenschein and Genesereth 1985), and (3) rationality—an agent will act in order to achieve its goals (Galliers 1988).

Intelligence can be defined as the ability to learn, the ability to adapt, the ability to improve one's performance over time, the ability to reason and make decisions, the ability to act without having to be instructed, the ability to plan and execute a complicated task that involves collaborations from other agents or human users, the ability to traverse a network and perform assigned tasks at each planned stop, and so on. We see an intelligent agent as a piece of program that is mobile, autonomous, reactive, and communicative. Of course, other attributes such as those mentioned above can be structured into the behavior of the program code to increase its intelligence. In our opinion, intelligent agents come in a spectrum of various degrees of intelligence, just like actual life forms in our world that range from single-purpose, single-cell life forms to extremely complex life forms like humans.

Moreover, in our discussion, we do not require agents to be intelligent to exhibit intelligent behavior collectively; that is, single-purpose agents performing a collectively intelligent task are considered, as a whole, an intelligent multiagent system. This is known as swarm intelligence. We discuss this field of intelligent agents in greater detail later in Section 19.4.1.3 of this chapter.

Traditionally, Distributed Artificial Intelligence (DAI) is defined as the research into systems composed of multiple agents, and this field has been divided into Distributed Problem Solving (DPS) and Multiagent Systems (MAS). DPS considers how a particular problem can be solved by a number of modules (nodes), which cooperate in dividing and sharing knowledge about the problem and its evolving solutions. A system of agents is a multi-agent system (Lesser 1995) or an agency (Agre and Rosenschein 1996). In contrast to DPS, research in MAS is concerned with the behavior of a collection of possibly preexisting autonomous agents aiming at solving a given problem. A multiagent system can be defined as a loosely coupled network of problem solvers that work together to solve problems that are beyond the individual capabilities of knowledge of each problem solver. The characteristics of MAS are: (1) incomplete viewpoint of the world for each agent, (2) lack of global system control, (3) decentralization of data, and (4) asynchronous computation (Jennings et al. 1998). Thus, we see that MAS is useful to telecommunication networks due to the implied efficiency, robustness, ability for integration of various systems, and flexibility.

A multiagent system inherits most of the advantages of distributing intelligence over centralized, sequential processing, since it is: (1) reliable—a multiagent system is more fault-tolerant and robust, (2) modular—agents can be added and deleted without greatly disrupting the system; a multiagent system has better scalability, (3) adaptive—agents can re-configure themselves to suit system changes such as noise, resource reallocation, and faults, (4) concurrent—agents can reason and perform tasks in parallel and asynchronously, resulting in faster and flexible execution of tasks, and (5) dynamic—agents can collaborate to form dynamic groups to solve specific problems, pooling together resources, and disband after the problems are solved, releasing resources to local usage.

One of the most important infrastructure for a multiagent system has to do with the communications: protocols and messages, communication channels, communication handling, and mobility of agents. In all implementations of multiagent systems in which agents communicate and travel, these issues must be resolved first before the development of the individual agents. The telecommunication networks currently in place are natural environments for intelligent agents to populate—agents can communicate, be mobile, and perceive detectable environmental cues to react to them. Some of these networks have already in place the fundamentals of the gateways and highways required for intelligent agents to grow and thrive.

19.3 Intelligent Agents and Telecommunication Networks

One of the most important research areas in telecommunication networks is network management. Network management determines the allocation of network resources (such as data storage, processing power, and memory), fault diagnosis and repair, system administration, routing (switching, bridging, etc.), communications among network components, etc. One of the goals is to have balanced loading and reliable loading on the network such that connections in the network can be established quickly without noise, delay, or numerous trials. In addition, network management also aims at house-keeping the networks so that they work efficiently and effectively, adapt to changes, and respond to problems such as traffic patterns. Thus, in this section and also in this chapter, we will emphasize network management for telecommunication networks.

Management systems for telecommunication networks usually come in four architectures: (1) centralized network management, (2) hierarchical network management, (3) peer network management, and (4) distributed network management. In a centralized architecture, a single manager handles the housekeeping of the whole network. It checks the network components regularly to ensure the smooth working of the network. It also responds to any warnings and errors issued by the network components. The information and data regarding the network components are stored at a centralized database warehoused by the manager. Thus, the central manager coordinates all network responsibilities from top to bottom. Once a network becomes larger and more complicated, a single centralized management is sometimes not sufficient or efficient. Hence, the hierarchical network management approach is used. In this strategy, managers are arranged in a hierarchy. On top of the hierarchy sits the central manager that administrates a group of assistant managers. Each assistant manager in turn over-looks a group of assistant-assistant managers and so on. In addition, each assistant manager communicates only to its parent manager, with no or minimal same-level interactions. Each manager maintains a localized database with the higher level managers having access to lower level databases. Note that usually these assistant managers are not application- or task-specific; instead, they manage a region of the whole network. Thus, in this architecture, each parent manager coordinates and delegates tasks among its children managers, providing an indirectly cooperative environment among the lower-level managers. In a peer-based network, there exist individual managers that are able to communicate among themselves. Each manager administrates a different domain of the network, interacts with its neighboring managers for information and data, and controls its own database. Hence, there is an increasing sense of cooperation. Finally, in a distributed network, the individual managers are application- or task-specific in terms of their specialties and responsibilities, and each has its own knowledge and databases. Therefore, this architecture requires the highest amount of cooperation from its managers. An extensive survey of network management approaches can be found in (Martin-Flatin and Znaty 1997).

Since intelligent agents come in handy whenever there is a system of coordination and cooperation, the intelligent agent technologies lend readily to the field of telecommunication networks. Intelligent agents in telecommunication networks are becoming more necessary and feasible because of the current research and development trends that are focused toward decentralization and cooperation. This is because today's networks are no longer manageable using just centralized or highly coordinated management strategies because of their size and heterogeneity.

From another viewpoint, the goal of network management technologies is to reduce the risks and cost exposure associated with operations of enterprise systems (Yemini 1993). Administrators or vendors equip their network components with agent software to monitor and collect operational data, such as error statistics, into databases, or to detect unexpected events such as connections overflow or overcrowded traffic. In (Yemini 1993), management platform workstations poll device data, or respond to event alerts sent by the network components. This management paradigm, after the Open Systems Interconnection (OSI)[1] management model, is platform centered where management applications are centralized at each platform, separated from the database and network components. Each managing platform is using the common management protocol to access managed information provided by an agent residing at a network node. The agent maintains a management information tree database that models a hub using managed objects to represent LANs, interfaces, and ports. A platform can use the protocol to create, delete, retrieve, or change managed objects in the tree model, invoke actions, or receive event notifications. Therefore, a network management system should be able to reconfigure the managed network and respond timely and effectively to monitored events.

According to Sahai et al. (1997) a network management system contains four types of components: (1) network management systems, (2) network management programs running on managed nodes, (3) management protocols, and (4) management information. A network management system uses the management protocol to communicate with agents running on the managed nodes. The information passed between the system and the agents is defined by a Management Information Base (MIB). The management standards that have emerged are the Simple Network Management Protocol (SNMP) and the OSI management system that utilizes the Common Management Information Protocol (CMIP). Thus, a network management has the following additional functions: behavior monitoring, installation of components, resource auditing (accounting), information retrieval, and health monitoring.

In (Meyer et al. 1995), network management is seen as capable of scripting and delegating agents to remote sites where they are incorporated into the local network management program and used for intelligent tasks such as management information base filtering. This application brings mobile agents into network monitoring and network control. Indeed, facilitating the migration or traversal of mobile agents in a telecommunication network allows asynchronous and cooperative processing of tasks, specialization of services, network configuration, decentralization of management, active

---

[1] OSI was created by the International Organization for Standardization (ISO) to develop standards for data networking to facilitate multivendor equipment interoperability.

service usage, intelligent communications such as negotiations among agents, and dynamic information flow.

With a centralized management system, telecommunication networks do not scale gracefully and lack flexibility. This is because of the difficulty in acquiring and maintaining an efficient control and computation with the centralized management system, the requirement that the system communicating to every component of the network, and the distributed and heterogeneous nature of today's network. So, from the standpoint of designing, building, and maintaining a telecommunication network, multiagent approaches are certainly more feasible.

Some standards regarding intelligent telecommunication networks have been proposed: the mobile agent framework of the Open Management Group (OMG) (Cheng and Covaci 1997), the Open System Interconnection (OSI) protocols defined in the 10040 Systems Management Overview by the International Organization for Standardization for data networking, the Telecommunications Management Network defined in M.3010 and the Intelligent Networks defined in Q.1200 by the International Telecommunications Union. These standards have called for more decentralized, distributed network management approaches, with specialized modules such as agents and treating network components as individual entities.

In conclusion, incorporating agent technologies and intelligence into telecommunication networks encourages extensive use of information processing modules, better management and usage of network components and resources, modularization and accounting of network components, flexible configuration of network components (such as self-adapting and self-configuring architecture with plug-and-play capability), and customization and specialization of services.

Some concerns have been raised for using mobile agents in telecommunication networks since these agents influence the performance of the networks. Baldi and Picco (1998) have performed tradeoff experiments regarding mobile agents in telecommunication networks, and the authors discovered that with the particular goal of optimizing network traffic, the tradeoffs depended on the characteristics of the network being managed (costs, number of nodes, protocols) and of the management task (possibility of local/global semantic compression, expected frequency of invocation, complexity of the task, dimension of replies). In addition, the characteristics of the technology supporting the implementation played a part according to the management protocols (overhead) and the mobile code system used (expressiveness of the language, formats used for transfer, overhead). Hence, according to the authors, whether to use a mobile agent design in place of a traditional client-server architecture requires (1) a model of the management functionality to be implemented, together with the information about the managed network, and (2) a precise quantitative characterization of the management protocols and the mobile code system to be used for the implementation.

19.4 Applications

In this section, we present several agent technologies in telecommunication networks and mention several projects and research activities that employ intelligent agents in network management architecture, network diagnosis, traffic control and routing, network mobility platform, network configuration, and network monitoring and accounting. These technologies include mobile agents or mobile computing, intelligent interface between agent and human users, swarm intelligence, and economic modeling. Mobile agents play an important role in *spreading* intelligence across networks when they travel. The mobility allows them to be created, deployed, and terminated without disrupting the network configuration. Interface agents model human managers and learn from them how to manage networks. This area of research has not been applied to telecommunication networks directly, but has the potential to automate or assist in the tasks of system administration and network management. Swarm intelligence stems from the work of artificial life in which unintelligent agents work independently or with relative small amount of collaboration to achieve a greater goal that requires intelligence. Then, we briefly touch on designing network management systems after economic models.

19.4.1   Agent Technologies in Telecommunication Networks

19.4.1.1   Mobile Agents

Mobile agents are sometimes known as mobile codes. On the least intelligent definition basis, mobile agents are programs that can be sent to and executed at a remote site. Some agents are deployed to gather information from a remote site, return with the information, and terminate. Some mobile agents, once launched, are able to plan, reason, and carry out tasks on their own and further decide their course of travel across the networks. Some mobile agents travel around the networks and perform itinerary housekeeping. Some mobile agents are thus also known as itinerant agents (Chess et al. 1995). Some agents such as *deglets* (Bieszczad and Pagurek 1998) are created with a sole task of sending the identifier of a visited node to the creator and travel by the means of the implemented migration patterns. These agents have also been identified as discovery agents—for finding and establishing connections.

In telecommunication networks, mobile agents are an exciting and increasingly important field. Joshi and Ramesh (1998) pointed out that the advantages of having a facility of mobile agents include mobile computing, remote search and filtering, and real-time production control. Mobile agents enable disconnected operation for clients where persistence of connection is not required for communication between the remote clients and servers. The client generates an agent for performing a remote operation, establishes a connection session, launches it, and terminates the session. When the client reconnects, the agent returns with the outcome of the operation. This reduces the network traffic and also releases the client from the connection, allowing it to establish connections with other sites. In computer-controlled manufacturing and production systems where machines are network nodes that are to be constantly monitored and reconfigured to perform different manufacturing tasks, run time control of machines require real-time management and communication latency is not acceptable. In such case, agents could be created with the control itinerary and dispatched across the network to remotely control the machines in real-time.

Baldi et al. (1997) further pointed out that the approaches proposed in the Internet Engineering Task Force and the International Organization for Standardization are of low degrees of flexibility and reconfigurability even though these research directions advocated agents for management. This is because the agents are direct extensions of the centralized management and do not have mobility. Baldi et al. (1997) showed that agent or code mobility is essential in enabling a better use of bandwidth resources and a higher degree of flexibility and reconfigurability in telecommunication networks.

In general, in order to manifest agent mobility for network management, a system has to incorporate a mobility framework (Bieszczad et al. 1998b). The framework has to support various agent models such as the life-cycle model, computational model, security model, communication model, and navigation model. For the life-cycle model, services have to be provided to allow the creation, termination, deployment, and suspension of agents. In our opinion, in addition to the above activities, one should consider the reconfiguration of agents as part of the life cycle, enabling agents to learn and adapt as they *live*. This would give the agents more intelligence and encourage knowledge flow, either among agents, or between agents and network managers. The computational model refers to the computational capabilities of an agent, which include data manipulation and thread-control primitives. The security model describes the ways in which agents can access network resources, as well as the ways of accessing the internals of the agents from the network. The communication model defines the communication between agents and between an agent and other network components. Finally, the navigation model defines how an agent should travel or reside, or how an agent should be delegated by its launcher.

Other important design issues in telecommunication networks housing mobile agents include transfer protocols, resource allocation, network component access, and user mobility.

Mobile agent technologies are sometimes known as mobile code technologies. An extensive survey and insightful analysis of mobile code technologies such as mobility mechanisms, including mobile code design paradigms and applications, can be found in (Fuggetta et al. 1998).

Hierarchical Agents—Appleby and Steward (1994) of British Telecom devised a distributed control mechanism for mobile agents to optimize routing in telecommunication networks according to a least cost criterion. There are two hierarchical levels of agents: load management agents and parent agents. A load management agent is launched on a particular node. The agent visits every node in the network efficiently using an adaptation of Dijkstra's shortest path algorithm, records the current spare capacity, and amends the routing tables. Hence, it collects the necessary information to determine the optimal routes from all other nodes to that particular node on which the agent has been launched. The parent agents control the next management level. They travel around the network and launch load agents where network management is needed. The decision to launch a load agent is made on the basis of information gathered, and a set of heuristic rules.

19.4.1.2 Intelligent Interface between Agents and Human Users

Intelligent interface agents interface with humans or other agents and learn through solicitation or vigilance (Chin 1991, Dent et al. 1992, Maes and Kozierok 1993, Sheth and Maes 1993, Lashkari et al. 1994). Intelligent interface agents can learn by (1) monitoring and imitating the human user, (2) receiving direct commands from the user, and (3) receiving feedback from the user. Sometimes, agents can learn from each other through knowledge transfer as well. We think that intelligent interface agent technology is an important aspect of knowledge-transfer between human system administrators or network engineers and software agents. It helps model the critical, diagnostic reasoning capabilities of humans in managing telecommunication networks.

There is little work using interface agents in telecommunications, but we foresee this as a very viable area of future development. Some work on using interface agents for network supervision is described in (Esfandiari et al. 1996).

19.4.1.3 Swarm Intelligence

The field of artificial life has given rise to swarm intelligence (Beni and Wang 1989), in which a group of unintelligent agents of limited capabilities exhibiting collectively intelligent behavior (White and Pagurek 1998, Bonabeau et al. 1999). Swarm agents are mobile agents, but with much less computational capability and intelligence—they traverse the network carrying out simple-purpose tasks with no explicit knowledge of the ultimate goal or network scenario, and without direct communications or contacts with other agents. However, the collective work of swarm agents can have global and intelligent impact on the whole telecommunication network. Usually these agents are distributed and highly adaptive to changes in the network and traffic patterns and can adapt to the network topology, the call probabilities of the nodes, and temporary situations caused by the randomness of the call patterns.

White et al. (1998a) described how multiple interacting swarms of adaptive mobile agents could be used to locate faults in networks. The authors proposed the use of mobile agents for fault finding in order to address the issues of client/server approaches to network management and control, such as scalability and the difficulties associated with maintaining an accurate view of the network. The authors further defined three principal types of mobile agents: servlets, deglets, and netlets. Servlets are extensions or upgrades to servers that stay resident as integral parts of those servers. Mobile agents constituting servlets are sent from one component to another and are installed as code extensions at the destination component. Deglets are mobile agents that are delegated to perform a specific task and generally migrate within a limited region of the network for a short period of time, for example, to undertake a provisioning activity on a network component. Netlets are mobile agents that provide predefined functionality on a permanent basis and circulate within the network continuously. These agents are small and mobile, and communication is top-down, instead of peer-based.

Ants—The intelligent behavior resulted from indirect communication between agents is called stigmergy (White et al. 1998a) and there are two forms of it. Sematectonic stigmergy involves a change in the physical characteristics of the environment. Ant nest building is an example of this form of communication in that an ant observes a structure developing and adds to it. The second form of stigmergy is sign-based. Here, something is deposited in the environment that makes no direct contribution to the task being undertaken but is used to influence subsequent task related behavior such as the foraging behavior of ants. Thus, some swarm systems in telecommunication networks are also known as ant systems.

Schoonderwoerd et al. (1996, 1997) designed an ant system for telecommunication networks after the trail laying ability of ants. Ants lay chemicals (pheromones) as they traverse trails. By creating small, mobile agents that are capable of traversing the network themselves and leaving behind simulated pheromones, agents coming after them can gather vital information about the paths and the network as a whole. In this research work, a simulated network models a typical distribution of calls between nodes; nodes carrying an excess of traffic can become congested, causing calls to be lost. The network supports a group of ants that have simple-purposed tasks, with no direct communication capability or explicit knowledge of the global goal. The ants move across the network between randomly chosen pairs of nodes; as they move they deposit simulated pheromones as a function of two variables: (1) the distance from their source node and (2) the congestion encountered on their journey. They select their path at each intermediate node according to the distribution of simulated pheromones at that node. Calls between nodes are routed as a function of the pheromone distributions at each intermediate node. The ant-based control system was shown to result in fewer call failures than other methods such as fixed shortest-path routing and algorithmic mobile agent (instruction-specific), while exhibiting many attractive features of distributed control.

Other work in ants in telecommunication networks for routing and load balancing has concentrated on improving the path-finding intelligence of ants using dynamic programming (Bonabeau et al. 1998), genetic algorithms (White et al. 1998c), and Q-learning algorithms (Gambardella and Dorigo 1995).

19.4.1.4 Economic Models

Economic models can serve as the motivation and navigation guidance of intelligent agents traversing a telecommunication networks. The distributed and emergent behaviors subscribed by the models are comparable to those of today's network components. Therefore we foresee many more such models being used in the future when designing multiagent systems in telecommunication networks.

Ferguson et al. (1996) used an economic model for intelligent agents in telecommunication networks. According to the authors, computer networks are being used by a growing and increasingly heterogeneous set of network components (computers, channels, and users) that have diverse Quality of Service (QoS) requirements. To support the diversity in large distributed networks, the tasks of efficient service provisioning and optimal resource allocation become very complex but can be accomplished via decentralization. In an economy, decentralization is provided by the fact that economic models consist of agents which selfishly attempt to achieve their goals. There are two types of economic agents, suppliers and consumers. A consumer attempts to optimize its individual performance criteria by obtaining the resources it requires, and is not concerned with system-wide performance. A supplier allocates its individual resources to consumers. A supplier's sole goal is to optimize its individual satisfaction (profit) derived from its choice of resource allocations to consumers. So in a provisioning telecommunication network, the economic model lends insights to the demand and supply balance, pricing strategies and mechanisms, decision making in choosing suppliers, system economy, and resource allocation.

19.4.2   Research Areas in Telecommunication Networks for Intelligent Agents

The following discussion is by no means inclusive of all research projects that apply intelligent agents to telecommunication networks. Most of these applications perform more than one single network management task. Therefore, the categorization is not exclusive either. We introduce the categorization to present several important areas in telecommunication networks where intelligent agent technologies are active.

19.4.2.1 Network Management Architecture

In Sahai et al. (1997) and Sahai and Morin (1998), a mobile agent environment for network management was described. The MAGENTA (Mobile AGENT for Administration) environment was designed to achieve dynamic and decentralized management of a distributed system comprising of heterogeneous machines running varied operating systems connected by LANs. The environment introduces the idea of a Mobile Network Manager (Sahai et al. 1997), equipped with autonomy, reactivity, proactivity, and communication skills. The environment also provides in its architecture *lieus* as places or locations where an agent can originate, reside, execute, and interact with the system as well as other agents. A lieu is a static program that checks the security, authenticates the suitability for execution, allows communications with other agents, provides residence, and keeps track of its agents. Thus, in this management architecture, lieus are static agents and they are capable of spawning and servicing mobile agents that traverse across the network.

Frei and Faltings (1998) proposed an abstraction approach to represent the original network as a hierarchy of simplified graphs. Each node of the graph abstracts a part of the network inside which routing of demands requiring a given amount of bandwidth is possible. The management of this virtual architecture and the routing of demands can then be distributed to intelligent agents.

19.4.2.2 Network Diagnosis

Mobile agents can be delegated to measure utilization and efficiency of nodes throughout the network. If the measurements violate some established network rules or fall below some expected network performance criteria, then the mobile agents have located a fault. Mobile agents can then perform a diagnosis to locate the source of the fault, report the problematic area to human managers, repair the fault itself, or transfer the responsibility to other agents. White et al. (1998a) used ants to perform network diagnosis by populating the network with a group of small agents to collect and tally observations which can be checked against diagnostic rules to determine faults.

19.4.2.3 Network Traffic Control and Routing

El-Darieby and Bieszczad (1999) proposed a system of intelligent mobile agents to reduce management traffic on the network (no bandwidth-exhaustive Client/Server message exchanges), enable more robust response to problems (timely, intelligent problem-solving behavior of the agents), reduce administrative overhead and cost due to function delegation (agents perform tasks autonomously), and allow low-level problems to be dealt with locally at the network component (reducing both the processing load on the manager computer and the traffic carrying instructions and data regarding the problems).

Other routing works include White et al. (1998b) in connection management and Schoonderwoerd et al. (1996, 1997) in network load balancing.

19.4.2.4 Network Mobility Platform (Mobile Code Languages)

Some research has examined guiding and enabling the navigation or migration of mobile agents in telecommunication networks (Acharya et al. 1997, Peine and Stolpmann 1997, Peine 1997, Gray 1996, Hylton et al. 1996, Straßer et al. 1996, White 1996, Lingnau et al. 1995). Issues integral to the mobility of agents in heterogeneous networks include code mobility technologies such as portability, execution, security, and resource access.

For example, Ara (Agents for Remote Action) is a platform for mobile agents designed for the portable and secure execution of mobile agents written in various interpreted languages on top of a common run-time core in an attempt to achieve location transparency. It allows agents to migrate at any point in their execution, fully preserving their states, and exchange messages with other agents. A network system can contain many virtual places, each establishing a domain of logically related services under a common security policy governing all agents at this place. Agents are equipped with allowances limiting their resource accesses, both globally per agent lifetime and locally per place. The Ara project emphasizes the system support for general mobile agents with minimal features for applications and behaviors of agents (Peine and Stolpmann 1997, Peine 1997).

19.4.2.5 Network Configuration

When network components are added, or suspended (for repair), or deleted from the network, the network reconfigures. Mobile agents can be used to reconfigure the networks and update the network topology. For example, agents can be designed to implement plug-and-play network components (Bieszczad et al. 1998a, Yemini 1993) in which components can either announce their existence by self-bootstrapping (sending out agents to inform others) or issue signals to be discovered by network-patrolling agents. New network services can also be provisioned using agents (Csurgay et al. 1997, Barr et al. 1993).

Pagurek et al. (1998) introduced an alternative to existing Asynchronous Transfer Mode (ATM) configuration management solutions. The authors designed and implemented a generic model based on mobile agents, that performs permanent virtual connections (PVCs) as configuration management functions in multi-vendor ATM networks. With the mobile code approach one operator was required to enter the end-to-end PVC configuration requirements (such as port connections for neighboring switches and bandwidth) and the PVC configuration application then sent a PVC agent to the network to conduct the configuration task, with the user requirements carried by the agent. The agent configured the switch at the source, migrated to the carrier's switch and configured it, and finished the task at the destination switch. In contrast to manager-to-manager PVC configuration management and CORBA-based PVC configuration management, the mobile agent approach has several advantages: there is no manager-to-manager software integration; understanding and knowledge about different switches are not required; a PVC agent automates the connectivity procedures and bypasses human decision making; and, PVC agents are task-specific and can be implemented more efficiently.

19.4.2.6 Network Monitoring and Accounting

Network monitoring and accounting are necessary to measure the performance of networks so that networks can be designed and improved over time for performance, efficiency, and cost. However, due to network delays the use of a centralized server makes measurements inaccurate or not timely, leading to network performance measurements that are not reliable or are difficult to collect. Instead of remotely polling network components, mobile agents can be dispatched to perform local analyses remotely (Bieszczad and Pagurek 1998). This improves the accuracy of the information, does not consume local resources permanently because of the agents' mobility, and makes the maintenance of the monitoring and accounting system easier since mobile agents are smaller and distributed in design and implementation.

Bieszczad and Pagurek (1998) also pointed out that mobile agents can be used to implement hot-swapping technologies (to keep stationary network monitoring agents up to date) and server migration (information analysis such as service demand, network load, and failure rate). The authors subsequently warned against agent flooding, that is, uncontrolled mobile agents populating the network and taking over a large proportion of its resources, affecting the network performance that is to be measured.

19.4.2.7 Service Management and Provisioning

As pointed out in Plu (1998), the goal of service provisioning in telecommunications is to allow corporations to provide more and more sophisticated information and electronic commerce services to many customers. Plu (1998) went on to describe several attributes of a service-based agent such as autonomy, trustworthiness, distributed, social ability, persistence and the ability to act asynchronously, data encapsulation, interoperability, and flexibility. In addition, the author prescribed policies to govern performative actions such as obligation, permission, and prohibition.

Other agent-based systems in service management include service maintenance agents and customer agents (Weihmayer and Tan 1992, Busuoic and Griffiths 1993, Weihmayer and Velthuijsen 1998) and operator assistance agents (Garijo and Hoffmann 1992).

19.5 Negotiating Agents

Cooperative and self-interested agents within a community communicate with each other to achieve a mutually acceptable agreement towards certain goals through negotiations. Negotiation protocols allow network components to conduct intelligent communications to either selfishly increase the performance of each component, or altruistically improve the overall effectiveness and efficiency of the network. The process can be seen as a distributed search through a space of potential agreements (Laasri et al. 1992). Traditionally, negotiating agents have used one of four techniques: game theory (Osborne and Rubinstein 1994, Rosenchein and Genesereth 1985), approaches inspired by operations research (Kraus et al. 1995, Sandholm and Lesser 1995), probabilistic/Bayesian techniques (Zeng and Sycara 1996), and heuristic models (Kraus et al. 1991).

Game-theoretic approaches usually make assumptions, such as shared prior probability estimates, common knowledge of agents' preferences, and other aspects of the game, and perfect rationality. Operation research approaches usually involve decentralization of computational algorithms. Bayesian approaches update the knowledge and belief that each agent has about the environment and other agents and naturally models the iterative negotiation behavior. Heuristic negotiation techniques such as case-based reasoning in PERSUADER (Sycara 1990a; 1990b) allow the agent to use any set of previously successful algorithms, without having to ascribe to one methodology.

Negotiations are important in multiagent systems in terms of cost reduction, performance optimization, cost budgeting, and autonomy. Negotiating agents have been used to detect and resolve certain kinds of feature interactions that occur in telecommunication systems (Griffeth and Velthuijsen 1993). This research area has the potential to contribute to intelligent agents in telecommunication networks. Readers are pointed to (Rosenschein and Zlotkin 1994) for further discussion on negotiation among agents.

19.6 A Design Example of an Intelligent Agent Structure

There are numerous designs of agent structure in the literature. This is due to various functions that the agent technologies apply in the area of telecommunication networks, as described above. In this section, we present a design of one such intelligent agent in the context of telecommunication networks. The following discussion is aimed to provide a general idea on how one such agent structure can be designed and, therefore, some issues related to telecommunication networks are only briefly touched upon.

Figure 19.1 shows a general design of an intelligent agent. There are six integral modules within the agent: (1) Event Monitor, (2) House Keeping, (3) Adaptive Learning Mechanism, (4) Message Interface, (5) Task Interface, and (6) Manager. The network support for the creation of an agent is provided through the Network Management module.

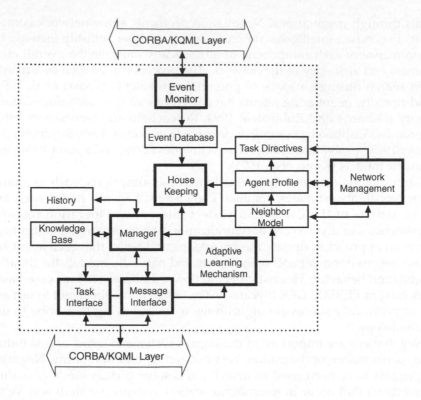

FIGURE 19.1
A design example of an intelligent agent structure for telecommunication networks.

At the creation of an agent, the Network Management module supplies the agent with its profile, its model of neighboring agents, and a set of task directives. The Agent Profile includes a set of parameterized attributes such as *name, origin, priority, attitude, category, class, role,* etc. These attributes will later be used in guiding the behavior of the agent in task planning, decision making, and interactions with neighboring agents. The Neighbor Model includes a list of agents known to the agent. Each neighbor model will include the *name, category, class, role,* etc. Note that we can either provide each agent with a set of complete knowledge of its neighbors, or a partial set. The more complete the knowledge, the more is required of the agent to maintain its knowledge base for decision making. The Task Directives describe the responsibilities of the agent. The *role* of an agent can be a Message Router, a Network Manager, a Resource Allocation Manager (such as computer, data, and information resources), a Navigator, a Network Maintainer, or a Network Housekeeper.

At the creation of an agent, the Network Management module will specify the role of the agent. Along with that specification is a set of task derivatives that outline the set of responsibilities of the agent. The Derivatives will specify the mobility (or residency), navigation plan, deployment plan, and capabilities of the agent, and, most of all, the tasks to be carried out by the agent. For example, a Navigator agent may be created to obtain the fastest route between a network node A in New York and a network node B in Lawrence, Kansas. Thus, this agent is mobile. The agent's deployment plan includes the times that it must be deployed and returned. The capabilities specify what resources the agent has in carrying out its tasks: query about a particular node's security and consistency, measurement or estimation of a node's traffic activity, logging of a node's activity, message sending to communicate with the originator of the agent, etc. The tasks, in this case, will describe several requirements: (1) whether the fastest route is for data flow (and its degree) or computer resource usage (and its degree), (2) the weights on the routes (speed, security, reliability, etc.), (3) whether the selected route is for short or long term usage, (4) a set of steps specifying the tasks. The set of task steps will include what procedures the agent must take when arriving at, say, a node. For example: (a) request for N computational resources, (b) measure node activity, (c) if the node activity is below a certain level, then accept it, (d) select another node down the route (based on the agent's current status, history, and knowledge base), and (e) move to the selected node.

The Event Monitor module interfaces with the Network Environments. This includes alerts or messages sent to the agent by users or system administrators of the networks: for example, a network shutdown of a particular node, network congestion, or even a temporary exclusive usage of a network area. The Event Monitor will log these events in the Event Database, which it will maintain and update for consistency and accuracy.

The House Keeping module fuses the events, tasks derivatives, agent profile, and the neighbor model together to provide the Manager module a modified but consistent set of information, making the Manager temporal and situational aware. The House Keeping module is essentially the information guru of the agent.

The Adaptive Learning Mechanism module allows the agent to learn from its interactions with other network elements. So, as the agent performs tasks during its life cycle, it will be able to learn the behavior of the neighboring agents or network nodes that it encounters. Of course, if the agent is short-lived, then we might want to tune down the learning ability. On the other hand, if the agent is an information-gathering agent, or has a long life cycle (during which it might encounter the same neighbor more than once), then this learning mechanism allows the agent to adapt. The Manager will thus be able to make decisions on more updated and accurate information.

The Message Interface module has two basic responsibilities: (1) send a message to a destination, and (2) receive a message. It will thus know how to compose a message and parse a message. The could be sent across a CORBA layer.

The Task Interface module performs the task. In the Navigator example, this module might have to ask for permission from the local node to perform a task. If the permission is granted, then the module runs its procedures to, say, collect information from the node. In another example, suppose that the agent is deployed and resides at a local node, it may have access to the local computer resources. Thus, in this case, the Task Interface module simply actuates the steps of task in the agent's directives.

The Manager module will be responsible for the decision making and planning of the agent. In order to be intelligent, it has a knowledge base. This knowledge base may consist of rules, features, weights, vectors, etc. and provide the reasoning process to the Manager. The Manager will decide whether, when, and how to carry out a task. It will also update its History base where past behaviors and observations are stored. For example, in the Navigator example, if the Manager realizes that it has found a fastest but also least reliable path, it might have to compute the utility of selecting this path based on its knowledge and information. If there is a rule that says "If the success of the transfer of data is of the HIGHEST priority, then it might be better to choose a more reliable albeit slower path", then the Manager may choose another safer path.

Note that even though we have used an example in which the agent is a Navigator, deployed to map out a fastest path between two nodes, we can extend the above discussion to other intelligent agent roles in a telecommunication networks such as (1) a Message Router—an agent that resides at a single node and distributes and dispatches message efficiently, (2) a Resource Manager—a resident agent that allocates its computer resource or data resource to other network functions or agents, (3) a Network Maintainer—a resident agent that analyzes processes on a node and decides which to terminate (such as hung processes) and which to maintain, and (4) a Network Housekeeper—an agent deployed to collect data on the activity and health of a particular network and terminated once it reports the findings.

19.7 Conclusions

We have presented an overview of the application of intelligent agents to telecommunication networks. We have discussed the properties of an agent and its intelligence. Again, we have not attempted to define what constitutes an agent and its intelligence. We have talked about how intelligent agents can be of help in managing telecommunication networks in various aspects such as reliability, security, control, routing, and fault diagnosis. In addition, we have discussed agent technologies and research areas in telecommunication networks and cited a number of contemporary research activities around the world. Moreover, we have included agent negotiations as closely related disciplines in intelligent agents to telecommunication networks. We have recommended this branch of technologies be involved in telecommunication networks because we believe that negotiations are key ingredients in designing cooperative and competitive intelligent agents. We have also presented a design example for an intelligent agent in the telecommunication networks context.

To conclude, we see intelligent agents in telecommunication networks as a research area that has great potentials in fusing previous works and experiences of other disciplines such as AI (e.g., DAI and MAS), network engineering, distributed processing, economy, social studies, information technology, data mining, knowledge engineering, system control and management, software engineering, computer languages, operations research, artificial life, psychology, and cognitive modeling. From the AI perspective, we believe that if we inject enough intelligence into telecommunication networks, then they will become self-sustaining, self-organizing, self-managing intelligent communities in the future.

References

A. Acharya, M. Ranganathan, and J. Saltz. 1997. Sumatra: A Language for Resource-Aware Mobile Programs, in *Mobile Object Systems: Towards the Programmable Internet*, J. Vitek and C. Tschudin (eds.), Springer-Verlag Lecture Notes in Computer Science, 111–130.

P.E. Agre and S.J. Rosenschein (eds.). 1996. *Computational Theories of Interaction and Agency*, Boston, Massachusetts: MIT Press.

S. Appleby and S. Steward. 1994. Mobile Software Agents for Control in Telecommunications Networks, *BT Technology Journal*, **12**(2):104–113.

M. Baldi S. Gai, and G.P. Picco. 1997. Exploiting Code Mobility in Decentralized and Flexible Network Management, in *Proceedings of First International Workshop on Mobile Agents (MA'97)*, April, Berlin, Germany, 13–26.

M. Baldi and G.P. Picco. 1998. Evaluating the Tradeoffs of Mobile Code Paradigms in Network Management Applications, in *Proceedings of the 20th International Conference on Software Engineering (ICSE'98)*, April, Kyoto, Japan, 146–155.

W.J. Barr, T. Boyd, and Y. Inoue. 1993. The TINA Initiative, *IEEE Communications Magazine*, March, 70–76.

J. Bates. 1994. The Role of Emotion in Believable Characters, *Communications of the ACM*, **37**(7):122–125.

G. Beni and J. Wang. 1989. Swarm Intelligence in Cellular Robotic Systems, in *Proceedings of the NATO Advanced Workshop on Robots and Biological Systems*, Italy.

A. Bieszczad and B. Pagurek. 1998. Network Management Application-Oriented Taxonomy of Mobile Code, in *Proceedings of the EEE/IFIP Network Operations and Management Symposium (NOMS'98)*, February 15–20, New Orleans, LA, 659-669.

A. Bieszczad, S.K. Raza, B. Pagurek, and T. White. 1998a. Agent-Based Schemes for Plug-and-Play Network Components, in *Proceedings of the Third International Workshop on Agents in Telecommunications Applications (IATA'98)*, July 4–7, Paris, France, 89-101.

A. Bieszczad, T. White, and B. Pagurek. 1998b. Mobile Agents for Network Management, *IEEE Communications Survey*, **1**(1):http://www.comsoc.org/pubs/surveys/4q98issue/bies.html.

E. Bonabeau, M. Dorigo, and G. Theraulaz. 1999. *Swarm Intelligence: from Natural to Artificial System*, New York: Oxford University Press.

E. Bonabeau, F. Henaux, S. Guerin, D. Snyers, P. Kuntz, and G. Theraulaz. 1998. Routing in Telecommunication Networks with "Smart" Ant-Like Agents, in *Proceedings of the Second International Workshop on Intelligent Agents for Telecommunication Applications (IATA'98)*, July 4–7, Paris, France, 60–71.

M. Busuoic and D. Griffiths. 1993. Cooperating Intelligent Agents for Service Management in Communications Networks, in *Proceedings of the 1993 Workshop on Cooperating Knowledge Based Systems (CKBS'93)*, September, University of Keele, UK, 213–226.

D.T. Cheng and S. Covaci. 1997. The OMG Mobile Agent Facility: A Submission, in *Proceedings of the First International Workshop on Mobile Agents (MA'97)*, April 7–8, Berlin, 98–110.

D. Chess, B. Grosof, C. Harrison, D. Levine, and C. Parris. 1995. Itinerant Agents for Mobile Computing, *IEEE Personal Communications*, 2(5):34–49.

D. Chin. 1991. Intelligent Interfaces as Agents, in J. Sullivan and S. Tyler (eds.), *Intelligent User Interfaces*, New York, ACM Press, 177–206.

P. Csurgay, A. Oesleboe, and F. A. Aagesen. 1997. Teleservices and Internet Application Technology, in *Proceedings of the IFIP TC6 International Symposium on Network Information Processing Systems*, October 14–16, Sofia, Bulgaria.

L. Dent, J. Boticario, J. McDermott, T. Mitchell, and D. Zabowski. 1992. A Personal Learning Apprentice, in *Proceedings of the Tenth National Conference on Artificial Intelligence (AAAI'92)*, July 12–16, San Jose, CA, 96–103.

P.S. Eaton, E.C. Freuder, and R. J. Wallace. 1998. Constraints and Agents: Confronting Ignorance, *AI Magazine*, 19(2):51–65.

M. El-Darieby and A. Bieszczad. 1999. Intelligent Mobile Agents: Towards Network Fault Management Automation, in *Proceedings of the 6th IFIP/IEEE Integrated Symposium on Integrated Network Management (IM'99)*, May 24–28, Boston, MA, 611–622.

B. Esfandiari, G. Deflandre, J. Quinqueton, and C. Dony. 1996. Agent-Oriented Techniques for Network Supervision, *Annals of Telecommunications*, 51(9/10):521–529.

O. Etzioni, H.M. Levy, R. B. Segal, and C. A. Thekkath (1995). The Softbot Approach to OS Interfaces, *IEEE Software*, 12(4):42–51.

O. Etzioni and D. Weld. 1994. A Softbot-Based Interface to the Internet, *Communications of the ACM*, 37(7):72–76.

D.F. Ferguson, C. Nikolaou, J. Sairamesh, and Y. Yemini. 1996. Economic Models for Allocating Resources in Computer Systems, in *Market-Based Control: A Paradigm for Distributed Resource Allocations*, S. Clearwater (ed.), New Jersey: World Scientific Press, 156–183.

C. Frei, and B. Faltings. 1998. A Dynamic Hierarchy of Intelligent Agents for Network Management, in *Proceedings of the Second International Workshop on Intelligent Agents for Telecom Applications (IATA'98)*, July 4–7, Paris, France, 9–16.

A. Fuggetta, G.P. Picco, and G. Vigna. 1998. Understanding Code Mobility, *IEEE Transactions on Software Engineering*, 24(5):342–361.

J.R. Galliers. 1988. A Strategic Framework for Multi-Agent Cooperative Dialogue, in *Proceedings of the Eighth European Conference on Artificial Intelligence (ECAI'88)*, August, Munich, Germany, 415–420.

L.M. Gambardella, and M. Dorigo. 1995. Ant-Q: A Reinforcement Learning Approach to the Traveling Salesman Problem, in *Proceedings of the Twelfth International Conference on Machine Learning (ML'95)*, July, Tahoe City, CA, 252–260.

F.J. Garijo and D. Hoffmann. 1992. MAITE: An Operator Assistance Expert System for Troubleshooting Telecommunications Networks, in *Proceedings of the 3rd International Conference of Database and Expert Systems Applications (DEXA'92)*, Valencia, Spain, 14–19.

R. Gray. 1996. Agent-Tcl: A Flexible and Secure Mobile Agent System, in *Proceedings of the 4th Annual Tcl/Tk Workshop*, July, Monterey, CA, 9–23.

N.D. Griffeth and H. Velthuijsen. 1993. Win/Win Negotiation Among Autonomous Agents, in *Proceedings of the 12th International Workshop on Distributed Artificial Intelligence*, Hidden Valley, PA, 187–202.

J. Hylton, K. Manheimer, F. Drake, B. Warsaw, R. Masse, and G. van Rossum. 1996. Knowbot Programming: System Support for Mobile Agents, in *Proceedings of the Fifth IEEE International Workshop on Object Orientation in Operating Systems (IWOOOS'96)*, October 27–28, Seattle, WA, 8–13.

J. Hylton and G. van Rossum. 1997. Using the Knowbot Operating Environment in a Wide-Area Network, in *the 3rd ECOOP Workshop on Mobile Object Systems (MOS'97)*, June 9–10, Jyväskylä, Finland.

N.R. Jennings, K. Sycara, and M. Wooldridge. 1998. A Roadmap of Agent Research and Development, *Autonomous Agents and Multi-Agent Systems*, 1:275–306.

N. Joshi and V.C. Ramesh. 1998. On Mobile Agent Architectures, *Technical Report*, ECE Department, Illinois Institute of Technology.

S. Kraus, E. Ephrati, and D. Lehmann. 1991. Negotiation in a Non-Cooperative Environment, *Journal of Experimental and Theoretical Artificial Intelligence*, 3(4):255–282.

S. Kraus, J. Wilkenfeld, and G. Zlotkin. 1995. Multiagent Negotiation Under Time Constraints, *Artificial Intelligence*, 75(2):297–345.

J.A. King. (1995). Intelligent Agents: Bringing Good Things to Life, *AI Expert*, 10(2):17–19.

B. Laasri, H. Laasri, S. Lander, and V. Lesser. 1992. A Generic Model for Intelligent Negotiating Agents, *International Journal on Intelligent Cooperative Information Systems*, 1(2):291–317.

Y. Lashkari, M. Metral, and P. Maes. 1994. Collaborative Interface Agents, in *Proceedings of the Twelfth National Conference on Artificial Intelligence*, August 1–4, Seattle, WA, 444–449.

V.R. Lesser, (1995). Multiagent Systems: An Emerging Subdiscipline of AI, *ACM Computing Surveys*, 27(3):340–342.

V.R. Lesser, J. Pavlin, and E. H. Durfee. 1988. Approximate Processing in Real Time Problem Solving, *AI Magazine*, 9(1):49–61.

H. Lieberman. 1995. Letizia: An Agent that Assists Web Browsing, in *Proceedings of the Fourteenth International Joint Conference on Artificial Intelligence (IJCAI'95)*, August 20–25, Montreal, Quebec, Canada, 924–929.

A. Lingnau, O. Drobnik, and P. Dömel. 1995. An HTTP-based Infrastructure for Mobile Agents, in *Proceedings of the 4th International WWW Conference*, December 11–14, Boston, MA, 461–471.

P. Maes. 1994. Agents that Reduce Work and Information Overload, *Communications of ACM*, 37(7):31–40.

P. Maes and R. Kozierok. 1993. Learning Interface Agents, in *Proceedings of the Eleventh National Conference on Artificial Intelligence*, August 18–20, Washington, D.C., 459–465.

T. Magedanz, K. Rothermel, and S. Krause. 1996. Intelligent Agents: An Emerging Technology for Next Generation Telecommunications?, in *Proceedings of INFO-COMM'96*, March 24–28, San Francisco, CA, 464–472.

J.P. Martin-Flatin, and S. Znaty. 1997. A Simple Typology of Distributed Network Management Paradigms, in *Proceedings of the 8th IFIP/IEEE International Workshop on Distributed Systems: Operations and Management (DSOM'97)*, October 21–23, Sydney, Australia, 13–24.

K. Meyer, M. Erlinger, J. Betzer, C. Sunshine, G. Goldszmith, and Y. Yemini. 1995. Decentralizing Control and Intelligence in Network Management, in *Proceedings of the Fourth International Symposium on Integrated Network Management*, Santa Barbara, California, 4–16.

H.S. Nwana. 1996. Software Agents: An Overview, *Knowledge Engineering Review*, **11**(3):1–40.

M.J. Osborne, and A. Rubinstein. 1994. *A Course in Game Theory*, Boston, MA: MIT Press.

B. Pagurek, Y. Li, A. Bieszczad, and G. Susilo. 1998. Network Configuration Management in Heterogeneous ATM Environments, in *Intelligent Agents for Telecommunications Applications*, S. Albayrak and F. J. Garijo (eds.), Berlin: Springer-Verlag, 72–88.

M.P. Papazoglou, S. C. Laufman, and T. K. Sellis. 1992. An Organizational Framework for Cooperating Intelligent Information Systems, *Journal of Intelligent and Cooperative Information Systems*, **1**(1):169–202.

H. Peine. 1997. Ara—Agents for Remote Action, in *Mobile Agents: Explanations and Examples*, W. Cockayne and M. Zyda (eds.), Greenwich, CT: Manning, 96–163.

H. Peine and T. Stolpmann. 1997. The Architecture of the Ara Platform for Mobile Agents, in *Proceedings of the First International Workshop on Mobile Agents (MA'97)*, April 7–8, Berlin, 50–61.

M. Plu. 1998. Software Technologies for Building Agent Based Systems in Telecommunications Networks, in *Agent Technology: Foundations, Applications, and Markets*, N.R. Jennings and M.J. Wooldridge (eds.), Berlin: Springer-Verlag, 241–266.

A.S. Rao and M.P. Georgeff. 1995. BDI Agents: From Theory to Practice, in *Proceedings of the First International Conference on Multi-Agent Systems (ICMAS'95)*, June 12–14, San Francisco, CA, 312–319.

J.S. Rosenschein and M.R. Genesereth. 1985. Deals among Rational Agents, in *Proceedings of the Ninth International Joint Conference on Artificial Intelligence*, Los Angeles, CA, 91–99.

J.S. Rosenschein and G. Zlotkin. 1994. *Rules of Encounter: Designing Conventions for Automated Negotiation Among Computers*, Boston, MA: MIT Press.

A. Sahai and C. Morin. 1998. Enabling a Mobile Network Manager through Mobile Agents, in *Proceedings of the Second International Workshop on Mobile Agents (MA'98)*, September, Stuttgart, Germany, 249–260.

A. Sahai, C. Morin, and S. Billiart. 1997. Intelligent Agents for a Mobile Network Manager (MNM), in *Proceedings of the IFIP/IEEE International Conference on Intelligent Networks and Intelligence in Networks (2IN'97)*, September 2–5, Paris, France, 449–463.

T. Sandholm and V. Lesser. 1995. Coalition Formation among Bounded Rational Agents, in *Proceedings of the 14th International Joint Conference on Artificial Intelligence (IJCAI-95)*, Montreal, Canada, 662–669.

R. Schoonderwoerd, O. Holland, J. Bruten, and L. Rothkrantz. 1996. Ant-Based Load Balancing in Telecommunications Networks, *Journal of Adaptive Behavior*, 5(2):169–207.

R. Schoonderwoerd, O. Holland, and J. Bruten. 1997. Ant-Like Agents for Load Balancing in Telecommunications Networks, in *Proceedings of the First International Conference on Autonomous Agents (Agents'97)*, February 5–8, Marina del Rey, CA, 209–216.

B. Sheth and P. Maes. 1993. Evolving Agents for Personalized Information Filtering, in *Proceedings of the Ninth Conference on Artificial Intelligence for Applications*, Orlando, Florida, IEEE Computer Society Press, 345–352.

Y. Shoham. 1993. Agent-Oriented Programming, *Artificial Intelligence*, 60(1):51–92.

M. Straβer, J. Baumann, and F. Hohl. 1996. Mole—A Java based Mobile Agent System, in *Proceedings of the Second ECOOP Workshop on Mobile Object Systems (ECOOP'96)*, July 8–9, University of Linz, Austria, 301–308.

K. Sycara. 1990a. Persuasive Argumentation in Negotiation, *Theory and Decision*, 28(3):203–242.

K. Sycara. 1990b. Negotiation Planning: An AI Approach, *European Journal of Operational Research*, 46(2):216–234.

R. Weihmayer and M. Tan. 1992. Modeling Co-Operative Agents for Customer Network Control Using Planning and Agent-Oriented Programming, in *Proceedings of the IEEE Global Telecommunications Conference (Globecom'92)*, December 6–9, Orlando, FL, 537–543.

R. Weihmayer and H. Velthuijsen. 1998. Intelligent Agents in Telecommunications, in *Agent Technology: Foundations, Applications, and Markets*, N. R. Jennings and M. J. Wooldridge (eds.), Berlin: Springer-Verlag, 241–266.

J.E. White. 1996. Mobile Agents, in *Software Agents*, J. Bradshaw (ed.), AAAI Press/MIT Press, 437–472.

T. White, A. Bieszczad, and B. Pagurek. 1998a. Distributed Fault Location in Networks Using Mobile Agents, in *Intelligent Agents for Telecommunications Applications*, S. Albayrak and F. J. Garijo (eds.), Berlin: Springer-Verlag, 130–141.

T. White and B. Pagurek. 1998. Towards Multi-Swarm Problem Solving in Networks, in *Proceedings of the Third International Conference on Multi-Agent Systems (IC-MAS'98)*, July 2–8, Paris, France, 333–340.

T. White, B. Pagurek, and F. Oppacher. 1998b. Connection Management using Adaptive Agents, in *Proceedings of the 1998 International Conference on Parallel and Distributed Processing Techniques and Applications (PDPTA'98)*, July 13–16, Las Vegas, NV, 802–809.

T. White, B. Pagurek, and F. Oppacher. 1998c. ASGA: Improving the Ant System by Integration with Genetic Algorithms, in *Proceedings of the 1998 Conference on Genetic Programming (GP'98)*, July 22–25, Madison, WI, 610–617.

M. Woodridge and N. Jennings. 1995. Intelligent Agents: Theory and Practice, *The Knowledge Engineering Review*, 10(2):114–152.

Y. Yemini. 1993. The OSI Network Management Model, *IEEE Communications Magazine*, 31(5):20–29.

Y. Yemini and S. da Silva. 1996. Towards Programmable Networks, in *Proceedings of the 1996 IFIP/IEEE International Workshop on Distributed Systems: Operations and Management (DSOM'96)*, October 28–30, L'Aquila, Italy.

D. Zeng and K. Sycara. 1996. Bayesian Learning in Negotiation, in *Working Notes of the AAAI Spring Symposium on Adaptation, Co-Evolution and Learning in Multi-agent Systems*, Stanford, CA.

Exercise Problems

1. Why do today's telecommunication networks require better handling of data?

2. What has motivated the multiagent systems (MAS) in telecommunication networks? What are the advantages of agent technologies and intelligence in telecommunication networks?

3. What areas in telecommunication networks can intelligent agents be of help and why?

4. What are some of the attributes sufficient or necessary for defining an agent (list at least 6)?

5. What makes an agent *intelligent*?

6. List some advantages of a multiagent system over a centralized system.

7. Describe four architectures of management systems for telecommunication networks and identify the differences.

8. Describe some mobile agents and their applications in telecommunication networks.

9. How can intelligent interface agents be useful in telecommunication networks?

10. What is swarm intelligence? What is an ant system? How can an ant system be useful in telecommunication networks?

11. What is a lieu in the environment of a Mobile Network Manager (Sahai et al. 1997)?

12. How is service management and provisioning related to telecommunications?

Index

I

K

L